OCR AS

Geography

Jane Dove | Paul Guinness | Chris Martin | Garrett Nagle | David Payne | Michael Witherick

Education
Library Service
LEARNING RESOURCES FOR SCHOOLS

R 29/4/10

1 9 MAR 2010
READING SCHOOL LIBRARY

KT-148-090

www.heinemann.co.uk

✓ Free online support
✓ Useful weblinks
✓ 24 hour online ordering

01865 888080

OCR
RECOGNISING ACHIEVEMENT

Heinemann

Official Publisher Partnership

Heinemann is an imprint of Pearson Education Limited,
a company incorporated in England and Wales, having its registered office at
Edinburgh Gate, Harlow, Essex, CM20 2JE. Registered company number: 872828

www.heinemann.co.uk

Heinemann is the registered trademark of Pearson Education Limited

Text © Pearson Education Limited 2008

First published 2008

12 11 10 09 08
10 9 8 7 6 5 4 3

British Library Cataloguing in Publication Data is available from the British Library on request.

ISBN 978 0 435357 53 5

Copyright notice
All rights reserved. No part of this publication may be reproduced in any form or by any means (including photocopying or storing it in any medium by electronic means and whether or not transiently or incidentally to some other use of this publication) without the written permission of the copyright owner, except in accordance with the provisions of the Copyright, Designs and Patents Act 1988 or under the terms of a licence issued by the Copyright Licensing Agency, Saffron House, 6–10 Kirby Street, London EC1N 8TS (www.cla.co.uk). Applications for the copyright owner's written permission should be addressed to the publisher.

Edited by Janice Baiton
Designed by Wooden Ark Studio
Typeset by Phoenix Photosetting, Lordswood, Chatham, Kent
Original illustrations © Pearson Education Limited 2008
Illustrated by Phoenix Photosetting
Cover design by Big Top Design
Picture research by Susi Paz
Cover photo/illustration © Bernhard Edmaier
Printed in China (SWTC/03)

Websites
The websites used in this book were correct and up to date at the time of publication. It is essential for tutors to preview each website before using it in class so as to ensure that the URL is still accurate, relevant and appropriate. We suggest that tutors bookmark useful websites and consider enabling students to access them through the school/college intranet.

Acknowledgements

We would like to thank Rob Bircher for his invaluable help in the development of this material. Garrett Nagle would like to thank Angela, Rosie, Patrick and Bethany for their help.

The author and publisher would like to thank the following individuals and organisations for permission to reproduce copyright material.

Maps and extracts

Pages 6, 7, 8, 9, 11, 13, 16, 20, 21, 25, 97, 103, 108, 232 Oxford University Press. Pages 15, 111, 112 Blackwell Publishing. Pages 19, 28, 93, 102, 106, 123, 228, 266, 269, 272, 282, 307, 310 Hodder. Pages 33, 79 Guardian. Page 34 Pearson Education. Pages 55, 57, 68 SCOPAC Standing Conference On Problems Associated with the Coastline. Page 57 Isle of Wight Centre for the Coastal Environment. Page 58 DEFRA. Page 66 pevensey-bay.com.uk. Page 71 The Wildlife Trusts. Page 77 Independent. Page 81 Soufriere Marine Management Association Inc. Page 91 Philip's. Pages 93, 94, 108, 195, 217, 227, 237 Nelson Thornes. Page 96 Longman. Pages 97, 100, 110, 191, 239 Ordnance Survey. Page 107 Price, R.J., Arctic and Alpine Research. Page 123 Ski-Zermatt.com. Pages 175, 178, 180, 181 Phillip Allan Updates. Page 177 UN Secretariat. Page 183 Cameron Dunn. Pages 197, 200 CartoonStock. Page 198 Environment Agency. Page 201 Thames Gateway. Page 204 Global Footprint Network. Page 205 Best Foot Forward. Page 234 Border Rural Committee. Page 237 Journal of Southern African studies. Pages 256, 257, 260, 261, 265, 267 BP plc. Page 262 Paul Mobbs, Mobbs' Environmental Investigations. Pages 263, 266, 277 National Geographic. Page 264 Natural Gas Supply Association. Page 268 Statistics for Education. Page 270 Royal Academy of Engineering. Page 281 Global Wind Energy Council. Page 282 Financial Times. Page 293, 294, 295, 296, 299, 303 World Tourism Organization (UNWTO). Page 301 National Statistics. Pages 304, 305 China National Tourist Office.

CD-ROM only – Figure 2.55 *Independent*. Figure 6.44 Woodstock Action Group.

Photographs

Garrett Nagle pp5–6, 8, 9, 10, 11, 12, 14, 16, 18, 23, 24, 25, 26, 27, 28, 37, 39, 88–89, 92, 96, 98, 99, 100, 101, 102, 104, 105, 106, 107, 108, 109, 110, 111, 112, 113, 114, 116, 117, 118, 119, 121, 122, 124, 125, 214, 219, 220, 221, 225, 226, 229, 230, 235, 240, 241, 242, 246, 247, 248. iStockPhotos/Steven Stone p32. Gunter Marx/Alamy pp46–47. Keith Shuttlewood/Alamy p48. Rolf Richardson/Alamy p50. John Giles/PA Archive/PA Photos p51. Alex Beaton/Alamy p52. John Farmar; Cordaiy Photo Library Ltd./Corbis UK Ltd p55. Simon Holdcroft/Alamy p60. Geoscience Features Picture Library p63. Brian Harris/Alamy p65. Marc Hill/Alamy p69.David Payne pp53, 68, 69. Travelshots.com/Alamy p75. NASA p77. Richard T. Nowitz/Corbis UK Ltd p79. Robert Harding Picture Library Ltd/Alamy p82. Jane Dove pp132–133, 141, 147, 148, 150, 155, 158, 159. Moodboard/Corbis UK Ltd p141. Photolibrary Group p142. Jim Goldstein/Alamy p143. Getty Images p152. Jonathan Blair/Corbis UK Ltd p154. Ann Johansson/Corbis UK Ltd. p157. Corbis/HarcourtEducation Ltd pp172–173. Michael Witherick p174. Digital Vision/HarcourtEducation Ltd pp80, 144, 174. Webbaviation.co.uk pp175, 182. geogphotos/Alamy p183. Goncalo Diniz/Alamy p186. David Lyons/Alamy p188. Nicholas Pitt/Alamy p189. Danita Delimont/Alamy p191. David Newham/Alamy p193. Charles Sturge/Alamy p193. David Burton/Alamy p196. Mark A. Johnson/Alamy p196. Chris Hondros/Getty Images News p201. Simmons Aerofilms/Hulton Archive/Getty p206. VIEW Pictures Ltd/Alamy p207. Evening Standard/Hulton Archive/Getty pp254–255. Joe McNally/Getty Images News p258. Reza/Getty Images News p261. neal and molly jansen/Alamy p264. STR/AP/PA Photos/Empics p267. Empics pp208, 267, 280. Ulrich Doering/Alamy p268. Mark Boulton/Alamy p270. Chris Hondros/Getty Images News p278. Carol and Mike Werner/Alamy p281. Images&Stories/Alamy p283. Stephen Saks Photography/Alamy p284. Stuart Franklin/MagnumPhotos.com pp290–291. Mary Evans Picture Library p292. NDP/Alamy p298. Paul Guinness pp298, 307, 309. David Levenson/Alamy p301. Catarina Leal/Alamy p302. John Henshall/Alamy p304. China Tourism Press/Riser/Getty p305. Photos.com p311. Gary Cook/Alamy p319. Norma Joseph/Alamy p320.

CD-ROM only – Garrett Nagle Figure 1.55, 3.54, 3.55, 6.45, 6.48. Daniel Badino/Alamy Figure 2.54.

Every effort has been made to contact copyright holders of material reproduced in this book. Any omissions will be rectified in subsequent printings if notice is given to the publishers

Contents

Unit 2 Managing change in human environments

Introduction to geographical skills – See CD-ROM

Glossary – See CD-ROM

Introduction

Student Book

Units 1 and 2

Your AS Geography course is divided into two units, Managing Physical Environments and Managing Change in Human Environments. This Student Book provides an exact match to the OCR specification and as well as teaching and learning material it includes activities, exam support (through our Exam Café sections) and extension opportunities.

Features of the book

◆ **Key terms/Key concepts**: throughout the text you will find definitions of important key terms that may be new to you. These are a useful reference source.

◆ **Activities**: there are a range of motivating activities to help you practise what you are learning including '**discussion point**' activities that provide opportunities for small-group discussions and '**theory into practice**' activities which encourage the application of concepts to real-life contexts.

◆ **Case studies**: new and up-to-date case studies provide real-world examples of the topics you are studying. Questions on the case studies will enable you to explore the topic further, understand the key issues and deepen your understanding of the topic.

◆ **Knowledge check**: questions at the end of each chapter will check that you have taken on board all the concepts within the chapter and that you can use your knowledge synoptically.

Additional material contained on the CD-ROM is clearly signposted in the text, for example:

'Take it further' activity 1.3 on CD-ROM

In our unique Exam Café you'll find lots of ideas to help you prepare for your Unit 1 and Unit 2 exams. You'll find the Exam Café at the end of each chapter. You can **Relax** because there's handy advice on getting started on your AS Geography course, **Refresh your memory** with summaries and checklists of the key ideas you need to revise and to **Get the result** through practising exam-style questions, accompanied by hints and tips from the examiners.

Student CD-ROM

LiveText

On the CD you will find an electronic version of the Student Book, powered by LiveText. As well as the Student Book and the LiveText tools there are:

◆ Additional case studies to support the main chapter content.

◆ 'Take it further' extension activities, these provide opportunities for you to undertake further work on a topic.

◆ An introduction to geographical skills

◆ A glossary of all the key terms in the Student Book.

Within the electronic version of the Student Book, you will also find the interactive Exam Café.

Immerse yourself in our contemporary interactive Exam Café environment! With a click of your mouse you can visit three separate areas in the café to **Relax**, **Refresh your memory** or **Get the result.** You'll find a wealth of material including:

◆ Revision timetable – a blank template for you to complete

◆ Revision Tips from students, Bytesize Concepts, Common mistakes and Examiner's Hints

◆ Language of the exam (an interactive activity)

◆ Revision help in the form of checklists

◆ Multiple choice quizzes

◆ A 'Thinking and planning' tool to help you create your own mind maps

◆ Sample exam questions (which you can try) with student answers and examiner comments

◆ Sample exam papers

◆ Up-to-date weblinks that direct you to extra exciting resources and encourage further research on topics.

Managing physical environments

To develop an understanding of:

- the physical processes and factors that produce features of the physical environment
- the impact of these processes upon human activity
- how human action has affected the physical processes and the resultant features.

Chapter 1

River environments

Rivers have become regular features in the news. In any given year, one or more rivers become notorious for their devastating floods. But for most of the time, rivers are very useful – they provide many people with their livelihoods, a place to live near and are a great source of recreation and leisure. Learning to live and cope with rivers would seem to be of increasing importance. And yet, many human activities seem to make the river environment, and the human use of it, more hazardous. Is it possible for humans to live harmoniously with rivers, while at the same time, benefit from them?

Questions for investigation

- What processes and factors are responsible for distinctive fluvial landforms?

- In what ways can river basins be a multi-use resource?

- What issues can arise from the development of river basins?

- What are the management challenges associated with the development of river landscapes?

Consider this

The number of people living on floodplains is accelerating every year.

Increasing human activities on floodplains are raising the risk of floods and flood impacts. So, why do people live on floodplains, and to what extent is it possible to manage human use of floodplains?

1.1 What processes and factors are responsible for distinctive fluvial landforms?

Slope processes

Slopes are defined as any part of the solid land surface where there is an incline (gradient). Processes operating on slopes have a major impact on fluvial (river) landscapes. Slope processes transfer material downslope to the river. In doing so they reduce the angle of the slope. The material carried downslope may be transported by the river and some of the larger material may be used to erode the river channel and bed. The features of erosion and deposition are detailed on pages 17–22. Mass movements are any large-scale movement of the earth's surface that are not accompanied by a moving agent such as a river, glacier or ocean wave. They include very small movements, such as soil creep, and fast movements, such as avalanches. They vary from dry movements, such as rock falls, to very fluid movements such as mud flows (Figure 1.1).

Slow movements

Two of the most important slope processes in temperate areas are soil creep (Figure 1.2) and rain-splash erosion (Figure 1.3).

Soil creep

Individual soil particles are pushed or heaved to the surface by wetting, heating or freezing of water (Figure

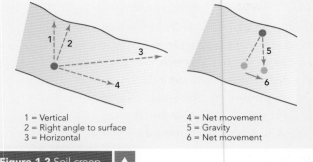

1 = Vertical
2 = Right angle to surface
3 = Horizontal
4 = Net movement
5 = Gravity
6 = Net movement

Figure 1.2 Soil creep ▲

1.2). They move at right angles to the surface (2) as it is the zone of least resistance. They fall under the influence of gravity (5) once the particles have dried, cooled or the water has thawed. Net movement is downslope (6). Rates are slow, 1mm/year in the UK and up to 5mm/year in the tropical rainforest. They form small scale terracettes such as the Manger in the Vale of the White Horse, Oxfordshire.

> **Key term**
>
> **Terracettes:** small terraces from a few centimetres to 0.5m across the face of a slope caused by soil creep or solifluction.

Rain-splash erosion

Figure 1.3 shows how on flat surfaces (1) raindrops compact the soil and dislodge particles equally in all directions. On steep slopes (2) the downward component (a) is more effective than the upward motion (b) due to gravity and so erosion downslope increases with slope angle. In contrast, solifluction is a form of accelerated soil creep that can produce braided rivers (see page 20). The term solifluction literally means 'flowing soil' and is affected by freezing and thawing in cold periglacial environments (see Chapter 3, page 91).

> **Key term**
>
> **Periglacial:** on the edge of glacial conditions, that is, an area which is covered by snow and ice for some of the year but also experiences a thaw for part of the year. Periglacial areas are commonly affected by intense freeze-thaw activity.

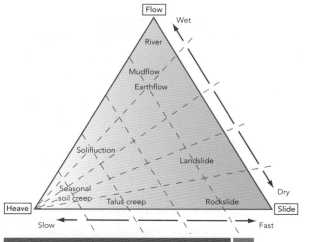

Figure 1.1 Classification of mass movement ▲

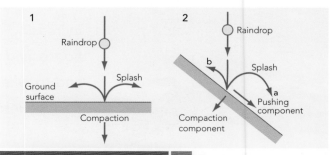

Figure 1.3 Rain-splash erosion ▲

Flow movements

Surface wash

This occurs when the soil's infiltration capacity is exceeded and can lead to the formation of gullies (see page 17). In Britain this commonly occurs in winter as water drains across saturated or frozen ground, following prolonged or heavy downpours or the melting of snow. It is also common in arid and semi-arid regions where particle size limits percolation.

Sheetwash

Sheetwash is the unchannelled flow of water over a soil surface. On most slopes, sheetwash is divided into areas of high velocity separated by areas of lower velocity. It is capable of transporting material dislodged by rain-splash. Sheetwash occurs in the UK on footpaths and moorlands. For example, during the Boscastle floods of 2004, sheetwash from the shallow moorland peat caused steep, narrow gulleys to form (see page 17). In the semi-arid areas of south-west USA, it lowers surfaces by 2–5mm/year compared with 0.01mm/year on vegetated British slopes.

Throughflow

Throughflow is water moving down through the soil. It is channelled into natural pipes (very small channels of water within the soil) in the soil. This gives it sufficient energy to transport material, and added to its solute load, may amount to a considerable volume.

Fast mass movements

Slides

Sliding material maintains its shape and cohesion (Figure 1.4) until it impacts at the bottom of a slope and leads to large, slumped terraces. Slides range from small-scale slides close to roads, to large-scale movements killing thousands of people; for example, the Vaiont Dam in Italy where over 2000 died on the 9 October 1963. In January 2006, prolonged heavy rain led to landslides in Central and East Java (Indonesia) killing 120 people.

Falls

Rock falls occur on steep slopes (>70°). The initial cause of the fall may be weathering; for example, freeze-thaw or disintegration, or erosion prising open lines of weakness. Once the rocks are detached they fall under the influence of gravity (Figure 1.5). If the fall is short, it produces relatively straight scree (also known as talus); if it is long, it forms concave scree. A good example of falls and scree is Wastwater in the Lake District. Falls lead to scree slopes and large slumped terraces. In upland areas falls and slides are an important sediment source for rivers.

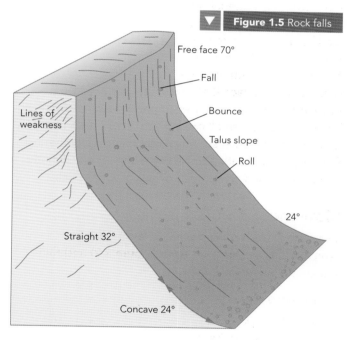

▼ **Figure 1.5** Rock falls

Free face 70°
Fall
Lines of weakness
Bounce
Talus slope
Roll
Straight 32°
24°
Concave 24°

Slumps

Slumps occur on weaker rocks, especially clay, and have a rotational movement along a curved slip plane (Figure 1.6). Clay absorbs water, becomes saturated, and exceeds its liquid limit. It then flows along a slip plane. Frequently the base of a cliff has been undercut and weakened by erosion thereby reducing its strength; for example, Folkestone Warren. Human activity can also intensify the condition by causing increased pressure on the rocks; for example, the Holbeck Hall Hotel, Scarborough.

Figure 1.4 Slides ▲

Scar
Slide plane
Detached block
Slide plane
Debris from an earlier slide

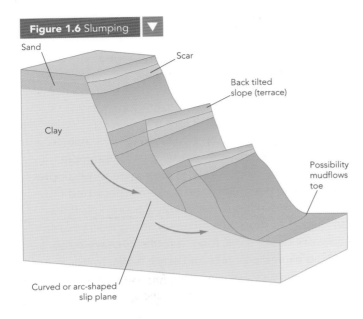

Figure 1.6 Slumping

Sand

Scar

Back tilted slope (terrace)

Clay

Possibility mudflows toe

Curved or arc-shaped slip plane

studies how different processes operate in different climatic zones and thereby produce different slope forms or shapes.

Rock type and structure

Slopes are also influenced by rock type. The Tees-Exe line is an imaginary line running from the River Tees to the River Exe (Figure 1.8). It divides Britain into hard and soft rock. To the north and west are old, hard, resistant rocks, such as granite, basalt and carboniferous limestone forming upland rugged areas. To the south and east are younger weaker rocks, such as chalk and clay, forming subdued low-lying landscapes.

Aspect

Aspect is another important factor. In the UK, north-facing slopes remain in the shade. During cold periglacial times, temperatures rarely rose above freezing. By contrast, south-facing slopes experienced many cycles of freeze-thaw. Solifluction and over-land runoff lower the level of the slope, and streams remove the debris from the valley. The result is an asymmetric slope such as the River Exe in Devon (Figure 1.9).

Key term

Aspect: the direction a slope faces; for example, north- and south-facing slopes.

'Take it further' activity 1.1 on CD-ROM

Factors affecting slopes

Climate

Slopes vary with climate. In general, slopes in temperate environments are rounder, due to chemical weathering, whereas slopes in arid environments are jagged or straight because of mechanical weathering and over-land runoff (Figures 1.7a and 1.7b). Climatic geomorphology is a branch of geography which

Figure 1.7a Rounded slopes in temperate environment

Figure 1.7b Rugged and straight slopes in semi-arid region

Figure 1.8 The Tees-Exe line ▼

N

River Tees

Older
resistant
rock

Younger
weaker
rock

River Exe

Weathering

Weathering is the decomposition and disintegration of rocks in situ. Decomposition refers to the chemical process and creates altered rock substances whereas disintegration or mechanical weathering produces smaller, angular fragments of the same rock. Weathering is important for landscape evolution as it breaks down rock and facilitates erosion and transport by rivers. The features of erosion and deposition are covered on pages 17–22.

> **Key terms**
>
> **Decomposition:** chemical weathering of rocks creating new materials.
>
> **Disintegration:** mechanical weathering of rocks resulting in smaller fragments of the same type.
>
> **In situ:** on the spot, that is, without any lateral (sideways) movement.

Mechanical (physical) weathering

The following are the four main types of mechanical weathering:

◆ *Freeze-thaw* This occurs when water in joints and cracks freezes at 0°C, expands by 10 per cent and then exerts pressure up to 2100kg/cm². Rocks can only withstand a maximum pressure of about 500kg/cm². It is most effective in environments where moisture is plentiful and there are frequent fluctuations above and below freezing point, such as periglacial and alpine regions (Figure 1.10).

◆ *Salt crystal growth* This occurs in two main ways: first, in areas where temperatures fluctuate around

U1

1

River environments

Figure 1.9 Asymmetric valley – River Exe in Devon ▼

▲ **Figure 1.10** The impact of freeze-thaw weathering on the Glydder Range, Snowdonia

26–28°C, sodium sulphate (Na_2SO_4) and sodium carbonate (Na_2CO_3) expand by 300 per cent; second, when water evaporates, salt crystals may be left behind to attack the structure (Figure 1.11). Both mechanics are frequent in hot desert regions.

◆ *Disintegration*　This is found in hot desert areas where there is a large diurnal temperature range. Rocks heat up by day and contract by night. As rock is a poor conductor of heat, stresses take place only in the outer layers and cause peeling or exfoliation to occur, showing that moisture is essential for this to happen.

Key term

Exfoliation: the 'peeling' of rocks on their outer edges as a result of repeated thermal expansion and contraction.

◆ *Pressure release*　This is the process where overlying rocks removed by erosion cause underlying ones to expand and fracture parallel to the surface. The removal of a great weight, such as a glacier, has the same effect.

Chemical weathering

The following are four main types of chemical weathering:

◆ *Carbonation solution*　This occurs on rocks with calcium carbonate; for example, chalk and limestone. Rainfall and dissolved carbon dioxide forms a weak carbonic acid. (Organic acids acidify water too.) Calcium carbonate reacts with the acid water and forms calcium bicarbonate, or calcium hydrogen carbonate, which is soluble and removed by percolating water.

◆ *Hydrolysis*　This occurs on rocks with orthoclase feldspar; for example, granite. Orthoclase reacts with acid water and forms kaolinite (or kaolin or china clay), silicic acid and potassium hydroxyl. The acid and hydroxyl are removed in the solution leaving china clay behind as the end product (Figure 1.12). Other minerals in the granite, such as quartz and mica, remain in the kaolin.

◆ *Hydration*　This is the process where certain minerals absorb water, expand and change; for example anhydrate is changed to gypsum during hydration.

◆ *Oxidation*　This occurs when iron compounds react with oxygen to produce a reddish brown coating.

Biological weathering

Biological weathering involves both mechanical impacts, such as the growth of roots, and chemical impacts, such as the release of organic acids.

▼ **Figure 1.11** Salt crystal growth on coral platform, Devil's Bridge, Antigua

Figure 1.12 The effect of hydrolysis on granite is to form china clay ▼

Factors affecting weathering

Climate

Rate of weathering varies with climate. Peltier's diagram (Figure 1.13) shows how weathering is related to moisture availability and average annual temperature. Frost shattering increases as the number of freeze-thaw cycles increases. Chemical weathering increases with moisture and heat. According to van't Hoff's Law, the rate of chemical weathering increases two to three times for every increase of temperature of 10°C (up to 60°C).

Geology

Rock type influences the rate and type of weathering in many ways due to:

- chemical composition
- the nature of cements in sedimentary rock
- joints and bedding planes.

For example, limestone consists of calcium carbonate and is therefore susceptible to carbonation solution, whereas granite with orthoclase feldspar is prone to hydrolysis. In sedimentary rocks, the nature of the cement is crucial: iron-oxide based cements are prone to oxidation whereas quartz cements are very resistant.

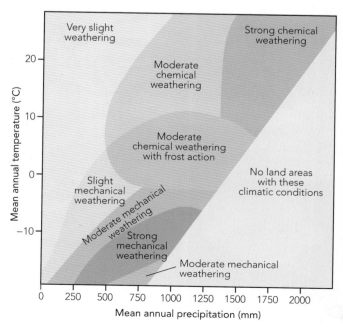

Figure 1.13 Peltier's diagram

1 Define the terms 'weathering' and 'mass movement'.

2 Briefly describe how:

  a freeze-thaw operates

  b carbonation solution operates.

3 Study Figure 1.13. Describe the relationship between chemical weathering and climate; then suggest reasons why there are two zones of moderate mechanical weathering.

4 Outline ways in which human activities might lead to an increase in the frequency and intensity of mass movements.

Factors affecting river flow and velocity

Types of flow

Water flow is subject to two main forces: gravity, which causes downstream flow; and frictional resistance with the bed and bank, which opposes the flow downstream. In addition, the volume of water within a channel and the shape of the channel affect the amount of energy a river has to move water and sediment. The type of flow in a river will affect the types of landforms produced by that river.

Turbulent flow

Water flow is not steady or uniform but turbulent, chaotic and eddying. Turbulence provides the upward motion in the flow that allows the lifting and support of fine particles. This will contribute to depositional landforms further down the river. The conditions necessary for turbulent flow to occur are:

◆ complex channel shapes such as meandering channels and alternating pools and riffles

◆ high velocities

◆ cavitation in which pockets of air explode under high pressure (Figure 1.14).

Key terms

Pools: the deep parts of a river located on the outer bend of a meander.

Riffles: the ridges – often formed of gravel – found in the straight part of a river between two meanders.

Cavitation: the explosion of air in joints and cracks as a result of air at very high pressure.

Laminar flow

By contrast, laminar flow is the movement of water in a series of sheets (or laminae). It is common in groundwater and in glaciers, but not in rivers, although it can occur in the bed in the lower course of a river. The best conditions for laminar flow are:

◆ shallow channels

◆ smooth, straight channels

◆ low velocities.

If laminar flow alone occurred in rivers, all the sediment would remain on the bed.

The highest velocity is found at a depth of $0.3 \times d$ (depth), but does vary depending on channel shape. The mean velocity of a stream is usually given as $0.8 \times$ surface velocity.

Channel shape

The efficiency of a stream's shape is measured by its hydraulic radius, which is the cross-sectional area divided by wetted perimeter (Figure 1.15).

Figure 1.14 Potholes caused by hydraulic action

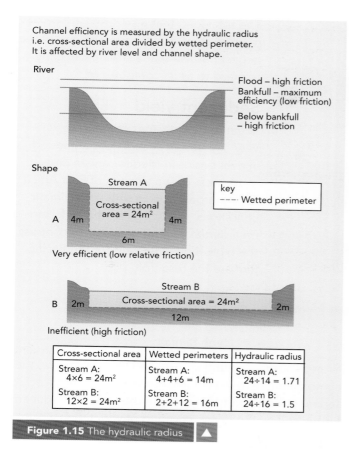

Channel efficiency is measured by the hydraulic radius i.e. cross-sectional area divided by wetted perimeter. It is affected by river level and channel shape.

River

Flood – high friction
Bankfull – maximum efficiency (low friction)
Below bankfull – high friction

Shape

Stream A

Cross-sectional area = 24m²

A 4m 4m

6m

key
--- Wetted perimeter

Very efficient (low relative friction)

Stream B

B 2m Cross-sectional area = 24m² 2m

12m

Inefficient (high friction)

Cross-sectional area	Wetted perimeters	Hydraulic radius
Stream A: 4×6 = 24m²	Stream A: 4+4+6 = 14m	Stream A: 24÷14 = 1.71
Stream B: 12×2 = 24m²	Stream B: 2+2+12 = 16m	Stream B: 24÷16 = 1.5

Figure 1.15 The hydraulic radius ▲

$$\text{Hydraulic radius} = \frac{\text{cross-sectional area}}{\text{wetted perimeter}}$$

The higher the ratio, the more efficient the stream and the smaller the frictional loss. The ideal channel shape is semi-circular.

Key terms

Hydraulic radius: the cross-sectional area of a river divided by the wetted perimeter.

Wetted perimeter: the total length of the bed and base of a river channel in contact with the water in a river.

Channel roughness

Channel roughness causes friction that slows down the velocity of water. Friction is caused by boulders, trees, vegetation, irregularities in the riverbed, and contact between the water and the bed and bank. Manning's formula below describes the relationship between channel roughness and velocity:

$$v = \frac{R^{\frac{2}{3}}S^{\frac{1}{2}}}{n}$$

where v = velocity, R = hydraulic radius, S = slope and n = roughness.

The higher value of n the rougher the bed as shown in Figure 1.16.

Bed profile	Sand and gravel	Coarse gravel	Boulders
Uniform	0.02	0.03	0.05
Undulating	0.05	0.06	0.07
Irregular	0.08	0.09	0.10

▲ **Figure 1.16** Factors influencing degree of roughness in a riverbed

Downstream changes in discharge and channel variables

Further downstream, the flow becomes less turbulent, the bed becomes smoother and consequently velocity increases despite appearances to the contrary (Figure 1.17).

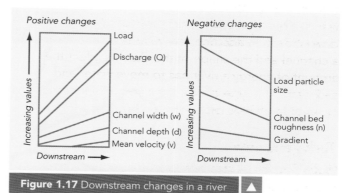

Figure 1.17 Downstream changes in a river ▲

Transport

Erosion by a river will provide loose material. This eroded material (plus other weathered material that has moved downslope from the upper valley sides) is carried by the river as its load. The load is transported downstream in a number of ways (Figure 1.18):

◆ *Suspension* – the smallest particles (silts and clays) are carried in suspension as the suspended load – this is especially important during times of flood (Figure 1.19).

◆ *Saltation* – larger particles (sands, gravels, very small stones) are transported in a series of 'hops' as the saltated load.

◆ *Traction* – pebbles are shunted along the bed as the bed or tracted load.

◆ *Solution* – in areas of calcareous rock, material is carried in solution as the dissolved load.

The material is moved by four main processes:

1 traction – larger pebbles and boulders roll or slide along the river bed. The largest may only be moved during times of severe flood.

2 saltation – bouncing or hopping movement of pebbles, sands and gravel along the river bed. The material is picked up for a short time before being dropped again when the current falls.

3 suspension – material, often the finer sand and silt, carried along in the flow. The suspended load is often the greatest proportion of the total load and in sufficient quantity makes the river look brown. The quantity and size of the material transported in suspension increases with:

• increased volume and velocity of the river

• distance from the source.

4 solution – minerals dissolved in the water such as those contained in limestone rocks.

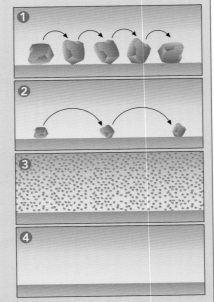

Figure 1.18 Different types of transport in a river ▲

The total load that a stream can carry is referred to as its capacity. The size of the largest particle that a stream can carry is referred to as its competence.

Sediment yield

Sediment yield varies considerably. Very high levels of sediment are found in river basins in Mediterranean areas, south-west USA and parts of East Africa. This is largely due to semi-arid climatic conditions, with irregular rainfall falling on a partial vegetation cover. In contrast, the high sediment levels for the Pacific Rim are the result of

◆ high relief

◆ tectonic activity

◆ high rainfall.

In areas such as central Eurasia and North America, the combination of low relief and resistant rocks reduced sediment levels. In tropical Africa the rates are reduced further still by the dense covering of vegetation which intercepts most of the rainfall.

Key term

Sediment yield: the amount of material (debris) carried away by water. Often measured in t/km/year.

Large amounts of sediment are transported in rivers in south-east Asia and Oceania (Figure 1.20). These areas account for just 15 per cent of the world's land surface but about 70 per cent of the sediment carried by rivers. The highest rates for suspended sediment came from the Huangfuchuan River in China, which had an average sediment yield of 53 000 t/km/year. The reasons for such a huge load are varied and include:

◆ erodable sediments (such as loess, glacial deposits, and volcanic ash)

◆ high relief

◆ tectonic activity

◆ erosive storms (tropical cyclones)

◆ limited vegetation cover

◆ intense human activity, removing vegetation and exposing ground surfaces.

Large amounts of sediment are also carried by mountain streams. In North America, for example, the loads of the Sustitna, Stekine and Yukon together carry more sediment than the Mississippi. Similarly, in

Figure 1.19 High rates of suspended sediment carried by a river in flood ▼

Country	River	Drainage area (km²)	Mean annual suspended sediment yield (t/km²/year)
China	Huangfuchaun	3199	53500
China	Dali	96	25600
Taiwan	Tsengwen	1000	28000
Kenya	Perkerra	1310	19520
Java	Cilutung	600	12000
Java	Cikeruh	250	11200
New Guinea	Aure	4360	11126
New Zealand	Waiapu	1378	19970
New Zealand	Waingaromia	175	17340
New Zealand	Hikuwai	307	13890
UK	Thames	c.50	13500

Source: adapted from A Goudie, *The Changing Earth* (1995)

Figure 1.20 Maximum values of mean annual specific suspended sediments yield for world rivers ▲

Europe, the Semani River of Albania, carries twice as much sediment as the Garonne, Loire, Seine, Rhine, Weser, Elbe and Oder combined. However, it is estimated that of the total sediment carried from land to sea each year – some 2.7–4.6km³ – more than 1.1km³ comes from the Himalayas.

River bank erosion is related to river discharge and width. Rates of erosion may be high enough (Figure 1.21) to cause concern and practical problems for riparian (riverside) communities. Riverbank erosion is an important source of suspended sediment in a river, and is an important part in the development of floodplains (caused by the migration of meanders, Figure 1.27).

Activities

1. Define the term 'hydraulic radius'.
2. Briefly describe the meaning of Manning's formula.
3. What are the main factors that affect global sediment levels in rivers. Use examples to support your answer.

Figure 1.21 Rates of riverbank erosion ▼

River	Dates of survey	Rate (m/year)
Cound, Shropshire	1972–4	0.64
Bollin, Cheshire	1872–1935	0.16
Exe, Devon	1840–1975	2.58
Creedy, Devon	1840–1975	0.52
Culm, Devon	1840–1975	0.51
Axe, Devon	1840–1975	1.0
Yatty, Devon	1840–1975	1.38
Coly, Devon	1840–1975	0.48
Hookamoor, Devon	1840–1975	0.19
Severn	1948–75	0.2–0.7
Rheidol	1951–71	1.75
Tywi	1905–71	2.65

Source: A S Goudie, *The landforms of England and Wales* (1990)

Theory into practice

For a small stream near your school, find out the size of the suspended load, dissolved load (use your school's chemistry department) and the size of the bedload. Compare this with another stream that you visit during fieldwork. Suggest reasons for the differences in the two streams.

Hjulstrom curves

The critical erosion velocity is the lowest velocity at which grains of a given size can be moved so the faster the stream, the larger the particles that can be picked up from the river bed and transported (Figure 1.22). The load of a river varies with discharge and velocity. The relationship between these variables is shown by means of a Hjulstrom curve. For example, sand can be moved more easily than silt or clay, as fine-grained particles tend to be more cohesive. High velocities are required to move gravel and cobbles because of their large size.

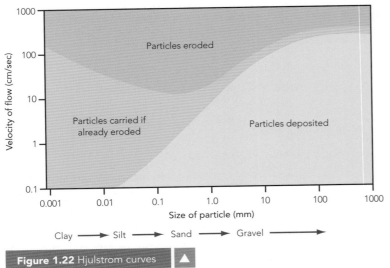

Figure 1.22 Hjulstrom curves

- *velocity* The greater the velocity, the greater the potential for erosion.

- *gradient* Increased gradient increases the rate of erosion.

- *geology* Weak, unconsolidated rocks such as sand and gravel are more easily eroded than consolidated rocks.

- *pH* Rates of solution are increased when the water is more acidic.

- *human impact* Deforestation, dams and bridges interfere with the natural flow of a river and frequently end up increasing the rate of erosion.

Key terms

Entrainment: picking up particles from the ground/bed of a river.

Unconsolidated rocks: loose, broken material that is often easily eroded.

Consolidated rocks: solid rocks or sediments that have been bonded together by a sedimentary cement.

There are three important features on Hjulstrom curves.

- The smallest and largest particles require high velocities to lift them. For example, particles between 0.1mm and 1mm require velocities of about 10cm/sec to be lifted up and carried compared with values of over 50cm/sec to lift clay (0.01mm) and gravel (over 2mm). Clay resists entrainment due to its cohesion; gravel due to its weight.

- Higher velocities are required for entrainment than for transport.

- When velocity falls below a certain level (settling or fall velocity) those particles are deposited.

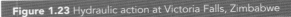

Activities

1 Briefly describe the main types of erosion in a river.

2 Explain the factors that affect the type and intensity of erosion in a river.

Erosion

There are a number of types of erosion in a river.

- *Abrasion* is the wearing away of the bed and bank by the load carried by a river.

- *Attrition* is the wearing away of the load carried by a river. It creates smaller, rounder particles. Hydraulic action is the force of air and water on the sides of rivers and in cracks.

- *Hydraulic action* occurs as a river tumbles over a waterfall and crashes onto and erodes the rocks below (Figure 1.23).

- *Solution* is the removal of chemical ions, especially calcium. It is a form of erosion especially effective on chalk and limestone.

There are a number of factors affecting rates of erosion. These include:

- *load* The heavier and sharper the load, the greater the potential for erosion.

Figure 1.23 Hydraulic action at Victoria Falls, Zimbabwe

Deposition

Deposition occurs when a river looses energy. This may happen because of:

◆ a shallowing of gradient which decreases velocity and energy

◆ a decrease in the volume of water in the river

◆ an increase in the friction between water and the channel

◆ human obstructions such as dams.

Landforms produced by erosion

V-shaped valleys

V-shaped valleys are formed by erosion from a river. Weathering and mass movements occur on the valley sides while the river erodes the base of the slopes. The angle of the V-shape depends on:

◆ the rate of downward erosion by the river

◆ the resistence of the rocks to weathering, mass movements and erosion

◆ climate

◆ the location along the course of the river – rivers in their upper course tend to have steeper V-shaped valleys than those in their lower course.

Graded stream profiles

In general, streams attempt to adopt (by erosion) a concave, upward, long profile, with steeper gradients found in the headwaters and gentler gradients as the stream approaches base level. If this theoretical shape is achieved, the stream would be regarded as having a graded profile (Figure 1.24).

Waterfalls and gorges

Major waterfalls frequently occur on the margins of horizontally bedded rocks. Resistant rock, underlain by weaker rock, is undercut by hydraulic action and

River environments

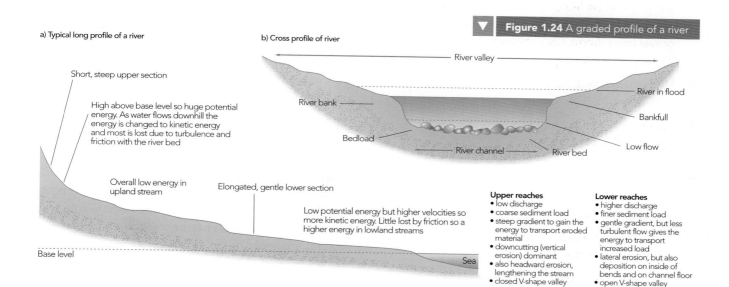

a) Typical long profile of a river

Short, steep upper section

High above base level so huge potential energy. As water flows downhill the energy is changed to kinetic energy and most is lost due to turbulence and friction with the river bed

Overall low energy in upland stream

Elongated, gentle lower section

Low potential energy but higher velocities so more kinetic energy. Little lost by friction so a higher energy in lowland streams

Base level

Sea

b) Cross profile of river

▼ Figure 1.24 A graded profile of a river

River valley

River bank

Bedload

River channel

River bed

River in flood

Bankfull

Low flow

Upper reaches
• low discharge
• coarse sediment load
• steep gradient to gain the energy to transport eroded material
• downcutting (vertical erosion) dominant
• also headward erosion, lengthening the stream
• closed V-shape valley

Lower reaches
• higher discharge
• finer sediment load
• gentle gradient, but less turbulent flow gives the energy to transport increased load
• lateral erosion, but also deposition on inside of bends and on channel floor
• open V-shape valley

Figure 1.25 High Force waterfall **▼**

① undercutting before collapse
② weight of water causes pressure on the unsupported Whin Sill
③ pieces of hard, igneous rock broken away from Whin Sill erode limestone
④ hydraulic action by force of falling water
⑤ organic-rich waters help dissolve the limestone
⑥ brown, peaty waters of the River Tees
⑦ softer limestone
⑧ Whin Sill hard igneous rock
⑨ flow of water over rocks
⑩ boulder-strewn gorge
⑪ waterfall gradually recedes

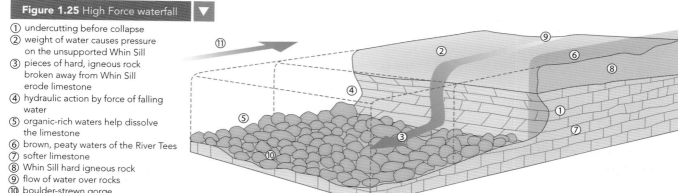

Figure 1.26 High Force waterfall ▲

abrasion in the process of enlarging the plunge pool at the base of the falls (Figures 1.25 and 1.26). Collapse and retreat then occurs. Some waterfalls may be initiated by faulting (e.g. Victoria Falls, Zimbabwe, Zambia border) whereas others are caused by a drop in sea level. Smaller waterfalls are common in alpine glaciated regions where tributaries fall into the main glaciated valley. Waterfalls move upstream as a result of the undercutting and collapse process and leave behind a gorge of recession. A gorge is a narrow, steep-sided valley which can be formed by the retreat of a waterfall (a gorge of recession) such as Niagara Falls on the US/Canada border, by the opening up of a fault line such as Victoria Falls on the Zambezi River, by superimposed drainage, or by the collapse of a cave system.

Waterfalls, gorges and human activity

Advantages

◆ The sudden change in gradient is ideal for the generation of hydro-electricity.

◆ They offer strong scenic and recreational attractions.

◆ The gorges of incised meanders provided safe defensive locations for settlement.

◆ The narrow valleys are relatively easily dammed.

Disadvantages

◆ Transport is severely affected – the valleys are often too narrow to accommodate roads or rail. Crossing gorges makes bridging necessary.

◆ Retreat of water falls limits the life of constructions.

Potholes

A pothole is a hole in the base of a stream or river. It is formed through abrasion by pebbles held up by turbulent flow in the eddies of the stream.

Landforms produced by erosion and deposition

As a stream approaches its base level, usually the sea, its gradient is reduced and it starts depositing its load. Erosion, if it occurs, is mainly lateral. Features commonly associated with this part of a stream include meanders, the flood plain and the delta.

Meanders

A meander is a bend in a river formed by a complex process (Figure 1.27). While meandering is the normal behaviour of fluids and gases in motion, the processes involved are complex. When stream velocity has increased and the load has become finer, in relatively straight channels, meanders develop as this is the only way the stream can use up the energy it now possesses equally throughout the channel reach.

Cross-section of a meander

Meanders have an asymmetric cross-section. They are deeper on the outside bank and shallower on the inside bank. In between, meanders are more symmetrical. They begin with the development of pools and riffles in a straight channel and the thalweg begins to flow from side to side. Helicoidal flow occurs where surface water flows towards the outer banks, while the bottom flow is towards the inner bank. This causes the variations in the cross-section and variations in erosion and deposition. These variations give rise to river cliffs on the outer bank and slip-off slopes (or point bars) on the inner bank.

Figure 1.27 Meanders ▼

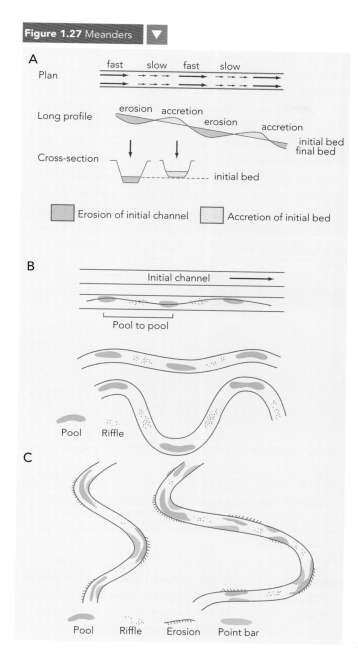

A

Plan — fast slow fast slow

Long profile — erosion accretion erosion accretion initial bed final bed

Cross-section — initial bed

☐ Erosion of initial channel   ☐ Accretion of initial bed

B

Initial channel

Pool to pool

Pool   Riffle

C

Pool   Riffle   Erosion   Point bar

Key terms

Thalweg: the line of maximum velocity in a river.

Helicoidal flow: the corkscrew-like motion that occurs in a meander.

River cliffs: the steep-sided slopes on the outer bend of the meander.

Slip-off slope: the gentle slope deposited on the inner bend of a meander.

Channel characteristics and meanders

◆ The curved part of the meander (the meander wavelength) is generally 6–10 times the width of the river channel and/or the discharge.

◆ Meandering is more pronounced when the bed load is varied.

◆ Meander wavelength increases in streams that carry coarse debris.

◆ Meandering best develops at or near the bankfull stage.

Key term

Bankfull stage: where the channel variables (depth, width and velocity) of a river are unable to cope with additional water.

Oxbow lakes

Oxbow lakes are the result of erosion and deposition (Figure 1.28). Lateral erosion, caused by centrifugal forces, is concentrated on the outer, deeper bank of a meander. During times of flooding, erosion increases causing the river to break through and create a new, steeper channel. In time, the old meander is closed off by deposition to form an oxbow lake.

Key term

Centrifugal: outward-moving force.

Figure 1.28 Oxbow lakes ▼

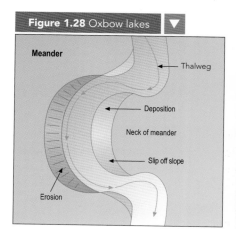

Meander — Thalweg — Deposition — Neck of meander — Slip off slope — Erosion

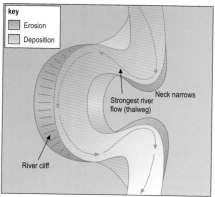

key
☐ Erosion
☐ Deposition

Neck narrows — Strongest river flow (thalweg) — River cliff

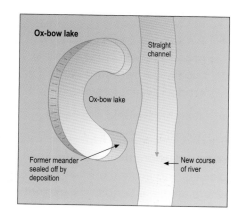

Ox-bow lake — Straight channel — Ox-bow lake — Former meander sealed off by deposition — New course of river

Activities

1 Describe how the net erosion curve (Figure 1.22 Hjulstrom curves) varies with (a) velocity and (b) particle size.

2 With the use of an annotated (labelled) diagram explain how waterfalls are formed.

3 Briefly explain how meanders are formed. You may use a diagram to help explain your answer.

Landforms produced by deposition

Braided channels

A river becomes braided when the main channel separates into a number of smaller interlocking channels (Figure 1.29). The channel is now highly unstable and its form changes constantly. Braiding tends to occur when the river does not have the capacity to transport its load in a single channel, be it straight or meandering.

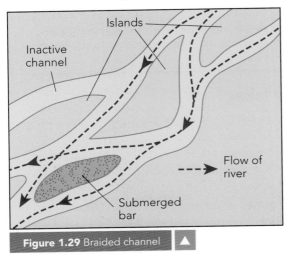

Figure 1.29 Braided channel

A number of conditions lead to braiding.

◆ A channel gradient that is slightly steeper than that of a meandering stream.

◆ A load that contains a large proportion of coarse material, usually derived from easily eroded bank material.

◆ A highly variable discharge.

Braided streams are especially common where they drain from glaciers and in semi-arid areas where all these conditions are met.

Floodplains

Increased discharge as a result of heavy precipitation or snow melt could lead to the bankfull stage. The channel variables (depth, width and velocity) become unable to cope with the additional water and the river spills out of the channel into the surrounding area. Deposition results as velocity drops to compensate for the increase in width. Repeated flooding and deposition leads to the build up of alluvium and forms the floodplain, a low-relief feature filling the original river valley (Figure 1.30). Because of their low gradients, meanders are a characteristic of rivers on floodplains. Floodplains are also formed by the erosion of bluffs (steep banks or cliffs).

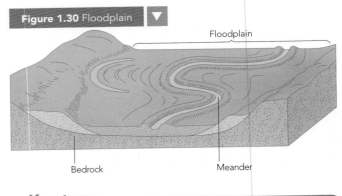

Figure 1.30 Floodplain

Key term

Alluvium: fine sediments of silt and clay transported in suspension.

Levees

Levees are raised ridges running parallel to the edge of the channel and formed by repeated flooding of a river (Figure 1.31). When the river floods, its velocity drops as the floodwaters spill out of the channel therefore, the heaviest and coarsest sediments are the first to be deposited and over time they will build up

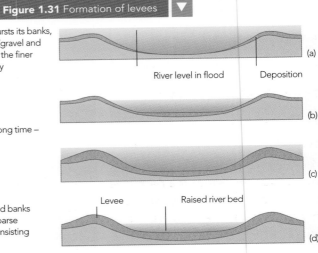

Figure 1.31 Formation of levees

(a) When the river floods, it bursts its banks, depositing the coarsest load (gravel and sand) closest to the bank and the finer load (silt and clay) further away

(b), (c) This continues over a long time – centuries

(d) The river has built up raised banks called levees, consisting of coarse material, and a floodplain, consisting of fine material

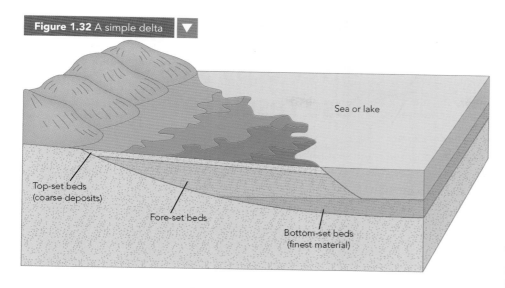

Figure 1.32 A simple delta ▼

Labels on figure:
Sea or lake
Top-set beds (coarse deposits)
Fore-set beds
Bottom-set beds (finest material)

can also be found along the course of a river (damming, small lakes, resistant layers of rock). The river, in its attempt to become graded (Figure 1.24), will develop low gradients as it approaches any base level.

Base levels, however, are not static and do change. Uplift or a fall in sea level will produce a negative change of base level, while coastal submergence produces a positive change in base level.

into levees. The levees, therefore, consist of coarse material, such as sand and gravel, while the floodplain consists of fine silt and clay.

Deltas

For deltas to be formed a river needs to:

◆ carry a large volume of sediment

◆ enter a still body of water such as a sea or lake – if there were strong currents, the sediments would be carried along the shoreline as longshore drift.

Deposition is increased if the water is salty, as this causes salt particles to group together, become heavier, and be deposited. This process is known as **flocculation**. Vegetation also increases the rate of deposition by slowing down the water.

The coarser material is deposited first and the finest material last and furthest away (Figure 1.32).

> ┌─ **Key term** ─────
>
> **Flocculation:** the process in which clay and silt particles combine when fresh water mixes with the salt water. The heavier particles sink more rapidly.

Features associated with negative changes in base level

As the potential energy of the river increases due to the greater difference in altitude between the source and the base level, the river responds by increasing its velocity and downcutting (a process sometimes referred to as rejuvenation). The following features arise from this process.

◆ *Intrenched/incised meanders*   Meanders that develop on a floodplain will maintain their form as the river cuts down towards its new base level. Where this process has continued for some time, the river will flow in a meander gorge.

◆ *River terraces*   When a river cuts down into its floodplain, the remnants of the floodplain form a terrace with a steep slope (Figure 1.33). Periodic changes in base level will form a sequence of

Effects of base level changes on the formation of features

The processes of erosion and deposition in a river are controlled by location relative to the base level of the river. This is usually the sea, but could be a lake in an area of inland drainage. Local base levels

▼ **Figure 1.33** Terraces formed by rejuvenation with knickpoints at K_1 and K_2

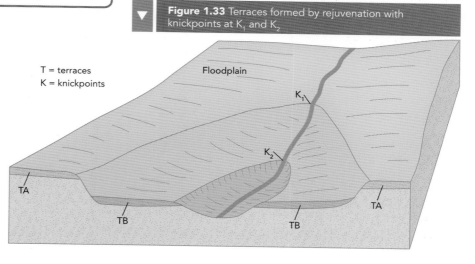

T = terraces
K = knickpoints

Labels on figure: Floodplain, K_1, K_2, TA, TB, TB, TA

terraces, with the oldest terraces at the top and the youngest being the lowest. These terraces may be paired (when downcutting keeps pace with the rate of change of base level) or unpaired (when there is an opportunity for lateral erosion by the river between incidences of base level change).

♦ *Knickpoints*   The long profile of a river is never smoothly concave, but shows many sudden changes in gradient. These knickpoints occur where the river changes from deposition to erosion and are caused by the negative changes in local base levels due to lithology or human actions such as damming or deforestation.

> **Key term**
>
> **Lithology:** the properties of rocks – mineral composition, cracks, joints.

Activities

1 What is a river terrace? How can terraces be formed?
2 How are deltas formed? Suggest why deltas are both attractive and hazardous as a place to live.
3 With the use of diagrams explain how levees are formed.

'Take it further' activity
1.2 on CD-ROM

1.2 In what ways can river basins be a multi-use resource?

Drainage basins

A drainage basin is the catchment area within which water supplied by precipitation or by underground sources is transferred out via a stream. Every stream has its own drainage basin, separated from an adjacent drainage basin by a watershed or divide. The drainage basin of a river with tributaries includes all the smaller drainage basins of the tributaries. Drainage basins are open systems. There are factors which affect them, processes which operate in them, and results (landforms) which are created.

> **Key terms**
>
> **Open systems:** systems with inputs and outputs.
>
> **Watershed** or **divide:** the 'line' dividing two drainage basins.

River basins are important to people for a variety of reasons. They provide:

♦ a source of water
♦ opportunities for industrial development
♦ opportunities for residential development
♦ fertile silt for agriculture
♦ a means of transportation
♦ a source of power
♦ a place for recreation and leisure
♦ a means of flood defence through floodplains
♦ a conservation value.

There are many examples of drainage basins which illustrate some, or all, of these uses and the conflicts they can lead to.

Case study | The Thames Basin

The Thames Region comprises the main drainage basins of the Thames and its tributaries. It is the most developed part of the UK, with a population of about 12 million, a fifth of the nation's population. It covers an area of over 13 000km², and includes 14 counties, 58 district councils and 33 local planning authorities. The region comprises the main drainage basin of the River Thames and its tributaries such as the Cole, Lee, Kennet, Wey and Loddon (Figure 1.34). Much of the region, particularly in the west, is rural in character, where the dominant land use is agriculture (Figure 1.35). In contrast, the lower reaches of the Thames Basin, in the east, are more urbanised and industrialised.

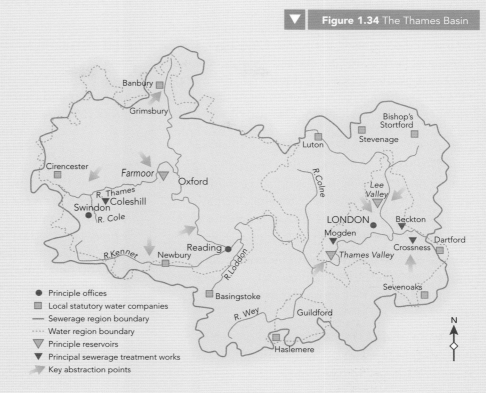

Figure 1.34 The Thames Basin

- ● Principle offices
- ▪ Local statutory water companies
- — Sewerage region boundary
- ⋯ Water region boundary
- ▽ Principle reservoirs
- ▼ Principal sewerage treatment works
- ➤ Key abstraction points

Figure 1.35 Rural landscape – the River Cole near Coleshill

Water supply

The Thames Region is one of the most intensively managed catchments in the world. Every day approximately 4700 million litres of water is abstracted from the region's rivers and groundwater. Rain falling in the Cotswolds can be used up to eight times before it reaches the Thames Estuary. Overuse of water has had a disastrous effect on some rivers. For example, there are five rivers with very low flow conditions in the Thames Region. In addition, the region has seen continued growth in housing and commercial development and in mineral extraction, increasing pressure on land use, water resources and the water environment generally.

Over the past 30 years, demand for public water supplies has increased by approximately 1.7 per cent each year. The key factors which have influenced demand are:

- the use of water in the home and garden
- losses through leakage from distribution systems and consumers' plumbing
- population growth and household size
- development pressure and economic activity.

Industrial development

Geologically, the area contains much chalk, limestone, sand and gravel, which creates opportunities for mineral extraction. However, the main use of rivers, as far as industry is concerned, is that they allow for the import and export of raw materials and finished goods. Many large-scale traditional heavy industries, such as car manufacturing and iron and steel works, located in the flat floodplains because land was cheap, there was water for cooling purposes, and the river for trade. The former industries in London Docklands are a good example from the past. The industries and refineries located at Tilbury and Canvey in the Thames Estuary are good current examples. On a much smaller scale, the river and floodplain at Osney in Oxford has attracted port industries, and modern industries such as warehousing, printing and publishing (Figure 1.36).

Residential development

Most major settlements are built by rivers, especially those in their lower course. This is due to the many varied benefits that rivers bring to settlements – a supply of drinking water, good potential for trade and communications, relatively flat sites for residential and

Figure 1.36 Industrial development near Osney Island, Oxford ▼

Figure 1.37 Aerial view of Oxford to show how the rivers have affected the growth of the city ▲

industrial development, and, in the past, fertile silt which is good for farming.

Many settlements are built on raised ground in order to reduce the risk of flooding. Oxford is an excellent example within the Thames Basin. Much of the floodplain of the Thames and the Cherwell has not been built upon (Figure 1.37). It has been left for farming and for recreational use. Housing and industry have tended to locate on the higher ground free from flooding (Figure 1.38).

However, there are some developments which appear to be at risk from flooding. The Thames Gateway initiative in the east of London and neighbouring counties is a national priority area for regeneration. There are 14 zones of change planned, with forecasts of housing growth from 2001 to 2016 of about 100 000 and employment growth of about 300 000. The proposed development in the Thames Gateway area is largely in areas which are at risk of tidal flooding.

Much of the conflict in the Thames Drainage Basin relates to the demand for more housing. The Thames Gateway is a case in point. The mudflats and salt marshes of the Thames Estuary are important ecological areas. The Thames floodplains provide important wetlands: nevertheless, there are plans for the building of many new settlements in the area.

▼ **Figure 1.38** Land use on Oxford's floodplains

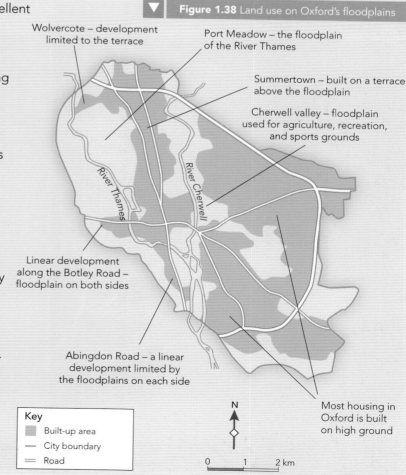

Wolvercote – development limited to the terrace

Port Meadow – the floodplain of the River Thames

Summertown – built on a terrace above the floodplain

Cherwell valley – floodplain used for agriculture, recreation, and sports grounds

River Thames

River Cherwell

Linear development along the Botley Road – floodplain on both sides

Abingdon Road – a linear development limited by the floodplains on each side

Most housing in Oxford is built on high ground

Key
- Built-up area
- City boundary
- Road

N

0   1   2 km

The new settlement will have 120000 homes below sea level. Among the schemes being considered to protect homes are raising homes on stilts; building roads on embankments above the flood level; and sacrificing large areas of Kent and Essex farmland to floodwaters during a tidal surge to prevent London defences being overwhelmed.

The reason for the development is the high demand for housing in south-east England. This is due to the large number of jobs in the London region, the continued in-migration of migrants from other regions of the UK and from overseas, and the desire to redevelop the former industrial part of the Thames Gateway.

Farming

The lower parts of river valleys are attractive to farmers due to the supply of water. In addition, the relatively gentle slopes allow for the use of machines. In contrast, in the upper regions of the drainage basin, conditions are generally cooler, wetter and steeper. Hence arable farming is less suited. Moreover, for areas that flood, livestock farming may be appropriate since animals can be led to dry ground during times of flood. In much of the Lower Thames Basin arable farming predominates, although in areas close to large, wealthy urban markets, market gardening is common.

Transportation and trade

Many of the world's great cities lie on important trade routes. London has the Thames, and Cairo has the Nile (at the confluence of the European–African and Asian trade routes). With trade there are jobs in transport, storage and logistics, and there are multiplier effects such as the demand for housing for workers and retail, education and health services. The more trade there is, the greater the number of jobs generated. For a city such as London, that was the centre of an Empire, the amount of trade it generated was out of all proportion to its already very large size.

Moreover, river valleys, especially those in the lower course, are relatively flat and make transport routes easy to build. The main railway line from Oxford to London, for example, follows the River Thames and passes through the cutting that the river makes through the Chilterns at Goring.

Figure 1.39 Punting in Oxford ▼

Energy development

The Thames is not used for hydroelectric power. This is because it is not steep enough and there is not a sufficient 'head' of water to drive turbines. However, many of the tributaries of the Thames have been used for water power in the past. The River Cole at Coleshill near Faringdon, for example, has a mill that was used for grinding grain. In London, the River Wandle, which flows from Beddington near Croydon northwards to the Thames, had over 30 mills.

The Thames is also used indirectly to generate power. Much of London's waste is carried by barge to incinerators in the east of London which convert municipal waste into electricity. The east of London has been described as the area where London 'generates its heat and buries its waste'.

Recreation and leisure

Rivers offer many opportunities for leisure and recreation, and the Thames is no exception. It is used for fishing, swimming, canoeing, rowing, sailing, cruising, guided tours, and walking and hiking. In the Oxford region, for example, the valleys of the Cherwell and Thames are used for sports grounds, farmland, the city's botanical gardens and allotments, while the rivers are used for punting (Figure 1.39) and water sports.

The Cotswold Water Park is an extremely important recreational resource within the Thames Basin. The park is made up of many lakes which are used for a wide range of water-based recreation and other forms of recreation attracting many visitors. The Thames Path runs through the area.

Flood defences

The Thames Barrier provides a high standard of protection to the 420 000 London properties at risk from Thames tidal flooding (Figure 1.40). Beyond 2030, upgrading of the Barrier and associated defences will be necessary at a currently estimated cost of £4 billion. The Thames Barrier became operational in October 1982. On average the Barrier has to close three times per year. However, in 2000/01 there were 24 Barrier closures.

The floodplains of the River Thames, and its numerous tributaries, provide an important flood alleviation and strategic facility which reduces discharges in the floodplain prior to reaching major urban areas downstream. One of the main flood alleviation 'facilities' for Oxford is Port Meadow (Figure 1.41). On a local scale, the restored meanders and floodplain of the River Cole act as an important local flood relief measure.

Figure 1.40 The Thames Barrier

Figure 1.41 Flooding on Port Meadow, Oxford ▲

Conservation

In the upper reaches of the Thames Basin, the Cotswold Water Park (CWP) (Figure 1.42) is considered a pressure point because of the variety of developments which focus on this environmentally sensitive area. Minerals extraction in particular is focused on the CWP. However, it must be realised that the Park would not exist without its history of mineral workings. The Park area also supports a wide diversity of wildlife and habitat features of acknowledged national and international importance. Water areas are ecologically important for wildfowl and a number of wet meadows have been designated as Sites of Special Scientific Interest (SSSIs).

Key term

Sites of Special Scientific Interest (SSSI): an area of land of special interest because of its flora, fauna, geological or physiographic characteristics. There are about 6000 SSSIs in Britain. They should be protected from development.

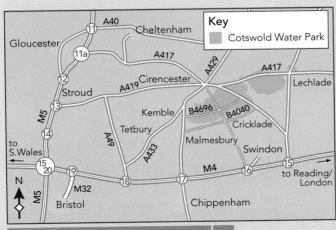

Figure 1.42 The Cotswold Water Park ▲

Other parts of the drainage basin that attract considerable attention are mudflats. These are seen by some as an ugly eyesore, ripe for industrial development. However, increasingly, as society reappraises the value of wetlands, mudflats and estuaries are seen as important 'hotspots' for biodiversity. The mudflats and salt marshes of the Thames Estuary are important ecological havens, including the Thames Estuary and Marshes Special Protection Area (SPA)/Ramsar site, and Benfleet and Southend Marshes SPA/Ramsar site (a Ramsar site is a wetland of international importance).

Key term

Biodiversity: the range of organisms present in a particular ecological community or system. It can be measured by the number of different species or the variation within or between species.

Activities

1 Using examples, suggest contrasting ways in which rivers can affect the location of settlements.

2 Comment on the potential conflicts in the use of river environments.

3 Outline the opportunities and constraints of utilising either mudflats or the upper course of a river.

4 Outline some of the ways in which the Thames has been useful as a resource to people living in the region.

Discussion point

Consider your home area. To what extent is it possible to balance the environmental concerns of river basin management with the economic needs of using the drainage basin?

Theory into practice

How is the drainage basin in which you live used by people? Build specific named examples that you could use in an exam.

The Mekong is south-east Asia's largest river (Figure 1.43) and the world's eighth largest. Its value as a resource is immense; from water supply to industrial and recreational uses. However there are six countries who share the Mekong and there are conflicts between them over the use of this resource. Unusually for such a large river, the largest city along it, Phnom Penh, has just 1.1m inhabitants. This also makes the river unusual in another respect: the pressure of a burgeoning population and fast economic growth is only just beginning to make an impact on the Mekong.

Transportation

Transport development along the Mekong is limited because the river is not navigable much beyond Phnom Penh. In the dry season, when the river is low, there are reefs and shifting sandbars. When the water level rises, the many rapids of Si Phan Don, or 'Four Thousand Islands', form an obstacle to shipping.

Industrial and economic development

Although industrial development in the region was limited in the early twentieth century, it has accelerated rapidly since. The Mekong river basin is located in the fast-emerging Pacific Rim, and the governments of China and Vietnam, in particular have been keen to develop their national economies.

The first dam on the river, at Man Wan, in China, was not completed until 1993. The first bridge across the lower Mekong (that is, outside China) was built in 1994, between Vientiane in Laos and Nong Khai in Thailand. The population of the area is growing rapidly but economic growth is increasing even faster (Figure 1.44).

Residential development

The Mekong river basin is home to some 55 million people and each year they have to contend with wide-spread flooding. In 2000, over 800 people died and many more people were displaced. The population is predicted to rise and this is going to place further demands on the water supply and infrastructure. The disposal of raw sewage is already a challenge with much of it finding its way into the river system.

Recreation and leisure

As one of the fastest growing tourist destinations in the world, the Mekong river basin is benefiting from economic growth. However the development associated with the expansion of tourism has the potential to conflict with requirements of the local residents and traditional activities such as fishing.

The region is receiving funding from the Asian Development bank to help conserve the natural environment and implement sustainable development strategies.

	Population growth 2007 (%)	Economic (GDP) growth 2006 (%)
Cambodia	1.73	7.2
China	0.61	10.7
Laos	2.37	7.4
Myanmar	0.81	3.1
Thailand	0.66	4.8
Vietnam	1.00	8.2

▲ **Figure 1.44** Population and economic growth in the Mekong region, 2005

▼ **Figure 1.43** The Mekong Basin

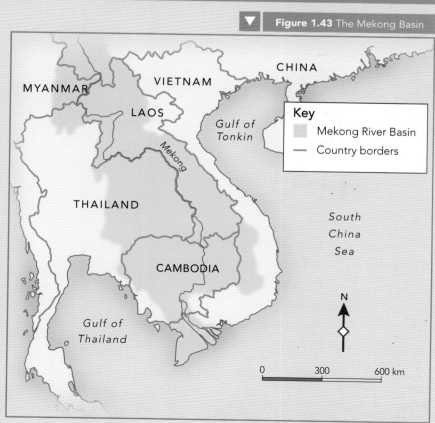

Energy development

The hydroelectric potential of the Mekong and its tributaries is considerable, mainly due to the steep relief of the area and the large volumes of water transported by the rivers, and its tributaries, which are mostly untapped. So far only 5 per cent (1600MW) of the Lower Mekong Basin's (LMB) hydroelectric potential of approximately 30000MW has been developed and the projects have all been on the tributaries.

These dams generate valuable electricity, aid irrigation and regulate flooding. However, in the process they have caused irreparable damage to what was, until recently, the Mekong's most valuable resource: its fisheries. In addition dams built further up-stream e.g. by China, siphon off water needed by countries such as Laos.

Farming and water supply

Farming development in the Mekong has been hampered by the large seasonal floods that occur each year. Nevertheless, about 80 per cent of rice production in the Lower Mekong Basin depends on water, silt and nutrients provided by the flooding of the Mekong. The Mekong also supports very rich fishing grounds. Dams on the upper Mekong affect water flow in the lower Mekong during the dry season, changing the natural cycle of the river. The dams could mean less frequent floods, adversely affecting farming and fishing.

Fishing and water supply

The Mekong and its tributaries yield more fish than any other river system. The annual harvest, including fish farms, amounts to about 2 million tonnes, or roughly twice the catch from the North Sea. The Mekong is home to over 1200 different species of fish, more than any other river save the Amazon and the Congo. Over 1 million people in Cambodia depend solely on fishing to make a living, while in Laos 70 per cent of rural households supplement their income by fishing.

The abundance of fish stems from the Mekong's seasonal ebb and flow. During the monsoon, the habitat for fish suddenly increases by as much as ten times. Moreover, much of the floodplain is forest, which provides leaves for the fish to feed on. They spawn at the end of the dry season, so that the coming floods can carry the fish to the floodplain. The bigger the flood, the greater the leaves on offer, and so the fatter and more numerous the fish.

More dams, however, mean smaller floods. Most hydroelectric plants aim to generate the same amount of energy all year round. That requires a consistent flow through the turbines, which in turn requires rainwater to be held in a reservoir for use in the dry season.

The rapid development of the Mekong river basin can be seen to be having a positive impact on energy production and industrial production, but is having a very negative impact on the two main traditional industries of the area: farming and fishing.

Country	Name	Location	Capacity (MW)	Output (GWh/year)	Date commissioned
China	Manwan	M	1500	7870	1993
	Dachaoshan	M	1350	5930	2001
Laos	Nam Ngum	T	150	900	1971–85
	Xeset	T	45	150	1991
	Theun Hinboun	T	210	1645	1998
	Houay Ho	T	150	600	1999
	Nam Leuk	T	60	184	2000
Thailand	Sirindhorn	T	36	115	1968
	Chulabhorn	T	15	62	1971
	Ubolratana	T	25	75	1966
	Pak Mun	T	136	462	1997
Vietnam	Dray Ling	T	13	70	1995
	Yaly	T	720	3642	2000

Key: M = Mekong; T = Tributary

Figure 1.45 Completed hydroelectric power stations on the Mekong

Activities

1 Suggest reasons why there is a need to develop the Mekong.

2 Explain why fishing is so important on the Mekong River.

3 With the use of an atlas, suggest why the Mekong has considerable potential for hydro electric power.

4 Comment on two reasons why it is more difficult to manage the Mekong than the Thames.

Discussion point

Are people more dependent on rivers in LEDCs or MEDCs?

1.3 What issues can arise from the development of river basins?

Case study | The 2007 floods in southern England

There are many reasons why drainage basins are vulnerable to flooding. Factors promoting flooding include:

- areas of intense convectional rainfall
- prolonged heavy rain
- seasonal snow melt
- small basins in which run-off is rapid
- steep slopes which lead to rapid overland flow
- areas where the natural vegetation has been replaced
- basins with large areas of impermeable surfaces due to urbanisation and industrial development.

The torrential downpours which hit central and western England in July 2007 were officially the worst in over 200 years. Rainfall was more than double the seasonal average. The flooding started in June, when a month's rain fell in a day in South Yorkshire and the Severn Valley. In East Yorkshire the entire average for June came down in an hour. Water overwhelmed drains, drowning three people in the initial onslaught. A Victorian dam cracked at Ulley near Rotherham, and only 48 hours of shoring and pumping saved three villages. Within two days vast runoff had burst banks on the Don, Hull and Severn. In Hull, 20 000 homes were damaged and an estimated 7000 people were still staying in hotels and emergency centres one month later. In Doncaster 5000 properties were damaged and in Sheffield there was £30 million worth of damage. Some 27 000 homes were damaged in total

across the region. Rain continued, leaving lakes of filthy water contaminated with drain refuse and sewage. The national death toll reached seven (Figure 1.46).

Discussion point

To what extent are people the victims and/or the causes of floods in the UK?

Gloucestershire and neighbouring areas experienced more than 320 per cent more rainfall than the average for the previous three decades, Figure 1.46 overleaf details the impacts of the flooding. Over 140 000 people were without water for ten days when the Mythe water treatment plant was damaged by the flooding.

Brize Norton in Oxfordshire received 121.2mm of rain in just 17 hours, three times the amount it would expect for the month of July. Across Oxfordshire more than 900 properties were flooded and 650 people forced to leave their homes. Over 90 residents were evacuated from Osney Island and taken to Oxford United's Kassam Stadium. The cost to repair the damage was put at about £43 million.

The unusual position of the jet stream has been blamed for the bad weather. It normally lies over the North Atlantic holding back bad weather while allowing warm weather to move up from Europe. But it moved several hundred miles to the south, possibly due to a La Nina event, which occurs when cool water surges from the bottom of the Pacific and cools the air above.

Figure 1.46 The impacts of the 2007 floods in Southern England ▶

£3 billion	Estimated cost of flood damage covered by insurance
£3 billion	Estimated cost of flood damage not covered by insurance – only 25 per cent of UK residents have home contents protection
500 000	Houses affected nationwide
10 000	Homes wrecked in Hull
340 000	People in Gloucestershire left without running water for over a week
387.6mm	Average rainfall across England and Wales from May to 22 July, more than twice the usual average
£600 million	Existing budget for flood protection
£14 million	Cut from the Environment Agency's flood protection budget in 2006
£1bn	Figure the Environment Agency says it needs per year
15 per cent	Real-terms cut in flood defence budget since 2000
25	Warning reports ignored by the government since 1997 on need for UK flood defences to be improved
30	How many more times costly it is on average to clear up your house after a flood than a burglary

Figure 1.47a The floods in numbers ▲

Activities

1 Outline the ways in which human activities within river basins may increase the risk and impact of flooding.

2 Using Figures 1.46 and 1.47a suggest reasons why the floods in 2007 were so devastating.

3 Briefly explain why flood hazards in large catchments differ from those in small catchments. Use examples to support your answer.

Activity

Study the information on the 2007 floods. Classify the impacts into environmental, social and economic.

Figure 1.47b Flooding in Gloucestershire ▶

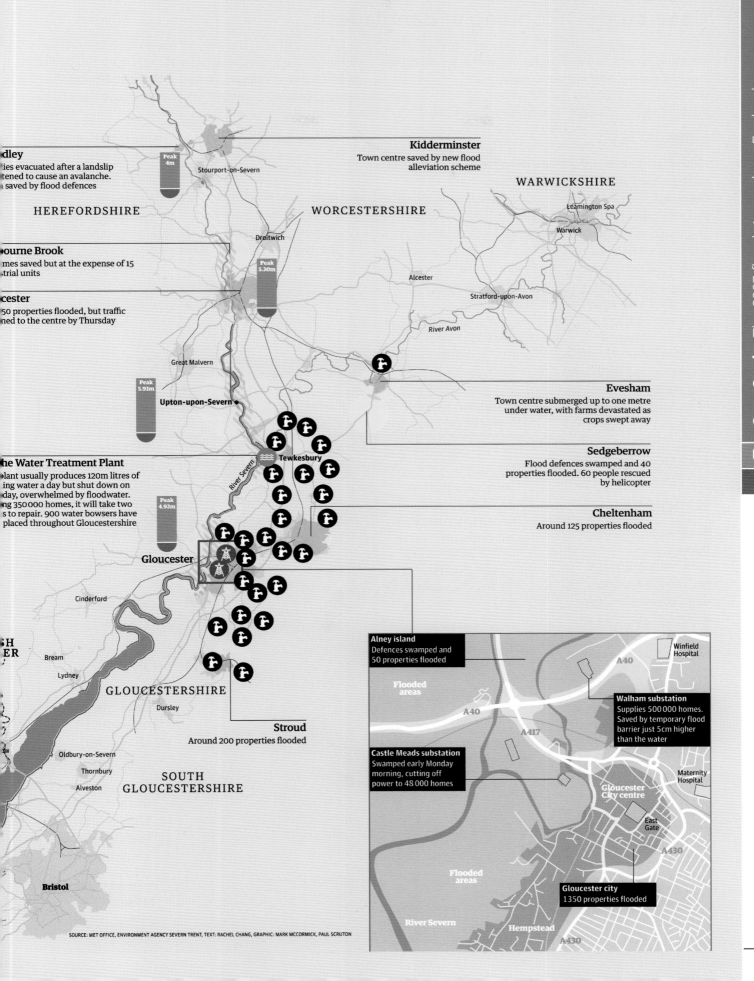

...dley
...ies evacuated after a landslip
...tened to cause an avalanche.
...a saved by flood defences

HEREFORDSHIRE

...ourne Brook
...mes saved but at the expense of 15
...trial units

...cester
...50 properties flooded, but traffic
...ned to the centre by Thursday

...he Water Treatment Plant
...lant usually produces 120m litres of
...ing water a day but shut down on
...day, overwhelmed by floodwater.
...ng 350 000 homes, it will take two
...s to repair. 900 water bowsers have
...placed throughout Gloucestershire

Gloucester

Cinderford

Bream

Lydney

GLOUCESTERSHIRE

Dursley

Oldbury-on-Severn

Thornbury

Alveston

SOUTH
GLOUCESTERSHIRE

Bristol

Peak
4m

Stourport-on-Severn

Droitwich

Peak
5.30m

Great Malvern

Peak
5.93m

Upton-upon-Severn ◆

Peak
4.92m

River Severn

Tewkesbury

Kidderminster
Town centre saved by new flood
alleviation scheme

WARWICKSHIRE

Leamington Spa

Warwick

WORCESTERSHIRE

Alcester

Stratford-upon-Avon

River Avon

Evesham
Town centre submerged up to one metre
under water, with farms devastated as
crops swept away

Sedgeberrow
Flood defences swamped and 40
properties flooded. 60 people rescued
by helicopter

Cheltenham
Around 125 properties flooded

Stroud
Around 200 properties flooded

Alney island
Defences swamped and
50 properties flooded

Flooded
areas

A40

A40

A417

Castle Meads substation
Swamped early Monday
morning, cutting off
power to 48 000 homes

Winfield
Hospital

A40

Walham substation
Supplies 500 000 homes.
Saved by temporary flood
barrier just 5cm higher
than the water

Maternity
Hospital

Gloucester
City centre

East
Gate

A430

Flooded
areas

Gloucester city
1350 properties flooded

River Severn

Hempstead

A430

SOURCE: MET OFFICE, ENVIRONMENT AGENCY SEVERN TRENT, TEXT: RACHEL CHANG, GRAPHIC: MARK MCCORMICK, PAUL SCRUTON

Case study | Flooding in Bangladesh

Flooding is common in Bangladesh and it is caused by a mix of natural and human factors.

Natural causes of flooding in Bangladesh

Almost all Bangladesh's rivers have their source outside the country. For example, the drainage basin of the Ganges and Brahmaputra cover 1.75 million km² and includes parts of the Himalayas and Tibetan Plateau and much of northern India (Figure 1.48). Furthermore, more than half the country is less than 5m above sea level. Total rainfall within the Brahmaputra–Ganges–Meghna catchment is very high and seasonal – 75 per cent of annual rainfall occurs in the monsoon between June and September. During the monsoon season these rivers burst their banks and flood the land. Bangladesh is also subject to coastal flooding; storm surges caused by intense low-pressure systems are funnelled up the Bay of Bengal.

Human causes of flooding in Bangladesh

Bangladesh is a very densely populated country (over 900 people per km²) and is experiencing rapid population growth (nearly 2.7 per cent per annum). It contains nearly twice as many people as the UK, but is only about half its size. Hence, the flat, low-lying delta is vital as a place to live as well as a place to grow food. Over three-quarters of the population live within the world's largest delta, an area of land formed by the meeting of the Ganges, the Brahmaputra and the Meghna rivers (Figure 1.48).

Although the main cause of flooding have traditionally been excessive rain associated with the monsoon season, a series of human actions has made the situation much worse.

◆ *Deforestation* The removal of many trees in the Himalayan uplands has led to more overland flow and less infiltration of water.

◆ *River diversion* India's River Ganges has been diverted for irrigation purposes. This has stopped the supply of river sediment. Effectively, Bangladesh is sinking each year.

Figure 1.48 The rivers of Bangladesh ▼

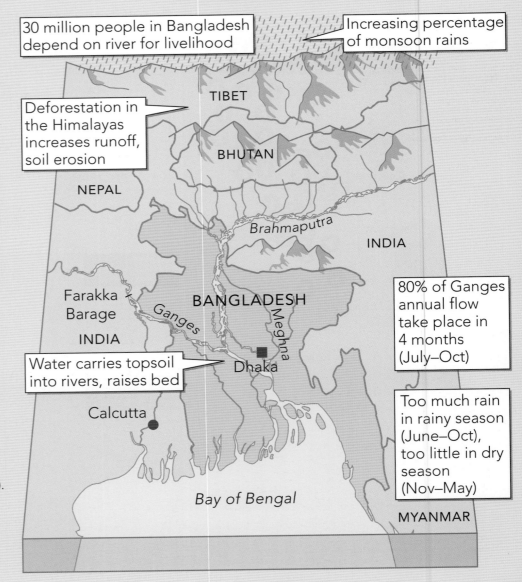

30 million people in Bangladesh depend on river for livelihood

Increasing percentage of monsoon rains

Deforestation in the Himalayas increases runoff, soil erosion

80% of Ganges annual flow take place in 4 months (July–Oct)

Water carries topsoil into rivers, raises bed

Too much rain in rainy season (June–Oct), too little in dry season (Nov–May)

- *Fresh water wells*   In the 1980s, more than 100 000 tube wells and 20 000 deep wells were sunk into the delta to provide drinking water. These wells have reduced the water table and added to the country's subsidence at a rate of 2.5cm per year.

- *International loans*   The Bangladeshi government often has to spend its money on repaying debts to other countries rather than on social schemes such as better flood protection.

- *Lack of international investment*   Few overseas companies can see any gain from investing in long-term social projects that would help Bangladesh.

- *Overseas pressure*   The Bangladeshi government is under pressure to invest in industry at the expense of social schemes.

- *Inadequate overseas aid*   Many of the flood schemes set up by the World Bank have been ineffective.

- *Corruption*   Funds for flood protection might have been diverted into the private funds of government officials.

Floods are vital to Bangladesh. They provide much-needed nutrients to the soil and also replenish the groundwater reserves (water in the underlying rocks); however, the benefits also have to be balanced with the many negative impacts of flooding as seen in 1998 and again in 2007.

1998 floods

In 1998, floods destroyed much of Bangladesh and its economy. Two-thirds of the country was submerged by water and Dhaka, the capital, was cut off. Over 23 million people were made homeless, more than 130 000 cattle were killed, at least 660 000 hectares of crops were badly damaged, and it is thought that the country's entire stock of rice was wiped out.

Almost one-quarter of children under the age of five were malnourished and widespread starvation was forecast. In Dhaka the sewage system collapsed and the drinking water supply was contaminated. Diarrhoea and dysentery affected the many people who came into contact with contaminated water.

Bangladesh's economy was also badly affected. The incomes of two of the largest industries, textiles and shrimp production, were about 20 per cent below their normal earnings. The textile industry employs more than 1.5 million people, mostly living in low-lying areas around the cities. After the floods, 400 factories were forced to close, leaving 166 000 workers (of whom 130 000 were women) jobless. More than 11 000km of roads were damaged in the floods, leaving communications in a state of chaos. More than 1000 schools were destroyed, and students lost all their books. Teachers believe that it will take years for them to replace learning materials.

International aid meant that the disaster did not become a catastrophe. The Bangladeshi government bought 350 000 tonnes of cereals from Asian countries and a further one million tonnes of international food was provided. The World Health Organisation launched an appeal for medicine and water purification tablets.

2007 floods

The monsoon rains of 2007 were particularly heavy – in Bangladesh 565 000 hectares of farmland and more than 10 000km of road were submerged. Over 300 people died, 35 from snakebites, the second biggest cause of death after drowning.

More than 7 million in Bangladesh have been affected by the floods. Health issues are of particular concern with reports of fever, acute respiratory infections, diarrhoea and snake bites. Floods inundated Dhaka, forcing many to flee to the homes of neighbours, relatives and friends. An estimated 300 000 people have been moved into refugee camps.

Activities

1 List the natural causes of flooding in Bangladesh.
2 List the human causes of flooding in Bangladesh.
3 Describe two environmental impacts of flooding in Bangladesh.
4 Outline the social and economic impacts of flooding in Bangladesh.

Discussion point

Is the Ganges a curse or a blessing to Bangladesh?

1.4 What are the management challenges associated with the development of river landscapes?

Managing river landscapes is often about balancing socio-economic and environmental needs. This requires detailed planning and management.

Managing development of river basins in Britain

Britain has 4.3 million people living in flood-risk areas, and one-third of the area earmarked for new housing development is on floodplain land. Some £240 billion worth of housing, economic assets and infrastructure is vulnerable to flooding.

One in ten new houses are being built in similarly risky places, including 31 000 planned for Ashford in Kent and 120 000 in London's Thames Gateway. A tenth of England's houses are on floodplains, and many of the 3 million new homes the government wants to see built by 2020 are to be in similarly risky areas. For example, villagers in Longford, near Gloucester, protested about the granting of planning permission for 650 new homes on land in the floodplain. However, there is little alternative: often floodplains are the only land left once metropolitan greenbelts and sites of outstanding beauty or interest are excepted. This is especially so in England's crowded south-east where demand for new homes is highest. These new homes are needed to accommodate a growing population.

According to the Association of British Insurers, storm and flood damage in the UK doubled to more than £6 billion between 1998 and 2003. Its fear is that if London is hit, the flood damage could cost £40 billion. Development plans for the Thames Gateway, where 91 per cent of new homes are planned for the floodplain, only add to the sense of impending gloom. Total insured losses for the 2007 floods are estimated to be between £2 billion and £3 billion. The full cost – including uninsured losses, foregone production and the bill for emergency relief – will be much more.

Managing development of river basins in Europe

Faced with similar problems, other countries in Europe have taken a much more interventionist approach to planning powers and obligations. In four of the main German regions, much tougher planning laws have been set. A planning application will not even be looked at if it does not include reservoir facilities. If soak-away land is going to be removed, its water holding capacity must be replaced on site. German local authorities can also specify that rainwater harvesting and recycling must be incorporated as design features for all new buildings. It is a provision as relevant to drought as to flood.

The Dutch are doing things on an even larger scale. Some 60 per cent of their land is below sea level. Already, the Dutch have the highest standard of flood protection in the world. Some €13 billion (£9 billion) has been invested in raising and strengthening the dykes over the past decade. But above a certain height, dykes become a problem not a solution partly because of the cost and partly because if they fail there will be huge devastation.

So a new strategy, 'living with water', has been formulated. It will reallocate 1.2 million acres of dry land as flooding zones. Rivers will be widened and new standards set for housing that has to be 'flood compatible'.

On the floodplain of Maasbommel, the Dutch are building permanently floating and amphibious homes. Anchored to mooring piles rather than fixed into foundations, the concrete-based homes rise and fall with floodwater levels. Wiring and sewage is ducted through the mooring piles. In the newest, changes in water level are used to generate electricity to make the houses energy self-sufficient. The Dutch see a future that has to accommodate the 'hydrometropolis' – major housing areas that partly float and may be surrounded by water.

Discussion point

Why do developers continue to develop floodplain areas when they know there is a risk of flooding?

Activity

Comment on the methods in the Netherlands to deal with flood risk.

Theory into practice

For the drainage basin that you live in, find out about the flood risk in the area. You may find the Environment Agency's website useful (www.environment-agency.gov.uk/). Click on the icon 'Are you at risk of flooding?

Case study | Managing the Thames at Oxford

The flood hazard

Oxford is a small city and there is not much land available for development. However, there is great pressure for new residential and industrial developments. The nature of these developments can cause conflict and increase the potential for flooding. In particular some of this new development is taking place on or near floodplains.

Oxford is prone to flooding for a number of reasons.

- It is low lying.
- Water from a large area, 2500km² drains into the two major rivers, the Thames and the Cherwell, and they meet in Oxford.
- The River Cherwell now floods more often than it used to in the 1970s and 1980s. This is largely the result of agricultural activities in the Cherwell valley. Trees and hedges are cut down reducing interception and increasing overland flow.

- Much of the area is covered by clay, which is fairly impermeable; hence, water does not sink into the rock but flows over the surface and into the rivers.
- Floodplain development, such as house building, has increased the amount of surface area covered by impermeable concrete. This results in less infiltration and an increased volume of water entering the storm sewers and drains at greater speed (Figure 1.49).
- Agricultural practices have led to a reduction in tree cover in the catchment and an increase in the amount of water flowing over the surface and into the rivers.
- Many bridges hold back water because debris gets trapped in the river channels making less room for the river to flow.

Figure 1.49 Flooded homes in Osney Island, Oxford

Current flood management strategies

The main ways in which the flood risk at Oxford (Figure 1.50) has been managed has been through a mix of land use zoning and flood relief schemes.

Land use zoning

In general, the areas close to the rivers that are most liable to flooding are given over to farming (Cherwell valley), recreation (Port Meadow), sports grounds (Magdalene College playing fields) and allotments (Figure 1.51). It causes economic disruption but it does not lead to a loss of life or a loss of property.

Flood relief schemes

Flood relief channels, channel dredging and flood prevention schemes have resulted in localised declines in peak flood levels and the length of flood periods (Figure 1.52). In addition, levees, channel scour and straightening have all been used. This has caused a large reduction in the average peak flow in floods in the Thames between Oxford and London. However, in some areas, deforestation for agricultural purposes has increased the height of flood peaks.

Figure 1.50 West Oxford ▼

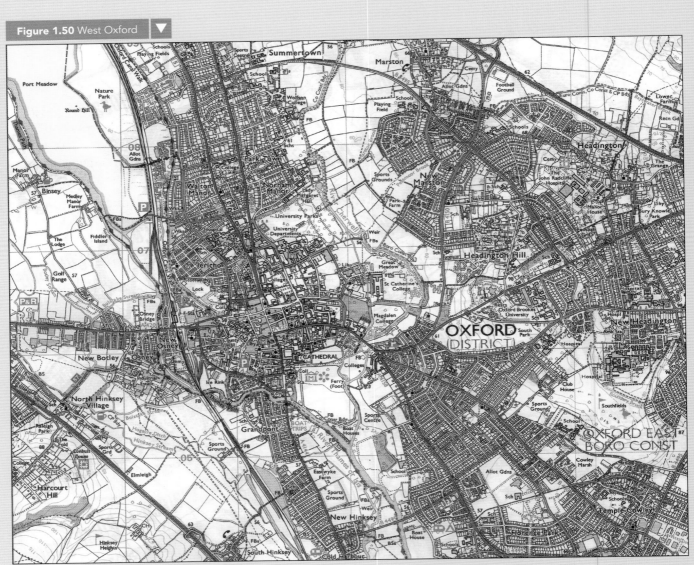

Figure 1.51 Land use in Oxford's floodplains ▲

Figure 1.52 Flood relief channels at Osney, Oxford ▼

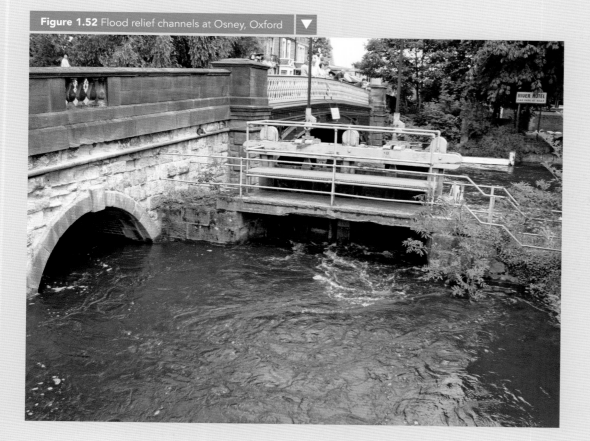

Urbanisation

Increasing urbanisation has increased the rate of runoff into the main river by increasing the proportion of impermeable surface and increasing the drainage density. But arguably this may also be seen in a positive light. Improved land drainage has the potential for increasing and decreasing flood levels. If runoff is increased, less water remains in the soil and the potential for the soil to become saturated is reduced.

Proposed flood management strategies

Flood relief channel

After years of study the Environment Agency (EA) concluded that the most effective solution to the problem of flooding in Oxford is an 8km long, 25m wide flood relief channel from the River Thames at Binsey to Sandford Lock. The £100 million flood relief scheme could see Oxford bypassed by a water channel of similar size to the Thames. However, any scheme incorporating a flood relief channel through the western corridor has the potential for impacts upon the Oxford Meadows Special Area of Conservation and Iffley Meadows Site of Special Scientific Interest. For this reason, and perhaps the cost of the project, the relief channel is no longer being considered by the EA.

Water storage areas

The EA is now investigating the creation of four large water storage areas beside the Thames and Cherwell north of Oxford. These would consist of large areas of farmland that would be allowed to flood. The EA has considered other possible solutions; for example, building above-ground reservoirs upstream of Oxford. It dismissed this as expensive and unworkable.

A small number of pioneering projects backed by the EA are showing the way forward. At Sherborne, Gloucestershire, the National Trust has restored extensive riverside water meadows, while the EA has reinstated meanders on stretches of the river Windrush. Likewise, on Otmoor, near Oxford, the RSPB has raised water levels on 267 hectares of formerly arable land, and also created a 22-hectare reedbed reservoir with the help of the EA. The reserve holds back upwards of half a million cubic metres of winter rain. Unfortunately, such initiatives have always been underfunded, relative to hard-engineered drainage and flood relief schemes.

'Take it further' activity 1.3 on CD-ROM

Activities

1 Study the map extract (Figure 1.50) and photo (Figure 1.51). How has the risk of flooding affected land-use in the Oxford area? Illustrate your answer with a sketch map and examples.

2 What are the main ways in which the flood risk in Oxford is managed and what are the associated challenges?

3 To what extent are these measures sustainable?

4 Is it worth trying to protect against floods?

Case study | Managing flooding in Bangladesh

The Flood Action Plan

It is impossible to prevent flooding in Bangladesh. The Flood Action Plan attempts to minimise the damage and maximise the benefits of flooding. The Plan relies upon huge embankments (levees) which run along the length of the main rivers. At an estimated cost of $10 billion they could take 100 years to build. Up to 8000km of levees are planned for the 16000km of river in Bangladesh. However, they are not able to withstand the most severe floods, for example in 1987, 1988 and 2007, but will provide some control of flooding. The embankments contain sluices which can be opened to reduce river flow and to control the damage caused by flooding.

The embankments are set back from the rivers. This protects them from the erosive power of the river, and has the added advantage of being cheap both to install and maintain. In addition, the area between the river and embankment can be used for cereal production.

Nevertheless, the Flood Action Plan is not without its critics. There are a number of negative impacts and potential land use conflicts of the scheme.

◆ Increased time of flooding, since embankments prevent back flow into the river.

◆ River channelisation by levees may increase the risk of flooding downstream and in the area between the levees.

◆ Channelisation will also increase deposition between the levees rather than on the floodplain, reducing the flood control capacity of the river channel.

- Not enough sluices have been built to control the levels of the floodwaters in the rivers – this means that there may be increased damage to the land by flooding if the embankments are breached, since the rapid nature of the breach is more harmful than gradual flooding.

- Sudden breaches of the embankments may also deposit deep layers of infertile sand reducing soil fertility and affecting agricultural practices.

- Compartmentalisation of the drainage basin may reduce the flushing effect of the floodwaters, increasing the concentration of pollutants from domestic effluents and agrochemicals.

- By preventing back flow to the river, areas of stagnant water will be created which may increase the likelihood of diseases such as cholera and malaria spreading amongst the residential areas close to the rivers.

- Embankments may cause some wetlands to dry out, leading to a loss of biodiversity.

- Decreased flooding will reduce the number of fish, which is a major source of protein, especially among the poor.

The rivers of Bangladesh are, in part, controlled by factors beyond the country. Moreover, the causes of floods in Bangladesh are not just related to rivers. There is a delicate balance between the disadvantages that the rivers create, such as death and destruction, and advantages that the rivers bestow, such as a basis for agriculture and export earnings. To date, there has been little agreement as to how to control the peak discharges of the rivers. The Flood Action Plan uses embankments to control the distribution and speed of flooding, although the embankments have, in turn, led to serious social, economic and environmental problems.

Activities

1 Outline the advantages of the Flood Action Plan.
2 Describe some of the disadvantages of the Flood Action Plan.
3 How would it be possible to manage the risk of flooding, and develop the Ganges delta, more successfully?

Discussion point

Are MEDCs more vulnerable to floods than LEDCs?

 Please refer to the CD-ROM for an additional case study on 'Flooding in London'

Knowledge check

1 Outline the ways in which weathering and mass movements can help form fluvial landscapes.

2 With the use of labelled diagrams, explain how one feature of erosion and one feature of deposition in a river environment are formed.

3 With the use of examples, explain how rivers can provide opportunities as well as be hazardous to people.

4 For a named river basin, describe and explain the physical and human factors that have led to a risk of flooding.

5 For a drainage basin that you have studied:
   a outline the impacts of flooding
   b examine how effective the measures to control flooding are.

A list of useful websites accompanying this chapter can be found in the Exam Café section on the **CD-ROM**

Exam Café
Relax, refresh, result!

Relax and prepare

What I wish I had known at the start of the year…

Alice

"Getting started is the most difficult stage in examination preparation. I try to start early but gently, as too much too soon leads to boredom and soon becomes a chore. When planning my revision programme, I build in slack and flexibility as unforeseen events crop up. I now know the way I learn best and the best place for learning so I try to find an area without too many interruptions. I try not to use my bedroom as this is the place I go to escape from revision."

Sanitta

"I find it easy to confuse the different types of mass movement that work on the sides of valleys. Creep is very slow, slides are fast and cohesive, flows are fast and wet and lack cohesion, slumps involve rotation and are wet, and avalanches and Free Falls are very fast and dry on very steep slopes. The term 'slip' may cover slides and slumps – that's the confusing part!"

Hot tips

Ian

"I've found that a 'study-buddy', someone I can trust to share problems with and act as a tester of my knowledge and understanding, helps me. It isn't always a geographer as sometimes another viewpoint can help. I also try to involve my family in my preparation as they have got to put up with me and they can sometimes help!"

Common mistakes – Nicky

▷ "In my last exam I didn't time my answers. Once I started writing I kept on until I finished the question. This meant that I only had 15 minutes to answer the last question. I had to rush and was unable to finish it.

▷ Now I work out the time the question before the exam required for each question. I look at past papers and divide them up according to the marks. So for AS papers it is three questions in 90 minutes for 75 marks. This means half an hour a question or roughly a mark a minute. This helps me pace myself over the shorter section questions, e.g. 9 marks equal 9 minutes maximum. I daren't take 12 minutes as this would upset the schedule. If I can answer it fully in under 9 minutes, then fine! Sometimes I have to stop before I have finished the question but a 'mark a minute' schedule does leave some slack to go back and complete unfinished sections."

1.1 What processes and factors are responsible for distinctive fluvial landforms?

Physical	Climate – evaporation, temperature, precipitation type and volume
	Relief – slope, altitude and base level
	Rock type – geology, structure, beds, porosity and tilt of rocks
	Vegetation – type and percentage cover
Human	Water supply – abstraction
	Channel work – dams, weirs, embankments, straightening, widening, deepening, dredging, flood prevention and meander management
	Drainage - soils, from industries and roads etc
	Agriculture – crops, deforestation, irrigation and drainage
	Urbanisation – impervious surfaces and channel controls
	Transport – canalisation, bridges and weirs
Time	Feature development, climatic changes and tectonic changes
Processes	Erosion – hydraulic action, attrition, corrasion solution
	Transport – traction, saltation, suspension, solution, floatation
	Deposition – heaviest is dropped first
Landforms	Erosion – waterfalls, potholes, valleys, meanders, caves, rapids
	Deposition – deltas, levees, ox bows, meanders

1.2 In what ways can river basins be a multi-use resource?

Residential development	Settlements are located on or near rivers because of the flat land and water supply, e.g. 90% of Egypt's population live by the Nile
Power source	Hydroelectric power e.g. the Aswan dam on the Nile
Industrial development	Fishing, tourism and heavy industry that requires flat land and a water supply
Minerals	Sediments, oil, gas, gravel e.g. gas in the Nile delta
Services	Tourism, recreation, waste tourism (cruises)
Agriculture	Fish farming and arable farming (fertile silts) e.g. cash and subsistence farming in Egypt depend upon irrigation from Nile
Transport	Ports, bulk cargo ports, cheap bulk transport
Conservation	Nature reserves

1.3 What issues can arise from the development of river basins?

Flooding	Pressure to develop on floodplains can make them increasingly vulnerable to flooding. Some areas are naturally vulnerable as flat and low lying e.g. Mississippi delta area
Flood prevention	To protect property, land, the transport infrastructure and conserve the local environment (historic and biotic). It can have negative impacts in other areas so you need to weigh up costs versus benefits
Methods	Hard and soft engineering, planning restrictions, planned retreat, do nothing
Where	Floodplains, channel, catchment, channel and valley sides
Types	Afforestation, banning building, embankments, diversion channels, dams, storage lakes, channel straightening, widening and deepening

1.4 What are the management challenges associated with the development of river landscapes?

Planning	Balancing environmental and socio-economic needs including costs (short term and long term); technology, political will, time, scale, knowledge of the issues and the wider impact
	Rivers often flow from one country or district to another and may cross different administrative boundaries. Many different types of planning authorities may have interest in any one river
	Rivers are difficult to manage as they change course, are fluid and are influenced by the climate

Top tips . . .

▷ Different approaches work for different people so you need to discover your preferred learning style before you start. Some people are visual learners and they learn best from pictures, maps, diagrams and by writing or drawing. Many such learners can recall where items are on a page or map. Others are auditory learners and learn by listening to songs, rhymes recordings and talking over revision with friends. Some people learn from doing – 'kinaesthetic' learners – and they respond best to making things such as patterns, models, card indexes or acting out a topic.

Get the result !

Explain the origin of waterfalls. [6 marks]

Examiner says

The student has clearly explained the basic concept of a waterfall and the structural cause for the formation of waterfalls. The student then elaborates on the formation and structure.

Examiner says

A diagram here would have been equally effective in illustrating the origin of waterfalls.

Student answer

Waterfalls occur when there is a sudden abrupt break in the long profile of the river. These may result from a change in rock type from hard to soft, with the soft being more easily eroded so allowing the river to 'fall'. Faults may bring hard and soft rocks together or a hard band of rock may be injected into softer rocks such as the Great Whin Sill that forms High Force waterfall on the Tees. In some cases it is two different processes working at different rates that cause the waterfall e.g. a glacial hanging valley or where the sea has cut back rapidly into the lower course of a river e.g. Osmington Mills in Dorset.

Examiner says

This is a good answer because two examples of a waterfall have been given as well as an alternative explanation for their formation. This answer would gain 6 marks.

Examiner's tips

One of the commonest errors candidates make is in not understanding what the question is asking. Understanding the meaning of terms used in questions is crucial in appreciating what the question and examiner want. A lot of time and trouble goes into deciding the exact phrasing of a question so every word is vital.

Chapter 2

Coastal environments

With over four billion people living in coastal areas and the numbers growing rapidly each year, the need to understand and manage these areas is critical. Coastal areas are one of the most dynamic environments on the earth and are being constantly re-shaped by both natural processes and human development. They have both economic and environmental value – and it is these conflicting demands that bring about the need for long-term, sustainable management strategies.

Questions for investigation

- What processes and factors are responsible for distinctive coastal landscapes?

- How can coasts be protected from the effects of natural processes?

- In what ways can coastal areas be a valuable economic and environmental resource?

- What are the management challenges associated with the development of coastal areas?

Consider this

The coast is the frontier between land and sea.

A frontier that will come under increasing pressure as global warming brings about changes to both sea level and the frequency of coastal storms.

2.1 What processes and factors are responsible for distinctive coastal landforms?

The dynamic nature of coasts

The coast is the frontier between land and sea and is one of the most dynamic and fragile environments on earth (Figure 2.1). It is constantly being re-shaped by waves, tides, **ocean currents** and the effects of the weather. Where rock structures are more resistant or sheltered from **prevailing wind** and waves, changes occur slowly. Where rock structures are less resistant and are open to storm conditions and heavy rainfall, sudden and dramatic changes can occur, reshaping the landscape in minutes in the case of coastal landslides or rockfalls.

Figure 2.1 Storm conditions ▼

> **Key terms**
>
> **Ocean current:** the flow of a large body of water in an ocean.
>
> **Prevailing wind:** the most frequently occurring wind direction in an area.

Waves – shaping the coast

As wind blows over the ocean, friction occurs and energy is transferred, creating waves. The stronger the wind, the greater the friction, making the waves higher and more powerful. Waves move across open ocean in a circular motion as ocean swell. As they move towards shallower, coastal waters the frictional drag of the seabed slows down the base of the wave making it increasingly elliptical in shape (Figure 2.2). This reduces the wavelength and increases wave height until the top of the wave breaks because it is moving faster than its base.

The force of the breaking wave pushes up the beach, creating the **swash**. When the wave has reached its furthest point up the beach, it runs back down the slope towards the sea as gravitational **backwash**. Wind

Figure 2.2 Breaking waves ▼

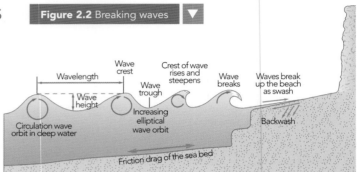

direction and strength, as well as fetch are the main factors determining the height and energy of breaking waves, fetch being the length of uninterrupted water over which waves can be generated.

> **Key terms**
>
> **Swash:** body of water pushing up a beach after a wave has broken.
>
> **Backwash:** the movement of water back down a beach after a wave has reached its highest point.

Constructive and destructive waves

Waves are the most important cause of changes to beaches and coastal landforms. There are two main classifications of wave types: constructive waves, which help to build beaches, and destructive waves, which remove sediment from beaches.

Constructive waves

Constructive waves are often called 'spilling' waves (Figure 2.3). Wavelengths are long and wave height often less than 1 metre. The breaking waves have low levels of energy and 'spill' on to the beach. The resulting swash is quickly absorbed by the beach. Sediment thrown up by the breaking waves accumulates in ridges or 'berms'. The backwash has little power to move the sediment back towards the sea so the beach gradually develops.

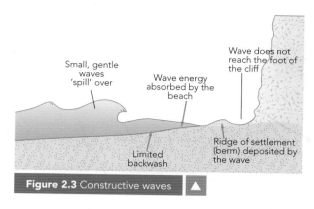

Figure 2.3 Constructive waves ▲

Destructive waves

Destructive waves are called 'plunging' waves and are often the result of storm activity (Figure 2.4). Wavelengths are short and waves high so the wave breaks from considerable height creating large amounts of energy which cannot easily be absorbed by the beach. Powerful waves run up the beach, the volume of water creating the opportunity for strong backwash to move sediment back down the beach and eventually out to sea.

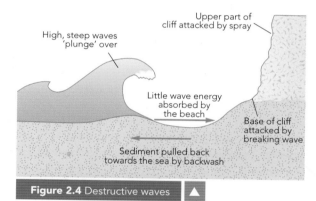

Figure 2.4 Destructive waves ▲

Classifying coastal environments

◆ Storm wave environments are characterised by frequent low pressure systems creating strong onshore winds which generate powerful waves. An example of this can be seen in north-west Europe.

◆ Swell wave environments are characterised by a less extreme pattern of wind and waves but a considerable swell built up over a long fetch. An example of this can be seen in west Africa.

◆ Tropical cyclone wave environments are characterised by extreme winds which allow huge waves to develop and push massive amounts of water towards coastal areas. This can be seen in south-east Asia and the Caribbean.

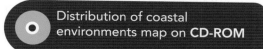

Distribution of coastal environments map on **CD-ROM**

Tides

A tide is the alternative rise and fall of the level of the sea. Tides are caused by the gravitational pull of the moon and, to a lesser extent, the earth. The moon pulls water towards it, causing a swell which creates a high tide. Twice each month the sun, moon and earth are in alignment; this creates the maximum gravitational pull and the highest tides. Tide levels are predictable and are published in tide tables. When high tides coincide with unpredictable events such as strong winds, large waves and heavy rainfall, the result can be catastrophic flooding events.

Storm surges

A **storm surge** is created when the following factors coincide:

◆ high tides

◆ strong onshore winds creating high levels of wave energy

◆ low pressure weather systems allowing the sea to expand.

The shape of the landscape can add to the intensity of a storm surge. Where the sea is 'pushed' into a narrow area between two land masses, it is forced to rise, flooding coastal areas.

> **Key term**
>
> **Storm surge:** high onshore winds and tides combine to give unusually high sea levels.

Discussion point

Using an atlas, consider which parts of the UK are most likely to be vulnerable to extreme wave conditions.

Activities

1 Explain how waves are formed.

2 Draw an annotated sketch to explain the causes and consequences of breaking waves. Include the use of appropriate terminology in your explanation.

3 Explain why waves are classified as 'constructive' and 'destructive'.

4 Why are waves often considered to be more significant than tides when considering threats to coastal areas?

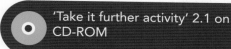

'Take it further activity' 2.1 on CD-ROM

Coastal processes

Coastal **landforms** are the result of the interaction of a number of processes. These include: marine **erosion**, **weathering**, human activity and mass movement/slumping.

> **Key terms**
>
> **Landform:** a physical feature on the earth's surface.
>
> **Erosion:** the wearing away of material by the action of water, ice or wind.
>
> **Weathering:** the disintegration of rocks by the action of the weather, plants, animals and chemical action.

Marine erosion

A number of **marine processes** are responsible for the landforms associated with coastal erosion.

> **Key term**
>
> **Marine processes:** the action of the sea on coastal landforms.

Hydraulic pressure

In areas where there is limited beach material to absorb the energy of breaking waves, cliff faces can be attacked. Breaking waves exert a force of up to 40 tonnes per square metre on cliff faces (Figure 2.5). They also force air into joints and cracks in the cliff surface. This compressed air has the power to loosen and break away pieces of rock.

Abrasion/corrasion

During storm conditions, waves have the energy to pick up sand particles and pebbles and hurl them at the cliff face. This 'sand blasting' effect is thought to be the most rapid process of coastal erosion in the United Kingdom.

Figure 2.5 Waves breaking against corroded limestone cliff face

Attrition

Rocks and pebbles are constantly colliding with each other as they are moved by waves. This action reduces the size of beach material and increases its 'roundness' by smoothing away rough edges.

Weathering

Corrosion/solution

All landscapes are affected by weathering processes. In coastal areas the proximity of sea water can speed up the effect of chemical weathering. Saltwater evaporation from sea-spray leads to the growth of salt crystals in rocks. As they develop, the salt crystals expand, forcing rocks to disintegrate. Particular types of rocks are susceptible to corrosion, especially if they contain limestone, which is dissolved by the carbonic acid in salt water (Figure 2.5).

Wetting/drying

Softer rocks such as clays and shales are very susceptible to wetting and drying. Where these rocks are in the coastal splash zone, they are constantly prone to expansion and contraction as they become wet and then dry out. This causes weaknesses in the rock which allows marine processes to attack and erode the rock away easily.

Human activity

The increasing use of coastal areas for leisure and recreation can put pressure on fragile coastal environments. The human erosion of cliff-top footpaths can cause weaknesses in rock structure and the removal of vegetation can leave rock surfaces more prone to weathering and erosion.

> **Discussion point**
>
> Consider the ways that human activities might make coastal areas more vulnerable to the natural processes of weathering and erosion.

Mass movement/slumping

Rockfalls and landslips are common features of cliff coastlines, often occurring as a result of the combination of wave action weakening the base of a cliff and **sub-aerial processes** attacking the upper part of the cliff face. Sub-aerial processes can include weathering processes and the effects of rainfall eroding cliff surfaces or weakening rocks by **percolation**. In more resistant rocks such as chalk, rockfalls are often the result of undercutting by the sea and the weakening of the rock by corrosion and ongoing wetting and drying. These processes can lead to individual fragments of rock falling or in extreme circumstances whole sections of cliff collapsing as a rockfall.

┌─ **Key terms** ──────────────────────────────

Sub-aerial processes: processes active on the face and top of cliffs.

Percolation: the rate at which water soaks into rocks. A shingle beach has a higher percolation rate than a sand beach.
└──

Landslides and slumping are often associated with weaker rocks such as clays and sands and can be triggered by periods of heavy rainfall. When the ground becomes saturated, the combination of extra weight, slope and increased lubrication can lead to slope failure and cause small-scale mudslides or landslides (Figure 2.6). Slumping is a result of a combination of marine processes undercutting the base of a cliff, heavy rainfall, and curved slipping planes where different rock types meet.

Figure 2.6 Landslide ▲

The impact of geology on coastal landforms

Both rock type and structure can have a significant impact on coastal landforms. More resistant rocks, such as chalk and limestone, erode more slowly and often produce spectacular cliff and headland features. Weaker rocks, such as clay and sands, have less structural, strength and are eroded easily, producing a lower cliff profile with mudslides and slumping.

The direction in which rocks occur in relation to the coast plays an important part in the resulting landforms. Coasts where the rock type runs parallel to the sea are called concordant coasts and often produce straighter coastlines. Coasts where the rock types outcrop at right angles to the sea are called discordant coasts and often produce headlands and bays. An example of both of these coastal types can be seen in the Purbeck coast of Dorset (Figure 2.7).

Activities

1 What is meant by:
 a marine processes
 b sub-aerial processes?
2 Explain how marine erosion and weathering help to shape the coast. Include the use of technical language in your explanation.
3 How does rock type and rock structure affect the rate of erosion?

⊙ 'Take it further' activity 2.2 on CD-ROM

Coastal environments

Figure 2.7 Purbeck coastline ▼

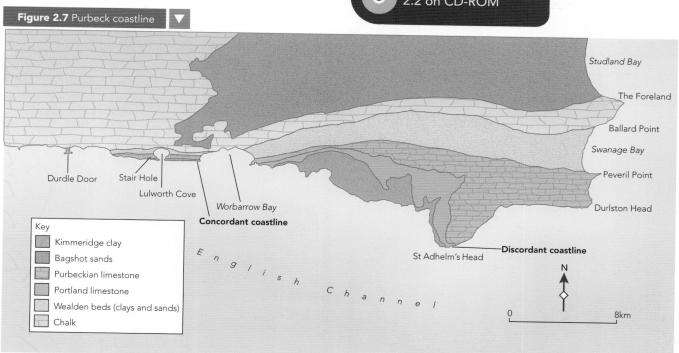

Durdle Door Stair Hole Lulworth Cove Worbarrow Bay **Concordant coastline**

St Adhelm's Head **Discordant coastline**

Studland Bay
The Foreland
Ballard Point
Swanage Bay
Peveril Point
Durlston Head

English Channel

N

0   8km

Key
■ Kimmeridge clay
■ Bagshot sands
▦ Purbeckian limestone
■ Portland limestone
□ Wealden beds (clays and sands)
▤ Chalk

Features of coastal erosion

There is a clear link between high-energy coastlines and rates of erosion. However, wave power is only one factor that explains both the rate of erosion and the resulting landforms in coastal areas. The characteristics of the rocks (**lithology**) found in coastal areas also play a significant part in the resulting features created by the processes of weathering and marine erosion. Some rock types are more coherent which means that they have well-connected individual particles and few lines of weakness – examples of this can be seen in igneous rocks but also in sedimentary rocks such as chalk and sandstone. The result of this is often a more solid coastline with a steep cliff profile and slow rates of erosional retreat. Other rock types are more incoherent, which means that they have poorly connected particles or a lot of cracks and joints, giving rise to a high level of weakness. Clay is an example of a poorly 'cemented' rock which is affected by both sub-aerial and marine processes, often resulting in a slumped or stepped cliff profile (Figure 2.8).

> **Key term**
>
> **Lithology:** the properties of rocks – mineral composition, cracks, joints.

The angle at which rock strata reaches the coast can also affect cliff profiles. Horizontally bedded strata usually forms steep cliffs, where strata is dipping towards or away from the sea usually forms a more gentle cliff profile is usually formed.

Landforms associated with resistant rock – headlands

Many coastal areas are made up of headlands of more resistant rock and bays where weaker rocks outcrop. Because of the **promontory** nature of headlands, they are under constant attack from marine processes on both sides. The result of this is a number of distinct landform features which can be seen in Figure 2.9 and Figure 2.10.

> **Key term**
>
> **Promontory:** highland area jutting out into the sea.

Landforms associated with weaker rock

All slopes are under stress because of the force of gravity. If the forces acting on a slope are greater than the resistant strength of the rocks, the slope will fail, resulting in a **landslide**.

> **Key term**
>
> **Landslide:** the general term for gravity controlled processes (mass movement). The three main types of landslide process are: falling, sliding and flowing.

Figure 2.8 Stepped/slumped cliff ▼

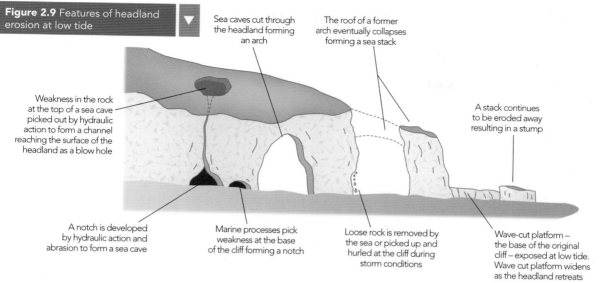

Figure 2.9 Features of headland erosion at low tide

Sea caves cut through the headland forming an arch

The roof of a former arch eventually collapses forming a sea stack

Weakness in the rock at the top of a sea cave picked out by hydraulic action to form a channel reaching the surface of the headland as a blow hole

A stack continues to be eroded away resulting in a stump

A notch is developed by hydraulic action and abrasion to form a sea cave

Marine processes pick weakness at the base of the cliff forming a notch

Loose rock is removed by the sea or picked up and hurled at the cliff during storm conditions

Wave-cut platform – the base of the original cliff – exposed at low tide. Wave cut platform widens as the headland retreats

Clay coastlines are particularly prone to landslides because clay is poorly consolidated and becomes very unstable when wet. Consequently, coastal areas that have a high level of clay content, or are made up of layers of rock which include clays, are especially vulnerable to landslide events (Figure 2.11).

Parts of the Dorset coast are affected by landslides because the geology contains a complex mixture of porous rocks, clays and mudstones, making the coast susceptible to wave attack, weathering and heavy rainfall. The area is more likely to have future landslide events because movement can occur along stress lines that already exist.

Figure 2.10 Chalk headland

Figure 2.11 Landslide features

Heavy rainfall, soaking through fissures reduces the friction which helps to hold the slope together. As the ground becomes saturated, the bands between the rock particles fail and more fissures develop

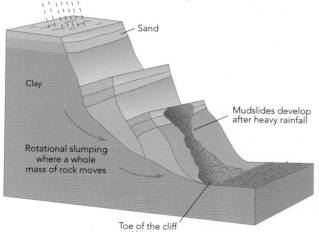

Sand

Clay

Mudslides develop after heavy rainfall

Rotational slumping where a whole mass of rock moves

Toe of the cliff removed by marine erosion which destabilises the cliff

Activities

1 Explain the relationship between geology and coastal processes.

2 **a** Draw sketches to show the profile of a chalk headland and clay cliff coastline.

  b Add annotations to your sketches to describe and explain the characteristics of chalk headland and clay coastlines.

'Take it further' activity 2.3 on CD-ROM

Features of coastal deposition

As waves enter shallow water and break, they pick up beach sediment and move it up the beach in the direction of the breaking wave (swash). The movement of the swash up the beach stops when the energy has been used. Any water that has not percolated into the beach flows back down the beach under the influence of gravity (backwash).

Types of beaches

There are two main types of beaches: swash-aligned beaches and drift-aligned beaches.

Swash-aligned beaches

This type of beach forms when waves approach the coastline parallel to the beach. The swash and backwash move sediments up and down the beach, often creating a stable, straight beach with an even, longitudinal profile. During storm conditions severe backwash can move sediment out to sea, creating sand or shingle bars on the seabed.

Drift-aligned beaches

This type of beach forms when waves approach the coastline at an angle and sediment is moved along the coast by the action of longshore drift (Figure 2.12). On some coasts the movement of material is slowed down by building wooden or concrete groynes. These allow the sediment to build up and create a wider beach on the up-drift side, but they often starve down-drift areas of sediment, making them lose their beaches and further adding to the possible problems of erosion.

The formation of spits, bars and tombolas

Spits and bars are ridges of sand or shingle that have been deposited by the action of the sea. They often form where there is an abrupt change in the direction of the coastline or a break in the coastline where a river enters the sea. At these points material that has been moved along the coast by longshore drift is deposited in the same direction as the coastline. The material is then shaped by wave action and ocean currents. A spit is joined to the land at one end and finishes in a recurved end in the open sea. A bar is formed when a spit extends across an opening and connects two separate headlands. Where a spit develops and connects to an offshore island, a **tombola** is formed. An example of this can be seen at Chesil Beach, which connects the mainland of Dorset with the Isle of Portland.

> **Key term**
>
> **Tombola:** where deposition of sediment joins an island to the mainland.

'Take if further activity' 2.4 on the CD-ROM

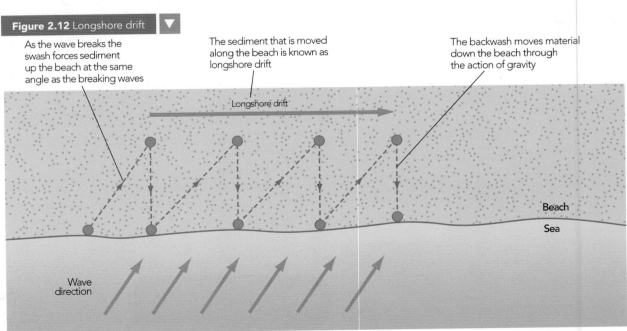

Figure 2.12 Longshore drift ▼

As the wave breaks the swash forces sediment up the beach at the same angle as the breaking waves

The sediment that is moved along the beach is known as longshore drift

The backwash moves material down the beach through the action of gravity

Longshore drift

Beach

Sea

Wave direction

Case study | Hurst Castle Spit – Hampshire

Hurst Castle Spit is a shingle ridge which has developed as a result of longshore drift in Christchurch Bay (Figure 2.13). The sudden change in the direction of the coast to the south-east of Milford-on-Sea allowed deposited shingle sediment to develop into a spit that reaches 2km out from the coastline. Strong winds and tidal currents have shaped the spit and created a distinct curved end (Figure 2.14).

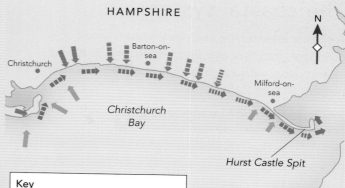

Figure 2.13 Sediment flows in Christchurch Bay ▶

Key

Sediment transport mechanism

➡ Littoral (beach) drift

➡ Cliff or coastal slope erosion

➡ Wave driven nearshore and zone transport

Volume of flow

➡ No quantitative data

➡ 3000–10 000 m³

➡ 10 000–20 000 m³

➡ >20 000m³

Figure 2.14 Hurst Castle spit and area behind ▼

1   Mudflats evident during low tide.

2   Colonisation of mudbanks by halophytic (salt loving) plants, trapping more sediment.

3   Ocean currents are stronger than the drift resulting in a recurved end to the spit.

4   Hurst Point, fed by gravel that drifts around the Hurst Coastal.

5   Keyhaven saltmarsh/Nature reserve.

6   Groynes built to maintain spit and Hurst Castle.

Activities

1   What combination of coastal characteristics might lead to the formation of a spit?

2   Describe the three depositional landforms identified on page 54.

3   Identify the main processes of sediment movement in Christchurch Bay.

4   **a**   Draw an outline sketch of Hurst Castle Spit.

     b   Annotate your spit to describe and explain the human processes and features of the spit.

The coastal system

A coastal area can be seen as a system which produces, transfers and deposits sediment. It is an open system with inputs, stores and transfers and outputs (Figure 2.15).

Figure 2.15 Parts of the coastal system ▼

- **Sediment inputs**
  - deposition by rivers
  - weathering and erosion of cliffs
- **Sediment transfer**
  - movement up and down the beach
  - movement along the coast by longshore drift
- **Sediment stores**
  - beach/sand dunes
  - sand/shingle onshore or offshore bars
- **Sediment outputs**
  - sediment lost to the open sea
  - sediment removed by human intervention

Where there is a balance between inputs and outputs, the system is said to be in equilibrium. A positive **sediment budget** suggests that beaches are developing and are relatively stable, while a negative sediment budget suggests a loss of beach material and the possibility of increasing wave action on cliffs. The following example describes parts of the Holderness coastal system on the Yorkshire coast (Figure 2.16).

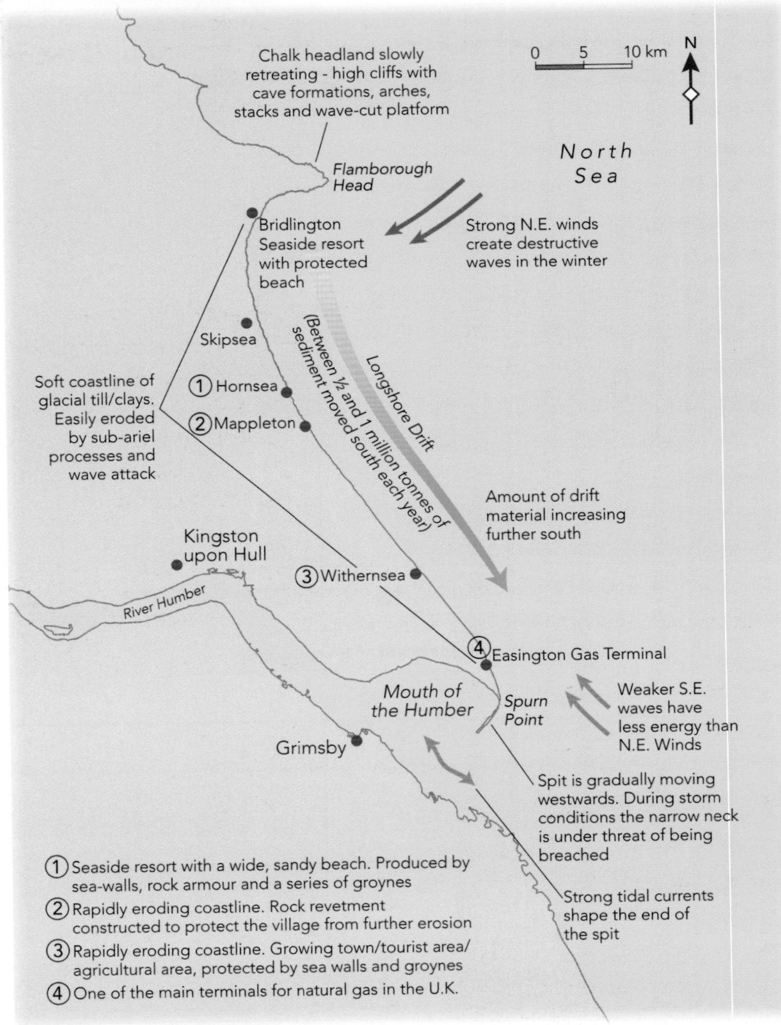

> **Key term**
>
> **Sediment budget:** the sediment budget is calculated by subtracting the volume of outputs from the volume of inputs.

Activities

1 Explain why a coastal area can be seen as a system.

2 Why is the idea of 'sediment budgets' important when considering the management of coastal areas?

Changing sea levels

Sea level is the relative position of the sea as it meets the land. There is clear evidence that sea levels have fluctuated considerably over the last 20 000 years. Rising sea levels are evident when looking at drowned river estuaries (rias) or glacial valleys (fjords) (**submergent coastlines**). Falling sea levels can be seen by the evidence of raised beaches or cliff formations now seen inland from current beach positions (**emergent coastlines**).

> **Key terms**
>
> **Submergent coastlines:** produced by rising sea levels.
>
> **Emergent coastlines:** produced by falling sea levels.

Figure 2.16 Holderness ▼

Chalk headland slowly retreating - high cliffs with cave formations, arches, stacks and wave-cut platform

North Sea

0 5 10 km N

Flamborough Head

Strong N.E. winds create destructive waves in the winter

Bridlington Seaside resort with protected beach

Skipsea

Soft coastline of glacial till/clays. Easily eroded by sub-ariel processes and wave attack

① Hornsea

② Mappleton

Longshore Drift (Between ½ and 1 million tonnes of sediment moved south each year)

Amount of drift material increasing further south

Kingston upon Hull

③ Withernsea

River Humber

Mouth of the Humber

Spurn Point

④ Easington Gas Terminal

Weaker S.E. waves have less energy than N.E. Winds

Grimsby

Spit is gradually moving westwards. During storm conditions the narrow neck is under threat of being breached

Strong tidal currents shape the end of the spit

① Seaside resort with a wide, sandy beach. Produced by sea-walls, rock armour and a series of groynes

② Rapidly eroding coastline. Rock revetment constructed to protect the village from further erosion

③ Rapidly eroding coastline. Growing town/tourist area/ agricultural area, protected by sea walls and groynes

④ One of the main terminals for natural gas in the U.K.

U1 2

AS Geography for OCR

Preparing for rising sea levels

There is growing evidence to support the view that global temperatures are changing and that this will result in a number of potential effects on coastal areas. With over 80 million people living in coastal areas in Europe alone it is clear that rising sea levels pose a significant threat to both life and property. The following potential affects of climate change have been identified by coastal scientists (Figure 2.17).

◆ Climate change is likely to result in changes to three key influential factors: rising sea levels, changes in wave patterns, increases in rainfall in winter.
◆ There may be an increase in the number of storm events.
◆ Flooding of developments built on lowlands is likely to intensify.
◆ Existing trends for erosion of salt marshes is likely to accelerate.
◆ Soft rock coastal cliffs are likely to retreat more rapidly because of increasing rainfall and rates of cliff top erosion. However, this may increase the supply of valuable sediments to nearby beaches.
◆ Protecting vulnerable coastlines will become increasingly difficult and more expensive.

Figure 2.17 The potential effect of climate change on coastal areas – adapted from *Coastal Defence – a non-technical guide* (SCOPAC) ▲

The Response Project

In order to plan for future sea-level changes, a number of European coastal scientists worked on the Response Project – which is about responding to the risk of climate change. They developed a response methodology based on a series of maps which gather existing information, explain what the information shows and uses the information to offer guidance for future management (Figures 2.18a and 2.18b). Each of the sequenced maps can be developed and stored within a Geographical Information System (GIS).

Gathering information about:	1	Existing landforms and processes
	2	Current coastal defences
	3	Existing coastal hazards
	4	Coastal assets/population
Interpretation of information	5	Coastal behaviour (soft/hard coasts/ rates of erosion)
	6	Potential coastal hazards
Guidance for future management	7	Potential future coastal risks
	8	Summary of potential coastal hazards
	9	Planning and guidance

Figure 2.18a The Response Project methodology ▲

Figure 2.18b Example of a response map ▼

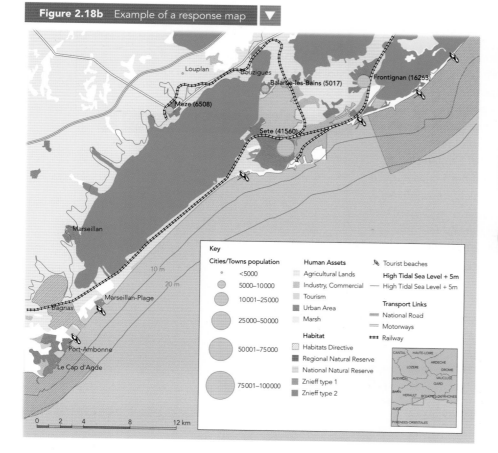

Activities

1 Describe and explain the evidence that shows historical changes to sea levels.
2 a What is the aim of the Response Project?
 b Explain the Response Project methodology.

⊙ 'Take it further' activity 2.5 on CD-ROM

Coastal environments

2.2 How can coasts be protected from the effects of natural processes?

Managing coastal areas from the effects of natural processes

Coastal environments are one of the most dynamic and fragile environments on earth. Consequently, managing them requires a detailed understanding of both natural and human influences. With over 4 billion people worldwide living within 40km of a coast, an ever-increasing demand for coastal leisure activities and the threat of increasing sea levels, the challenge is to find **sustainable** ways of managing the frontier between land and sea which offers protection from the forces of nature whilst being sensitive to the needs of the environment.

> **Key term**
> **Sustainable:** meeting the needs of the present time without compromising future needs.

How is the coastline managed in the United Kingdom?

In the UK the Department of the Environment, Food and Rural Affairs (DEFRA) has overall responsibility for the protection of the coastline from erosion and flooding. In order to have a more effective understanding of local coastal processes and to be able to implement management strategies, the coastline of England and Wales has been divided up into 11 sediment cells (Figure 2.19). These are lengths of coastline which are relatively self-contained in terms of the movement of sediment. The boundaries of the cells are often natural features such as headlands or estuaries which act as natural barriers to the movement of sediment. Each of the cells is divided into sub-cells. This division is based on knowledge of local processes within the whole cell and allows smaller areas to be managed more effectively. In order to fully understand and manage a sub-cell, a Shoreline Management Plan (SMP) is developed.

Discussion point

Why is the coastline sometimes called 'the frontier between land and sea'?

Why do some coastlines need to be managed?

This question was recently asked at a sixth form conference about coastal issues. Some of the responses are shown here:

'Lots of people live near the sea – without protection millions of pounds worth of property would be lost.'

'Coastal areas provide sand and gravel for building. The extraction of resources needs to be managed in case it affects sediment flows in an area.'

'Coastal environments like salt marshes, coral reefs and mangroves are becoming increasingly rare. Management is required to ensure the sustainability of these areas.'

'What about global warming? Rising sea levels and increasing storm activity will mean that more coastal areas will be vulnerable to flooding.'

'Coastal areas are important industrial areas with heavy industry and ports. This creates work as well as being important for national economies. If these areas were flooded, it would be disastrous.'

'Coastal tourism is a massive industry and the beach is an important amenity so it must be protected.'

'The fishing industry relies on natural processes in relation to breeding patterns. If these are disturbed, stocks of fish and shellfish would be affected.'

Figure 2.19 DEFRA sediment cells/sub-cells

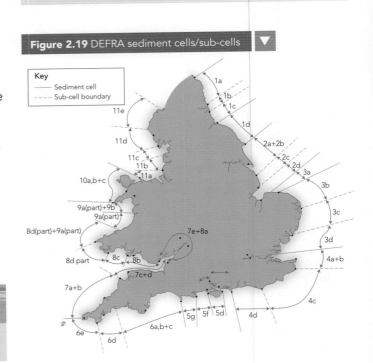

Key
— Sediment cell
--- Sub-cell boundary

AS Geography for OCR

These comments show that coastal areas need to be managed because they have economic, social and environmental value.

What is a Shoreline Management Plan (SMP)?

A Shoreline Management Plan is a document which examines the risks associated with coastal processes and presents a policy to manage those risks. It considers 'risks' to local people as well as the built and natural environment. The final policy document is drawn up after consultation with a wide variety of stakeholders, which includes local people and national groups such as the Environment Agency and English Nature.

The published SMP for a sub-cell will:

1 Divide the sub-cell into appropriate management units.

2 For each management unit:
   ◆ identify the economic and environmental assets that would be affected by flooding
   ◆ assess the issues and conflicts associated with the management of the unit
   ◆ propose a coastal defence plan for the next hundred years, divided into three time periods: 'present day' (next 20 years), 'medium term' (20–50 years) and 'long term' (50–100 years).

3 Use the DEFRA criteria for management which is:
   ◆ Hold the line – maintain existing defences.
   ◆ Advance the line – build new defences seaward of the existing line of defence.
   ◆ Managed realignment – allow the land to flood with careful monitoring and management further inland.
   ◆ No active intervention – no investment in providing any management.

Figure 2.20 shows a simplified version of part of the SMP for sub-cell 3b, The North East Norfolk Coast, with the suggested policy for the present day.

Integrated Coastal Zone Management (ICZM)

Integrated Coastal Zone Management (ICZM) is a method of managing not only the shoreline but also the whole of the coastal zone. Set up in 1996 by the European Union, ICZM sees management being 'environmentally sustainable, economically fair, socially responsible and culturally sensitive', and that 'uncoordinated coastal policies can lead to conflict or further deterioration of the coast'. A number of local authorities have drawn up ICZM policies to work alongside their SMP.

Activities

1 Construct and complete a table to show the economic, social and environmental value of coastal areas.

2 'Sediment cells are self-contained areas in terms of coastal processes.' Explain this statement.

3 Why is the coastline of the UK divided into cells and sub-cells for the purposes of management?

4 What is the difference between shoreline management and coastal zone management?

U1
2

Coastal environments

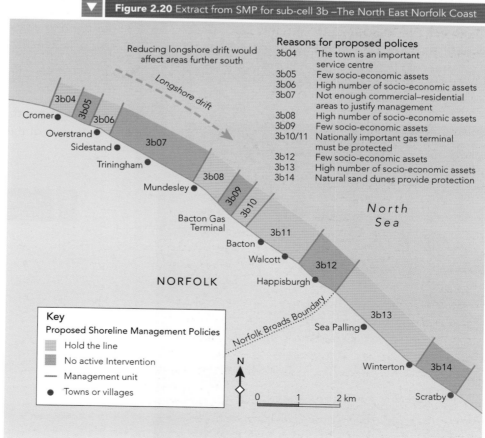

▼ **Figure 2.20** Extract from SMP for sub-cell 3b –The North East Norfolk Coast

Reducing longshore drift would affect areas further south

Longshore drift

Reasons for proposed polices
3b04 The town is an important service centre
3b05 Few socio-economic assets
3b06 High number of socio-economic assets
3b07 Not enough commercial–residential areas to justify management
3b08 High number of socio-economic assets
3b09 Few socio-economic assets
3b10/11 Nationally important gas terminal must be protected
3b12 Few socio-economic assets
3b13 High number of socio-economic assets
3b14 Natural sand dunes provide protection

Cromer • 3b04 3b05 3b06
Overstrand •
Sidestand • 3b07
Triningham •
Mundesley • 3b08
3b09 3b10
Bacton Gas Terminal
Bacton 3b11
Walcott •
NORFOLK 3b12
Happisburgh •

North Sea

Norfolk Broads Boundary
Sea Palling • 3b13

Winterton • 3b14
Scratby •

Key
Proposed Shoreline Management Policies
□ Hold the line
■ No active Intervention
— Management unit
● Towns or villages

N

0   1   2 km

59

Managing coastal erosion

With the increasing development of coastal areas the need for management has grown in order to protect residential and industrial areas as well as communication networks. This need is likely to increase in the future as climate change brings about increasing numbers of storms and rising sea levels.

There are a variety of coastal defence methods being used around our coast; they are usually divided into what is known as 'hard' or 'soft' engineering techniques.

Hard engineering techniques focus on reducing wave energy by putting large structures in place between the sea and the land (Figure 2.21a). Soft engineering techniques work with the existing natural processes rather than attempting to control them. In recent years a technique called '**managed retreat**' or 'coastal realignment' is being increasingly used. This is often seen as soft engineering and works by allowing existing defences to be breached and areas to flood up to their natural level. As sea levels rise, this method of coastal defence may be seen as increasingly economically and environmentally sustainable. Many coastal areas use a mixture of defence methods, often using hard engineering where there is a need to protect the built environment and softer methods where there are fewer built structures to protect. Before any method is put in place **Cost-Benefit Analysis (CBA)** and **Environmental Impact Assessment (EIA)** is carried out in order to assess their economic and environmental costs.

U1
2

AS Geography for OCR

Key terms

Managed retreat: also referred to as managed realignment or coastal realignment.

Cost-Benefit Analysis: a means of evaluating the economic/social and environmental costs and benefits of a project in order to reach a decision.

Environmental Impact Assessment: a means of assessing the environmental impact of a project so that damage can be minimised.

Hard engineering

The following are examples of hard engineering.

- *Sea walls*   Curved, straight or stepped reinforced concrete structures.

- *Groynes*   Wooden or concrete structures designed to break waves and slow down longshore drift.

- *Cliff drainage*   Piped drainage in cliffs to prevent landslides.

- *Rock bund*   A row of rocks along the beach.

- *Rock armour (rip-rap)*   Large rocks placed at the foot of sea walls to absorb wave energy.

- *Gabions*   Steel mesh cages filled with small rocks.

- *Revetments*   Wooden or concrete structures designed to absorb wave energy but still allow a flow of sediment.

- *Offshore bars*   Rows of boulders built up offshore to break waves before they reach the beach.

Figure 2.21b shows the current protection methods in place in an area of the south coast.

Figure 2.21a Hard engineering for coastal protection (seawall and groynes) ▼

▼ **Figure 2.21b** Shows the current protection methods in place in an area of the south coast

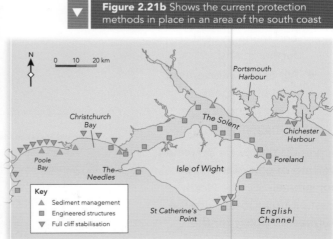

Hard engineering methods are a useful way of controlling erosion and protecting built up coastal areas. They can be expensive – usually between £3 million and £8 million per kilometre – although the cost of flooding might be much higher! Slowing down the movement of sediment by using groynes is very effective but it can reduce beach levels and increase erosion further down drift, causing a **promontory effect**. Also, hard engineering methods are not always visually attractive or very environmentally friendly.

Key terms

Promontory effect: where protected areas begin to protrude seawards as surrounding areas erode.

Soft engineering

The following are examples of soft engineering.

◆ *Beach replenishment* Pumping sand or shingle back onto the beach to replace eroded material.

◆ *Building bars* Underwater bars reduce wave energy.

◆ *Beach reprofiling* Changing the shape of the beach so that it absorbs more energy and reduces erosion.

◆ *Fencing/hedging* Preserves the beach by reducing the amount of sand being blown inland.

◆ *Replanting vegetation* Planting grasses or salt-resistant plants helps to stabilise low-lying areas.

◆ *Beach recycling* Moving material from one end of a beach to the other to counteract longshore drift.

Soft engineering methods try to work with natural processes in an area and are usually based on protecting or preserving the beach. Having a wide, gently sloping beach is excellent as a coastal defence because it absorbs so much wave energy, especially during storms. Soft engineering is usually less expensive and more environmentally friendly – also the resulting beach has amenity value, which is important in tourist areas. This type of coastal defence may not be suitable where areas are very developed and it does require on-going maintenance, which can be expensive.

Managed retreat

Managed retreat (also called coastal realignment, managed realignment or managed coastal realignment) involves abandoning existing coastal defences and allowing the sea to flood inland until it reaches higher land or a new line of defence (Figure 2.22).

Managed retreat is quite a new idea but may become increasingly used as sea levels rise. Allowing low-lying areas to flood and develop into salt marshes produces a natural defence against storms and increases the amount of salt marsh – an increasingly rare environment. It is a relatively inexpensive method of coastal defence where there are few buildings but is not suitable in highly developed areas or where there is high quality farmland near the coast.

Discussion point

Discuss the view that 'soft engineering works with the coasts while hard engineering tries to control it'.

Activities

1 Explain the importance of Cost-Benefit Analysis and Environmental Impact Assessment when considering the management of coastlines.

2 Describe and explain four examples of:
   a hard engineering techniques
   b soft engineering techniques.

3 Explain how managing one part of a coastline may cause problems in another coastal area.

4 Explain why different coastal management strategies may be suitable for different locations.

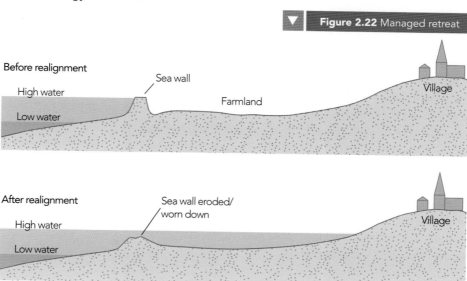

▼ **Figure 2.22** Managed retreat

Coastal management in north-east Norfolk

North-east Norfolk is part of sediment cell 3 (Figure 2.20). The cliffs in the area are largely made up of soft sands and clays, which makes them particularly vulnerable to wave attack as well as rotational slumping caused by rainwater saturation (Figure 2.11). The dominant winds in the area are north-easterly, creating a significant amount of sediment drift from north to south. This creates a challenge for coastal management because protecting one part of the coast may slow down rates of longshore drift and consequently starve beaches of sediment further south.

The proposed Shoreline Management Plan (SMP) for north-east Norfolk has recommended that places with a high level of socio-economic value should be protected while most places with limited socio-economic value should adopt a policy of managed retreat (Figure 2.22). This has caused considerable conflict as people in some places see their homes and villages being lost to the sea in the near future, while nearby towns are protected by expensive coastal engineering schemes.

The following examples describe the current situation and future policy proposals for the coastal settlements of Sea Palling and Happisburgh.

Coastal management at Sea Palling

In 1953 a storm surge broke through the sand dunes at Sea Palling and flooded large parts of north-east Norfolk. Since that time, coastal protection has been seen as increasingly necessary in this area in order to protect the inland area of the Norfolk Broads from flooding.

Following the 1953 floods, a sea wall was built in front of the sand dunes in order to ensure that the inland area was protected from any further storm surge activity. Further to the north at Eccles (Figure 2.23) a number of groynes were built in order to preserve local beaches by trapping sediment being moved southwards by longshore drift. This has proved very effective at Eccles in producing wide open beaches but has reduced the amount of sediment reaching beaches further south. By the 1990s the beach at Sea Palling had been reduced so much that during storms the sea often reached the sea wall, which was being gradually undermined. In 1991 rip-rap was put in front of the sea wall as a temporary measure until a more permanent solution could be put in place.

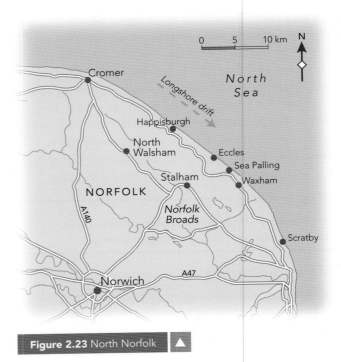

Figure 2.23 North Norfolk

What was done at Sea Palling?

In order to ensure long-term protection for this area, a 'beach management strategy' was put in place in 1992. Since that time the following have been implemented.

- Over 100 000 tonnes of boulders have been put in front of the existing sea wall to prevent it from collapsing.

- The beach has been replenished with one million cubic metres of sand, covering the boulders and widening the existing beach.

- A set of offshore bars, parallel to the coast, have been constructed. These were designed to break the waves before they reached the beach and also absorb wave energy during storm events. Gaps were left between the bars to allow the natural process of longshore drift to continue. A result of this has been a 'tombola' effect where sediment has built up behind the reefs and connected them to the beach, creating a series of bays (Figure 2.24). This has provided added protection and amenity value for Sea Palling but has reduced the flow of sediment further south. Consequently, beach replenishment has been required at Waxham.

- A second set of bars has been built to the south, using a different design to reduce the chance of the 'tombola' effect happening again.

- A third set of bars is planned to the south of Waxham.

U1

2

AS Geography for OCR

Figure 2.24 Sea Palling (north-east Norfolk)

What are the SMP recommendations for Sea Palling?

The following statement (Figure 2.25) is a simplified version of the current SMP statement for Sea Palling.

Policy

Due to the considerable assets at risk and the uncertainty of how the coastline could evolve, the policy is to hold the present line of defence. This policy is likely to involve maintenance of existing sea walls and reef structures, replacing groynes as necessary and continuing to re-nourish beaches with dredged sand. This policy will provide an appropriate standard of protection to all assets behind the present defence line, and, with the recharge, a beach will be maintained as well as a supply of sediment to downdrift areas.

Figure 2.25 Sea Palling SMP

The implications of the proposed plans are shown in Figure 2.26.

Activities

1 Explain why Sea Palling became increasingly vulnerable to coastal flooding.

2 Explain why managing the coast at Sea Palling is not only about protecting the shoreline area, but also the whole coastal zone.

3 Describe the main features of the beach management strategy at Sea Palling.

4 Discuss the view that 'coastal management has not only protected Sea Palling from destruction, it has also provided social and economic amenity for the local area'.

5 Summarise with reference to Fig 2.26 the current Shoreline Management Plan for Sea Palling. Then discuss the predicted implications.

 'Take it further' activity 2.6 on CD-ROM

Property and land use	Nature conservation	Landscape	Historic environment	Amenity and recreational use
No loss of property or land behind the existing defences.	No loss of dunes behind sea wall and beach maintained through recharge.	No change from present.	No loss of sites behind the existing defences.	Beach present (with recharge). Car parking/facilities maintained. Sea Palling lifeboat station maintained. No change to facilities behind existing defences.

Figure 2.26 Predicted implications of the SMP for Sea Palling

Coastal management at Happisburgh

Coastal management at Happisburgh has become a contentious issue in recent years as the soft cliffs continue to be eroded away and increasing number of homes and businesses are put at risk. The original timber defences, constructed in the late 1950s, have been destroyed by the force of the waves, leaving the cliff face open to wave attack during storm conditions (Figure 2.27). The local authority identified the need to replace the existing defences in the 1990s in order to protect the village of Happisburgh.

In 2001 the central government agency responsible for approving and funding major coastal defence works proposed a scheme which would involve the building of rock groynes. It was thought that this would encourage the accumulation of sediment and create a wider beach in front of the existing cliffs. This newly created beach would be able to absorb the wave energy during storm conditions, reducing the wave attack on the cliffs. There were a small number of objections to the proposal and it was eventually rejected on financial and technical grounds. In the meantime land and buildings continued to be lost to the sea.

In 2002, as an emergency measure, the North Norfolk District Council funded the provision of 4000 tonnes of boulders in order to create a rock bund at the base of the cliff. This was seen as a temporary measure to slow down the rate of erosion until a more permanent solution could be found.

The hopes of local people for a more permanent coastal defence scheme were dashed when the proposed Shoreline Management Plan (SMP) for the area was published in 2006. The plan proposed that the area should have 'no active intervention' and that the coast should be allowed to retreat (the possible implications of this are shown in Figures 28 and 29).

In 2007 the temporary **rock bund** was extended, aided by £50000 from the 'Buy a rock for Happisburgh' local appeal.

Summary of the shoreline management plan recommendations and justification

> *It would not be appropriate to defend Happisburgh due to the impact this would have on the shoreline as a whole, as the coastal retreat either side would result in the development of this area as a promontory making it impact significantly upon the sediment transport to downdrift areas. Although there are implications, such as loss of residential properties and amenities at Happisburgh, these are not sufficient to economically justify building new defences along this frontage. The existing rock bund will continue to have a limited effect on the retreat ratio in the short term (5–10 years), but will not prevent cliff erosion.*

What are people's views?

'All local people say they want is a clear strategy for the defence of Happisburgh which is going to be carried through.'

'The government makes millions of pounds from licences for offshore dredging but will not spend anything protecting this part of the coast.'

'Happisburgh has a number of important heritage sites which will be lost forever if the area is not protected.'

'When I bought this house there was another row of houses between us and the sea. Today the sea is only 20 metres away.'

'The problem is that if the beach is developed here, places further down the coast may be starved of beach material – so they will suffer in the future.'

'The recent SMP means that only larger towns will be protected. Villages like Happisburgh will be left to fall into the sea.'

'Happisburgh is a thriving community with local facilities and a worthwhile tourist industry so it is worth saving.'

'The value of local houses has fallen. Who would want to buy a house in this area?'

Discussion point

Why do the Shoreline Management Plan proposals at Sea Palling and Happisburgh suggest different strategies? What local conflicts might this create?

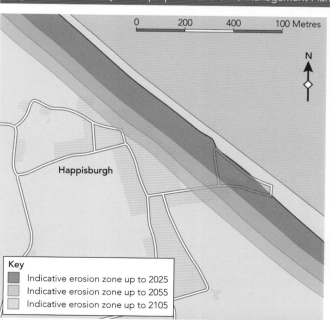

Figure 2.27 Happisburgh, North Norfolk ▼

Time period	Property and land use	Nature and historic environment conservation	Amenity and recreational use
By 2025	• Loss of less than 15 properties (commercial and residential) • Loss of cliff top caravan park land at Happisburgh • Loss of HM Coastguard Rescue facility • Loss of Grade 1 agricultural land	• Continued exposure of Happisburgh SSSI cliffs • No loss of cliff top heritage sites	• Little or no beach • Access may be maintained at Happisburgh
By 2055	• Cumulative loss of between circa 15 and 20 properties (commercial and residential) • Further loss of cliff top caravan park land at Happisburgh • Further loss of Grade 1 agricultural land	• Continued exposure of Happisburgh SSSI cliffs • Grade I St Mary's Church and Grade II Manor House at risk of erosion	• Beach present, but probably loss of existing access at Happisburgh
By 2105	• Cumulative loss of between circa 20 and 35 properties • Loss of cliff top caravan park land at Happisburgh • Total loss of up to approx. 45 hectares of Grade 1 agricultural land	• Continued exposure of Happisburgh SSSI cliffs • Probable loss of Grade I St Mary's Church and Grade II Manor House	• Beach present, but probably loss of existing access at Happisburgh

Figure 2.28 Summary of the proposed Shoreline Management Plan at Happisburgh ▲

0   200   400   100 Metres

N

Happisburgh

Key
- Indicative erosion zone up to 2025
- Indicative erosion zone up to 2055
- Indicative erosion zone up to 2105

▲ **Figure 2.29** Implications of the proposed SMP at Happisburgh

Activities

1 Why is the coast at Happisburgh particularly vulnerable to wave attack?

2 Describe the key features and impacts of the proposed Shoreline Management Plan for Happisburgh over the next 100 years.

3 Examine the conflicts created by the issue of coastal management at Happisburgh.

4 Do you think Happisburgh should be protected from coastal erosion? Explain your answer.

⊙ 'Take it further' activity 2.7 on CD-ROM

Case study | Soft engineering – the Pevensey Bay coastal defence scheme

The Pevensey Bay defence area stretches from Eastbourne in the west to Bexhill in the east, on the Sussex coast (Figure 2.30). The Shoreline Management Plan (SMP) recommendations for this area for the present, medium and long term is to 'hold the line' and protect the coast. There are a number of reasons for this.

- The immediate inland area has over 10000 properties.

- The area has a significant number of commercial premises, including several caravan parks.

- The main coastal road (A259) and railway network are close to the coast.

- The Pevensey levels, an environmentally sensitive area (Site of Special Scientific Interest and Ramsar Wetland Site of international importance) would be flooded with salt water if the coastal defences failed.

- The area has a large quantity of high-value agricultural land.

In order to manage this part of the coast a group of companies joined together to form a new single enterprise called Pevensey Coastal Defence Ltd. In June 2000 the new company signed a 25-year contract with DEFRA guaranteeing that the coastal defences of the area are maintained until 2025. One of the stipulations set by DEFRA was that any coastal management had to be environmentally sustainable and work with the existing coastal processes. Having

a 25-year contract gave the company the opportunity to experiment with new methods of coastal defence which could be fully monitored and evaluated.

How is the area being managed?

A number of techniques are being used to manage the area which include the following.

Beach replenishment and profiling

The dominant waves are from the south-west, which means that sediments are moved from west to east along the beach by longshore drift. This results in a loss of beach material of approximately 20000 cubic metres a year. Beach sediment is replaced by **dredging** it from the ocean floor and spraying it as a slurry back on to the beach. The sediment is then moved by bulldozer, to give the beach a natural profile. Replenishment creates a wide beach which is able to absorb wave energy during storm events. The beach is constantly monitored through a Global Positioning System (GPS), which shows when and where more beach material is required and when profiling needs to take place (Figure 2.31).

> **Key term**
>
> **Dredging:** taking sediment from a seabed or river bed.

Figure 2.30 Pevensey Bay (Eastbourne to Bexhill) ▼

EAST SUSSEX

A22 Hailsham
A259
Bexhill
Pevensey
Cooden Beach
A27
Pevensey Bay
Eastbourne

English Channel

Beachy Head

0   3   6 km

N

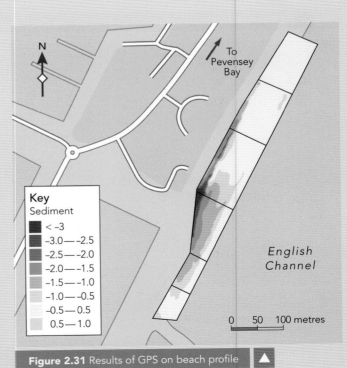

Key
Sediment
< −3
−3.0 – −2.5
−2.5 – −2.0
−2.0 – −1.5
−1.5 – −1.0
−1.0 – −0.5
−0.5 – 0.5
0.5 – 1.0

To Pevensey Bay

N

English Channel

0   50   100 metres

Figure 2.31 Results of GPS on beach profile ▲

Beach recycling

Two or three times a year beach material is moved from the eastern end of the coastal defence area and is deposited back to the western end to counteract the natural process of longshore drift. Infilling is also carried out where the beach has narrowed through natural processes.

Beach reprofiling

During winter storms beach material can be lost from the higher part of the beach and deposited on the lower part nearer the sea, where it can then be removed by wave action. In order to reduce this effect, the beach is regraded by moving material back up the beach. This reduces longshore drift and therefore helps to preserve the beach.

Groyne replacement

In 2000 there were 150 groynes in the coastal defence area, many of which were damaged. Most are being removed and replaced by a small number of more strategically placed groynes. This creates a more open and attractive beach which is important since the beach has high amenity value for the tourist industry.

Use of different materials

The defence scheme has been used to experiment and monitor the use of different materials and techniques including:

◆ recycled plastic, which is being used on some of the groynes in order to reduce the use of tropical hardwoods

◆ bales of old car tyres, which are buried in the upper part of the beach to provide additional bulk and stability to the beach material.

Wooden retaining walls

Wooden retaining walls have been built at the landward edge of the beach to allow a steeper shingle beach to be built and prevent beach material being blown inland onto local coastal roads.

Has the scheme been successful?

The scheme is seen as environmentally sustainable because it works with natural processes rather than using heavy engineering to control them. Maintaining and protecting the beach is seen as the most effective form of coastal defence in the area because a wide, gently sloping beach is able to absorb wave energy very efficiently.

The scheme is being run within the agreed budget and the company are working closely with both local people and DEFRA. It is too early to fully assess the scheme, but initial signs look promising in both economic and environmental terms.

Discussion point

Discuss the view that 'the best type of coastal defence is always a beach'.

Activities

1   Using the information from page 66 and Figure 2.29, explain why there is a need for coastal management at Pevensey Bay.

2   a   Explain what is meant by beach replenishment and beach profiling.

　　b   How do these methods help to reduce wave energy?

3   Explain how coastal management strategies at Pevensey Bay use information gathering technology and experimental materials and techniques.

4   a   Consider the Pevensey Bay coastal management scheme in terms of environmental impact assessment costs, benefit analysis.

　　b   Why is the management of Pevensey Bay referred to as 'soft engineering'?

Case study | A multi-engineered coastline – Ventnor to Bonchurch, Isle of Wight

Much of the coastline between Ventnor and Shanklin on the Isle of Wight lies within a landslide area known as the Undercliff (Figure 2.32). The geology in this area is a complex mixture of chalk, greensand and clay and has no real uniformity. Because of this the area has a range of landslide features including:

◆ rotational slumping, which creates linear 'benches' separated by steep slopes

◆ mudslides where clay is exposed and vulnerable to movement during wetter periods

◆ rock falls and coastal landslides, especially where exposed areas of cliff are not protected from wave attack.

Research in the area suggests that there are at least three different landslide systems operating and that the area becomes increasingly unstable after periods of heavy rainfall and during storm conditions.

The coastline is vulnerable to wave attack because of the following reasons.

◆ During storm conditions Atlantic waves from the south-south-west create a powerful storm surge over 1m higher than normal predicted levels (Figure 2.32).

◆ There is a limited supply of drift material in the area so in some places beaches are narrow or non-existent and offer little protection to the cliffs from wave attack.

◆ During storm conditions waves pick up gravel and attack the cliffs or existing coastal defences through shingle abrasion.

Why is coastal erosion an issue in this area?

This area is both a coastal resort and a residential area and has millions of pounds worth of commercial and residential property close to the coastline. The main east-west communications route passes close to the edge of the coast in some places. The town of Ventnor (Figure 2.33), originally developed largely in Victorian times, has a significant proportion of older buildings – many of which were not built with land movement protection in mind. It is estimated that serious slope failure on this coast could affect properties up to 200m inland, with enormous social and economic costs.

▲ **Figure 2.33** Ventnor town and coast, Isle of Wight

Coastal management between Ventnor and Bonchurch

Ventnor sea front and the coastline to the east of Ventnor have been protected from wave attack for many years but in the past the defences were often piecemeal and not well co-ordinated. By the 1970s many of the existing defences were suffering from corrosion and simply collapsing, leaving the area vulnerable to wave attack, especially during winter storms.

During the last 25 years a number of heavy engineering schemes have been put in place in order to protect the whole of the coastline between Ventnor and Bonchurch. The major example of engineering has been a continuous

Figure 2.32 Solent, Isle of Wight ▼

0   10   20 km
N

Southampton
Portsmouth Harbour
Christchurch Bay
The Solent
Chichester Harbour
Poole Bay
Hurst Spit
The Needles
Hengitsbury Head
Isle of Wight
Foreland
Shanklin
Bonchurch
Ventnor
St Catherine's Point
English Channel

Key
— High wave energy levels
— Moderate wave energy levels
— Low wave energy levels
— Cliff more than 15m in height
— Cliff less than 15m in height
→ Sites of cliff failure
▲ Major stack, arch and large cave
▬ Well developed shoreline platform
⋯⋯ Boulder-dominated beach

sea wall, the design of which varies according to when it was built (Figure 2.34).

Environmental Impact Assessment

Part of this area is recognised as having valuable ecological characteristics and priority habitats – consequently ecological assessments have been carried out. However, on-going geo-technical reports have made it clear that heavy engineering options are really the only viable option to protect the area from wave attack and sub-aerial processes.

Key terms

Habitat: a specific environment where plants and animals live.

Ecology: the study of the relationship between plants, animals and their environments.

Discussion point

Why is there a socio-economic need to protect the Ventnor–Bonchurch coastal zone?

Activities

1 Explain why this part of the Isle of Wight coastline is vulnerable to erosion.

2 Draw an annotated sketch to describe and explain the methods of coastal management used at Wheeler's Bay (Figure 2.35).

3 a Why are the management techniques used in the Ventnor area referred to as heavy coastal engineering?

   b Why was this type of coastal management chosen for the Ventnor area?

Slope stabilisation of previous landslide area

Straight-edged sea wall

Rip-rap – large boulders placed in front of sea wall

Metal reinforcement backed by concrete

Concrete tetrapods placed in front of sea wall

Cliff drainage

Promenade along top of sea wall

Lipped sea wall

Figure 2.35 Heavy engineering at Wheeler's Bay to east of Ventnor, Isle of Wight ◀

Case study | Managed retreat – Abbotts Hall Farm, Essex

Abbotts Hall Farm, at Great Wigborough in the Blackwater estuary (Figure 2.36), is one of the largest managed retreat schemes in Europe. It covers nearly 280 hectares of land on the north side of the **estuary**. The Blackwater estuary is the largest salt water estuary in East Anglia and has been recognised as having special importance for nature conservation. It has Special Protected Area (SPA) status (for the protection of migrating birds) and also contains Sites of Special Scientific Interest (SSSI).

> **Key term**
>
> **Estuary:** lowest part of a river which is affected by tides.

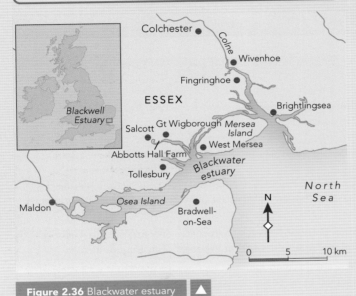

Figure 2.36 Blackwater estuary

Why was Abbotts Hall Farm chosen as an area for management retreat (realignment)?

During the 1990s the World Wildlife Fund (WWF) and Wildlife Trust presented evidence to the government which made it clear that:

- large areas of salt marsh and mudflats were being lost to coastal erosion and development
- the loss of salt marsh environments had a significant impact on flora and fauna
- well-established salt marsh can provide a cost-effective defence against storms and coastal flooding
- heavy engineering in environmentally sensitive areas could damage both economic and environmental opportunities.

In order to demonstrate the effectiveness of coastal realignment, Abbotts Hall Farm was bought by the Essex Wildlife Trust in 2000 with the aim of removing parts of the existing sea wall and allowing the land to be flooded by the incoming tide, creating a broader salt marsh environment. The scheme was supported by the Environment Agency, English Nature and the World Wildlife Fund.

This area was seen as a suitable site for managed realignment because there is only a narrow strip of land between the existing 3km sea wall and the 5m contour line. This means that when the land was flooded and fully developed as a salt marsh, both the vegetation and the natural gradient of the land would provide coastal defence against storm events.

What were the aims of the project?

- To support the work of the Essex Wildlife Trust.
- To provide leisure and recreation facilities.
- To show how farming and nature conservation can work together.
- To show how some coastal areas can be protected without the need for heavy engineering.
- To increase the nursery environment for shellfish and fish.

How was the coastal realignment process managed?

The low-lying parts of Abbotts Hall Farm had been protected by a sea wall for many years. This had effectively fixed the salt marsh between the sea and the sea wall and increasing storm activity had eroded away a large area of the salt marsh and mud flats, a process known as 'coastal squeeze' (Figure 2.37).

> **Key term**
>
> **Coastal squeeze:** reduction of an environment due to natural processes and development.

In order for the coast to be realigned, the sea wall had to be breached in a number of places. A number of conditions had to be satisfied to obtain the planning permission required in order to breach the existing sea wall. The conditions included were as follows.

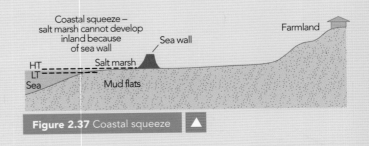

Figure 2.37 Coastal squeeze

- The construction of counter walls at both ends of the existing sea wall to protect neighbouring farms.

- The construction of **earth bunds** to protect freshwater areas from saltwater intrusion.

- An Environmental Impact Assessment (EIA) had to be carried out and the area monitored for three years prior to realignment to assess the movement of sediment and the flora and fauna.

- Archaeological surveys and excavations to identify evidence of ancient salt workings and habitation.

- Making sure the work was carried out with due care for the existing environment. This meant:

  - creating new environments for displaced animals

  - protecting existing freshwater ponds

  - not carrying out work during bird migrations.

Once these conditions were agreed and responded to, the sea wall was finally breached in October 2002.

> **Key term**
>
> **Earth bund:** mound of earth.

What are the effects of realignment?

A monitoring programme is being carried out to assess the effect of the realignment and provide guidance for future realignment schemes. Figure 2.38 shows what the area looks like today.

Discussion point

Discuss the view that 'managed retreat may not always be a viable option for managing coastal areas'.

Activities

1 Why is managed retreat seen as an increasingly important method of coastal management?

2 Explain why this part of the Blackwater estuary was a suitable location for a policy of managed retreat.

3 What is meant by 'coastal squeeze'?

4 Use an annotated sketch to describe the processes and impact of coastal realignment at Abbotts Hall Farm then construct a table to show the costs and benefits of the coastal realignment scheme at Abbotts Hall Farm.

Figure 2.38 Plan of Abbotts Hall Farm ▼

ABBOTTS HALL FARM, GREAT WIGBOROUGH

Abbotts Hall Farm, situated on the north side of the Blackwater estuary near Colchester, dates back to the Domesday survey of 1085 when it was "held" by Barking Abbey. In 1540, it was granted to Thomas Cromwell and was later owned by Thomas Howard, both of whom were beheaded! In 1810, the sale price was £23,500 – rather different from the £2.3 million asking price in 1999. In addition to the 283 hectares of land, Abbots Hall comprises farm buildings and three tied cottages

2.3 In what ways can coastal areas be a valuable economic and environmental resource?

Coastal areas can provide valuable economic and environmental opportunities, resulting in increasing numbers of people wanting to live near the coast. This trend has become more dynamic in recent years because of the increase in global trade, the development of the tourist industry and the increasing individual demand for coastal lifestyles.

Coastal areas are often placed under competing pressures and this can be a cause of conflict. The challenge in coastal areas is to find sustainable ways of managing an often complex set of demands.

A multi-use coastal area – the Solent

The map shows Southampton Water and the Solent, on the south coast of England (Figure 2.39). It describes some of the ways that this coastal area is used and highlights the main residential areas. A number of which have grown rapidly in recent years around small marinas. Some are used on a permanent basis and others as holiday homes.

Discussion point

Why did the proposal to build a container port at Dibden Bay create conflict?

Activities

1 Why is the Solent described as a multi-use coastal area?
2 Using examples, consider the reasons for the industrial, recreational and residential development in the Solent coastal area.

 'Take it further' activity 2.8 on CD-ROM

① **Dibden Bay**
An area of mud flats and grassland which attracts thousands of wild birds and is recognised as being an important coastal environment by the Royal Society for the Protection of Birds (RSPB). It also incorporates a number of Sites of Special Scientific Interest. Associated British Ports wanted to build a container port here, but after local opposition and a public enquiry the idea was rejected in 2004.

② **Container facility**
The container facility is part of the deepwater port and covers an area of over 200 acres. It has a frontage of 1350m and is next to the rail freight terminal which handles 25 per cent of the container traffic.

③ **Exxon Mobil oil refinery**
The largest refinery in the UK, covering an area of over 3000 acres with a terminal frontage of 1500m. The refinery handles over 2000 ship movements a year. The refinery takes oil from all parts of the world and turns it into fuels and petrochemicals. The refinery supplies about 15 per cent of all petroleum products in the UK. The total workforce is over 3000.

④ **Fawley chemical manufacturing plant**
The chemical plant produces over a million tonnes of chemical products a year. The main products are chemicals used in paints, adhesives and rubber for vehicle tyres. About 80 per cent of the chemicals produced are exported by ship from the deepwater terminal.

⑤ **Fawley Power Station**
Fawley is one of the largest oil-fired power stations in the UK.

⑥ **Calshot Activities Centre**
Calshot Activities Centre is one of the largest outdoor activity centres in Britain. It offers a range of water-based activities including windsurfing, canoeing, sailing, power-boating and water-skiing.

⑦ **Keyhaven Nature Reserve**
The nature reserve extends to over 2000 acres of salt marsh and mud flats. The area supports nationally and internationally important numbers of migrating birds and is important for its range of plants and animal life.

⑧ **Thorness Bay Holiday Village**
Holiday park overlooking the sea with a range of land and water-based facilities for family holidays.

⑨ **The City of Southampton**
With a resident population of over 250000 Southampton is the largest city in the south of England.

⑩ **Roll on/Roll off facility**
Nearly 750000 vehicles a year are delivered from Asia, North America and Africa. There is a vehicle storage and distribution compound which holds 6000 cars.

⑪ **Bulk terminal**
The bulk terminal has a 28000 tonne grain facility and a flour mill operated by RankHovis. It also has a fertiliser importing/processing plant. A glass reprocessing facility handles recycled glass brought in by coastal vessels.

⑫ **Cruise terminal**
Southampton is the UK's busiest cruise port handling over 200 cruise ships a year and 700000 passengers. It is the home of P&O and Cunard vessels and has three dedicated cruise terminals with reception areas and baggage handling facilities.

⑬ **The River Hamble**
Contains several large marinas and a number of boatyards which are used by local and international yachtspeople, both for recreational and competitive sailing. The area is increasingly used by recreational motor boat owners.

⑭ **Cowes**
Cowes is the Isle of Wight's main port and passenger ferry terminal. It has a large natural harbour used by recreational yachtspeople, fishing boats and powerboat owners. The town hosts the world famous 'Cowes Week' yachting regatta.

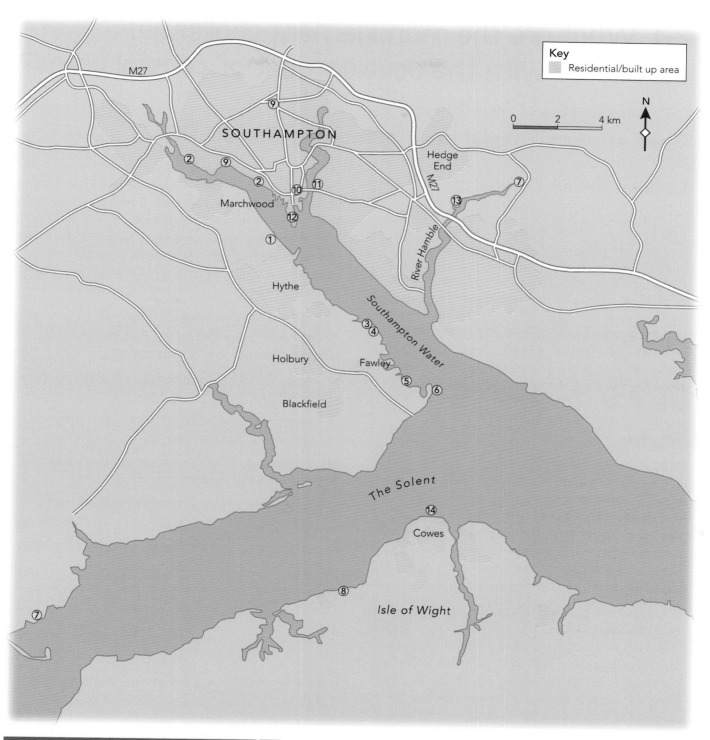

Figure 2.39 Southampton Water and the Solent ▲

⊙ **Additional case study on CD-ROM**: Using the coast as an economic resource: tourism development – Bahia, north-east Brazil

2.4 What are the management challenges associated with the development of coastal areas?

Case study | Coastal development in Florida, USA

Coastal development

In many parts of the world the coastal zone has become a core growth area while the inland area remains peripheral. Rapid development in coastal areas generates increasing socio-economic opportunities, creating the stimulus for population growth, which in itself creates a demand for on-going infrastructural development. Added to this is the attraction of coastal areas as both holiday destinations and locations for second home owners. The following case studies look at two parts of the world where there has been and continues to be rapid coastal development.

Florida

Florida is known as the 'sunshine state' in the USA because of its sub-tropical climate, which becomes increasingly tropical as you move southwards towards Florida Keys (Figure 2.40). The high temperatures can bring heavy rainfall and extreme climatic events such as lightning, storms, tornadoes and hurricanes. Sea breezes cool the coastal areas and make them attractive places for both residents and holidaymakers.

Florida has the fastest growing GDP (averaging 7 per cent between 2000 and 2006) of any US state and the third fastest rate of population growth, increasing from 9.7 million people in 1980 to 19 million in 2007. The comfortable climate, economic opportunities, high living standards and range of leisure activities are seen as major attractions for the working population and the area is increasingly seen as a magnet for second home owners and the retired.

Florida has a long history of tourist development and has a number of internationally known theme parks, including Walt Disney World Resort, Universal Orlando Resort, Busch Gardens and Sea World. The coastal region attracts millions of visitors each year, making tourism the largest sector of employment. Most of the development has been along the south-east coast, with the west coast becoming increasingly popular in more recent years.

Pressures on the coastal area

Economic development on the southeast coast has created a number of problems including:

- Algae blooms (called red tide) are linked to water pollution.
- Damage to fragile marine ecosystems linked to agriculture and tourism.
- Marine erosion in some areas creating a need for heavy coastal engineering.
- Sediment loss due to dredging for construction materials.
- The urbanisation of coastal areas with increasing demand for space (Figure 2.41).
- Increasing vulnerability to flooding as sea levels rise.

Activities

1 To what extent is the Florida coast a multi-use area?

2 Explain the reasons for the growth of tourist and residential populations in Florida.

3 Discuss the view that 'managing the Florida coast is about balancing socio-economic and environmental needs.'

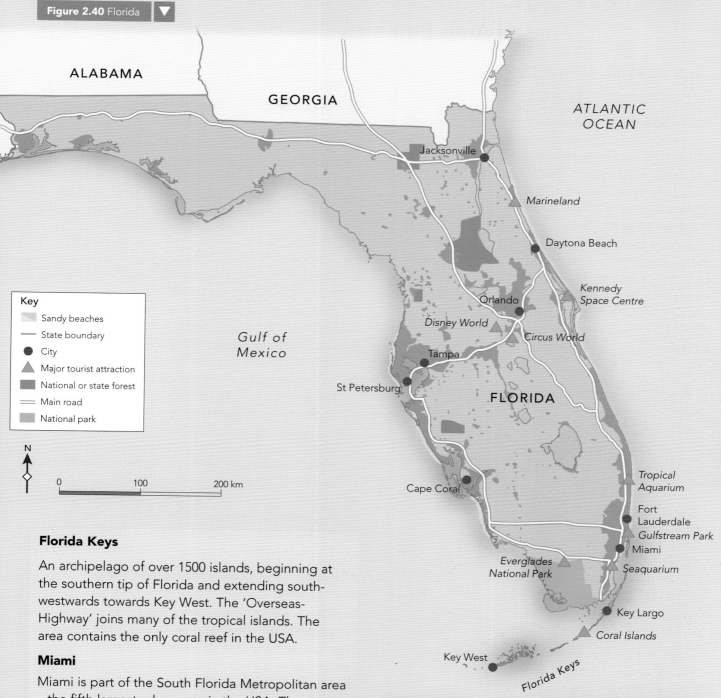

Figure 2.40 Florida ▼

ALABAMA

GEORGIA

ATLANTIC OCEAN

Jacksonville

Marineland

Daytona Beach

Kennedy Space Centre

Orlando

Disney World

Circus World

Gulf of Mexico

Tampa

St Petersburg

FLORIDA

Key
- Sandy beaches
- State boundary
- City
- Major tourist attraction
- National or state forest
- Main road
- National park

N

0   100   200 km

Cape Coral

Tropical Aquarium

Fort Lauderdale

Gulfstream Park

Miami

Everglades National Park

Seaquarium

Key Largo

Coral Islands

Key West

Florida Keys

Florida Keys

An archipelago of over 1500 islands, beginning at the southern tip of Florida and extending south-westwards towards Key West. The 'Overseas-Highway' joins many of the tropical islands. The area contains the only coral reef in the USA.

Miami

Miami is part of the South Florida Metropolitan area – the fifth largest urban area in the USA. The port of Miami has been called the 'cruise capital of the world' and the 'cargo gateway of the Americas'. It is the world's number one cruise liner port and the starting point for many Caribbean cruises. Miami is a linear city; inland growth being restricted by the Everglades National Park. It is a centre for finance and the headquarters for over 1000 multinational corporations including Disney, American Airlines, Exxon, Microsoft and Sony. Most of the built environment is less than 10m above sea level with the main part of the city on the shores of Biscayne Bay, including the world famous Miami Beach (Figure 2.41).

Figure 2.41 Miami Beach ▲

Case study | Coastal development in Dubai, United Arab Emirates

The world's largest coastal reclamation scheme

The largest coastal reclamation scheme in the world is currently under construction on the coast of Dubai, in the United Arab Emirates (Figure 2.42). In 2001 it was announced that the 'Eighth Wonder of the World' was going to be created in Dubai by building two artificial islands, the Palm Jumeirah (Figure 2.43) and the Palm Jebel Ali. The islands, built in the shape of date palm trees, are expected to be completed by the end of 2007. In 2003 it was announced that a third palm-shaped island, the Palm Deira and 'The World', a creation of over 300 islands built in the shape of the earth's continents, will be constructed. When completed, the three Palm developments will increase Dubai's shoreline by over 400km.

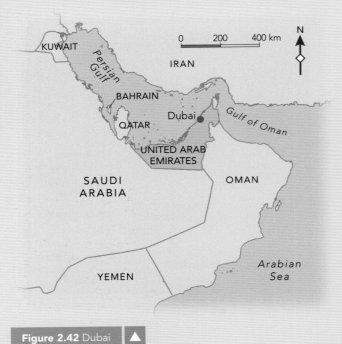

Figure 2.42 Dubai ▲

Dubai – Factfile

◆ Dubai is one of the world's fastest growing tourist locations with visitor numbers increasing from one million in 1990 to over six million in 2004.

◆ The 700km of coastline and sub-tropical climate make it an ideal holiday destination.

◆ It is also attracting increasing numbers of second home owners and foreign residents.

◆ It has recently opened a new cruise terminal and aims to build the tallest building in the world.

The following information is taken from www.nakheel.com, the developer's website.

The Palm Jumeirah, where the vision of Dubai gets built

Launched in 2001, the Palm Jumeirah (Figure 2.42) is the project that started it all. A man-made island in the shape of a palm. Jumeirah created a destination of world-class hotels, unprecedented retail, an array of home types and exceptional leisure and entertainment.

Project features

◆ A range of first class hotels.
◆ The Golden Mile retail destination offering more than 200 retail outlets and 860 residential units comprising townhouses, one-, two- and three-bedroom apartments and penthouses.
◆ The Village Centre featuring 68 000m² of retail and commercial offerings with over 1000 apartments and townhouses and a marina which offers approximately 750 berths for private vessels.
◆ Trump International Hotel and Tower: The Palm Jumeirah will be the first Trump property in the Middle East and will be the centrepiece of The Palm Jumeirah's trunk, featuring both a residential and hotel element.
◆ Shoreline Marina with approximately 400 berths for private vessels between 12m and 60m in length.
◆ Private beachside villas located on the fronds of The Palm will begin occupancy at the end of 2006.

Managing the environmental impacts of the development

In order to ensure environmental preservation and sustainability the developers created an environmental team consisting of over 20 international environmental institutions. A number of studies were carried out prior to starting the development and Environmental Impact Assessments (E.I.A.s) used to inform the planners. Specialist water quality studies were carried out and the information used to ensure that water circulation characteristics would be maintained after the development was completed. Innovative solutions to protect and enhance the environment were put in place, including state of art sewage treatment plants and curved breakwaters which provide a habitat for marine life. A marine biology laboratory is being created in order to monitor activities in the coastal area.

Despite these measures concerns have been expressed about the impact of the development on the marine environment. Some of these concerns are illustrated in Figure 2.44.

- Trump Int. Hotel and Tower
- Village Centre
- Nakheel Sales Centre
- The Fronds
- Shoreline Marina
- Atlantis
- Village Centre Marina
- The Crescent

Figure 2.43 The Palm Jumeirah development, seen from above

Activities

1 Describe the main features of the coastal development schemes in Dubai.

2 Discuss the view that 'developments in Dubai are more concerned with economic development than environmental management'.

AS OIL RESERVES run low, Dubai is building a series of exclusive resorts to attract the super-wealthy. But, as Severin Carrell hears from local experts, the natural world is already taking a battering.

When people stroll out of their luxury villas onto their private beaches on The Palm Jumeirah, it will seem as though they own a little slice of Eden.

However, naturalists claim this exclusive earthly paradise, shaped like a vast palm tree, is threatening the marine environment it exploits.

They argue that this retreat, now rising from the sea, is wrecking fragile coral reefs, devastating local fish stocks that support endangered sea birds, and destroying the seabed. The tiny oil state's last breeding ground for endangered hawksbill sea turtles is threatened by a second identical resort, the Palm Jebel Ali.

In Abu Dhabi there is great concern about the damage these resorts are causing. Abu Dhabi's Environmental Research and Wildlife Development Agency (ERWDA) revealed earlier this year that 'habitat degradation' such as

coastal building had helped cause 'major declines' of local fish populations by exacerbating other environmental problems. It said total stocks of 20 local species, such as the twobar seabream and silver pomfret, had fallen by more than 80 per cent since 1978. It has also raised the alarm about local hawksbill turtles – a globally endangered species – and green sea turtles.

However, they admit that these resorts will provide new homes for local sea life. Artificial reefs – reportedly including an old airliner – are being built to attract coral and fish and

will simulate 'famous dive destinations' such as the Maldives and Bali.

But independent experts fear the oil, noise and disturbance caused by yacht marinas, pleasure craft and jet skis will increase the pressure on the area.

Figure 2.44 Environmental concerns about the development – adapted from the *Independent* newspaper

Coastal areas and economic development – Rio de Janeiro

Many of the world's major cities and industrial areas are found in coastal areas because they offer advantages for industry and the development of international trade. People are attracted to coastal areas because of the economic and social advantages that they offer, including leisure and recreational possibilities. Rio de Janeiro (south-east Brazil) developed on Guanabara Bay (Figure 2.45), one of the largest natural harbours

many of the new companies. The area quickly became the centre for a wide range of industries, including vehicle manufacturing, chemicals, metal making, clothing and food processing. Industrial expansion has continued and includes the development of Brazil's major technology complex with over 200 national and international companies. There are currently discussions about the building of a new steel plant and another petro-chemical works in the area. In recent years, the warm climate and natural landscape of the area has attracted increasing numbers of tourists. Of particular attraction are the crescent-shaped bays of Copacabana and Ipanema, backed by mosaic promenades and expensive hotels.

The growth of industry in Rio de Janeiro has created enormous wealth for some of the inhabitants. This has encouraged the development of newer residential areas along the coast as wealthier people seek to move away from the city centre. Barra da Tijuca is an example of this type of development. Built on a sand dune coast in a stunning beach environment, it is a suburb of modern residential areas, shopping malls and extensive leisure facilities. The 18km-long sandy beach is one of the main attractions. Surfing is a popular past-time and there are a number of surf schools in the area. Barra is becoming an increasingly popular tourist resort and there are fears that it might become overdeveloped and face the pollution problems of some of Rio's other popular beaches. In order to protect the area from environmental destruction, nature reserves and wild bird sanctuaries have been set up and some beaches have been designated 'preserved beaches' with limited car access.

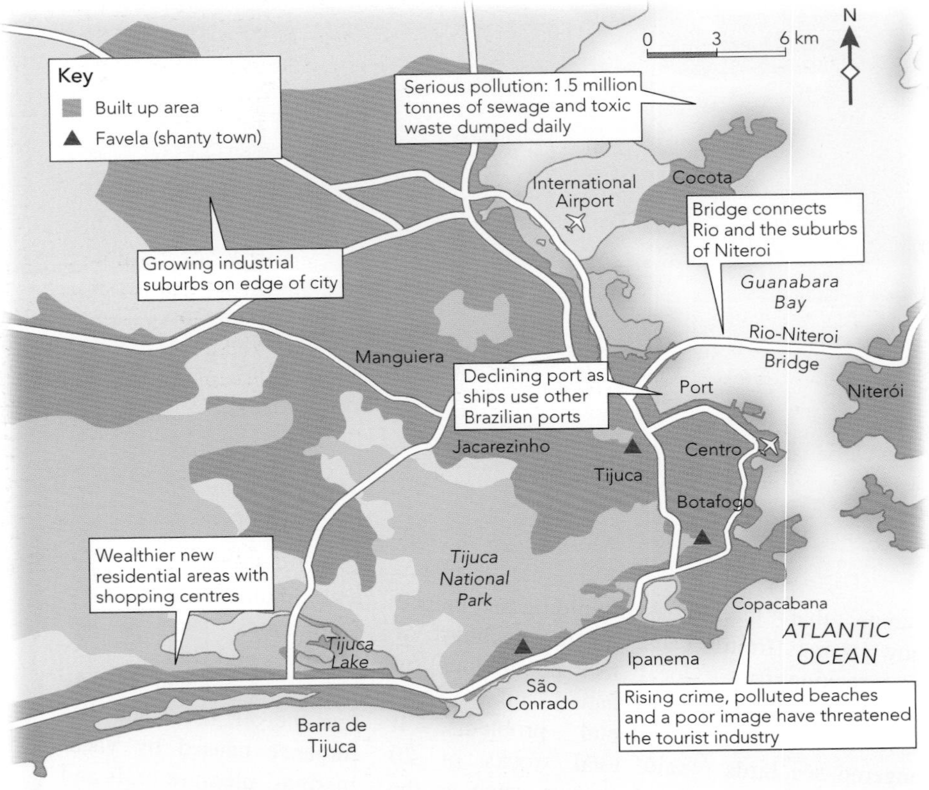

Figure 2.45 Rio de Janeiro

Key
- Built up area
- ▲ Favela (shanty town)

Serious pollution: 1.5 million tonnes of sewage and toxic waste dumped daily

Growing industrial suburbs on edge of city

Bridge connects Rio and the suburbs of Niteroi

Declining port as ships use other Brazilian ports

Wealthier new residential areas with shopping centres

Rising crime, polluted beaches and a poor image have threatened the tourist industry

in the world. The harbour is deep enough to handle the largest cargo ships, oil tankers, car transporters and cruise liners. The port has developed into the major centre for Brazil's imports and exports and is the distribution point for trade throughout the south-east, Brazil's industrial heartland.

In the 1940s large-scale industrial development started in this area with the building of the government-owned petrochemical works and steel-making plant. In the1950s and 1960s the area was seen as an industrial **growth pole** and **transnational companies** were encouraged to move in; the nearness of a large, sheltered port being seen as a significant advantage to

- **Key terms**

  Growth pole: an area within a country which is a target for economic development.

  Transnational corporations: a company operating in many different counties.

However, there remains a significant wealth gap in Rio, with a large proportion of the 12 million inhabitants

SEWAGE, ALGAE AND RUBBISH STAIN THE ONCE-PRISTINE WHITE BEACHES OF RIO

STANDING ON THE PUTRID banks of Rio's Cunha canal it is hard to believe that 30 years ago this was a favourite thoroughfare for dolphins, surrounded by pristine strips of sand.

These days the burnt-out chassis of an abandoned Volkswagon pokes through the surface of the black sludge and the air is permeated with the acidic stench of sewage, which flows into the water from the Complexo da Mare, a vast shantytown.

'There was white sand over there,' says Waldeck Monteiro, 42, a fisherman. 'You could swim in the water and there were fish all over the place. Now if you put your net in you'll probably end up pulling out a corpse.'

Pollution is nothing new to Rio de Janeiro, but environmentalists say many of

the city's waterways now represent a grave threat to public health and Rio's tourism industry.

In January a stretch of the Barra da Tijuca beach was cordoned off after toxic algae appeared in the water, and at the end of March authorities removed a tonne of dead fish from Guanabara bay.

Dark stains known as 'black tongues' periodically appear on Rio's beaches, and strips of white and yellow foam – the result of untreated sewage – have started to show up.

'This is now a question of public health,' said Mario Moscatelli, a biologist and prominent environmental activist, who says he has vaccinated his two young daughters against Hepatitis A and B so they can use the beaches.

Activists hope Rio's

Rio from the air

newly appointed environment secretary, the environmentalist Carlos Minc, can achieve some success. His deputy, Izabella Teixeira, said the state government planned to spend R$140m (£35m) to clean up Rio's beaches and lakes; further money would be put into sewage treatment projects.

Environmentalists have welcomed plans for a 'green revolution' in Rio. They warn, however, that

200 years of pollution cannot be undone in four years of government. 'In places where you could have tourism you have giant dustbins of litter and sewage. The city is still marvellous, but it is being trashed.'

Figure 2.46 Tourism development in Rio de Janeiro – adapted from the *Guardian* newspaper ▲

living in squatter areas throughout the city, often with a lack of basic facilities. Industrial and urban growth has put great pressures on the coastal zone. Some of these pressures are described in Figure 2.46 and illustrate the conflict between economic development and environmental protection in coastal areas.

Discussion point

Examine the conflict between economic development and environmental protection in the Rio de Janeiro coastal area.

Activities

1 Explain why the Rio de Janeiro coastal area has been called 'a valuable economic and environmental resource'.

2 Use the newspaper article (Figure 2.46) to describe how pollution is 'threatening the beauty of Rio de Janeiro'.

 'Take it further' activity 2.9 on CD-ROM

Coastal ecosystems under threat – coral reefs

Coral reefs have been called the 'gardens of the sea' because of the variety of plant and animal life they contain, including one-third of the world's fish species. They are one of the earth's most dynamic ecosystems, producing more plant material each year than tropical rainforests (Figure 2.47).

Ecosystem	Productivity (grams)
Coral reefs	2500
Tropical rainforests	2200
Temperate forests	1250
Savanna grasslands	900
Open sea	125
Semi-desert	90

Figure 2.47 The productivity of selected ecosystems (amount of plant material produced per m²/year)

How are coral reefs formed?

Coral reefs are formed from colonies of small animals called polyps and one of the world's most spectacular ecosystems (Figure 2.48). Particular conditions are needed for the growth of reef building corals, including:

◆ warm sea temperatures – reefs only develop where the average water temperature is above 18°C and are most productive in stable water temperatures between 22 and 25°C.

◆ light – required for photosynthesis to take place.

◆ high salt levels – close to the open sea.

◆ clear water – sediment-rich water will block the light and smother reefs.

Figure 2.49 A coral reef with tropical fish

These conditions are mainly found in shallow water, either side of the equator, generally away from freshwater estuaries (Figure 2.49).

Why are coral reefs important?

Coral reefs are important for a number of reasons.

◆ They provide the foundations for many islands.
◆ They provide a barrier against storm waves.
◆ They contain many rare species and chemicals, some of which have been used in the treatment of disease.
◆ They support the fishing industry, providing food and income for local communities.
◆ They are a major attraction for tourists – over 200 million tourists visit coral reefs each year.

Why are coral reefs under threat?

Coral reefs grow very slowly and are easily damaged by small changes to their environment. In many parts of the world coral reefs are increasingly under pressure because of:

◆ pollution caused by sewage outfalls, agricultural and industrial runoff
◆ damage by boats or divers
◆ collection of coral for souvenirs
◆ increasing sediment being deposited in rivers
◆ overfishing
◆ mining of the coral for building materials
◆ global warming – increasing the frequency of damaging storms.

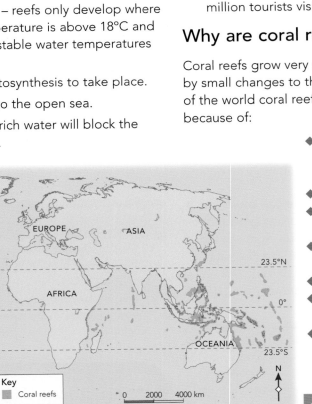

Figure 2.48 Location of coral reefs

Case study | Protecting coral reefs – the Soufriere Marine Management Area, St Lucia

Tourism is one of the fastest growing industries in the Caribbean, creating both economic opportunities and challenges for managing what are often fragile environments. In order to resolve conflicting demands and protect the most productive reefs on the west coast of St Lucia, The Soufriere Marine Management Area (SMMA) was formed in 1994 (Figure 2.50).

The SMMA was formed after consultation with local people, particularly those involved in the fisheries and tourism sectors. The SMMA agreement established five different zones within the area surrounding Soufriere. The aim of the zoning system is to cater for the wide range of users and encourage the economic and environmental sustainability of the area.

What does it do?

The SMMA is involved in a number of activities:

◆ the regular monitoring of coral reefs and water quality

◆ managed provision for different users

◆ public information and education

◆ the promotion of appropriate development

◆ the enforcement of rules and regulations

◆ conflict resolution.

Has it been successful?

Recent surveys show that the reserve area has had an increase in commercial fish biomass and biodiversity as well as a reduction in the level of damage to the reef by human activity.

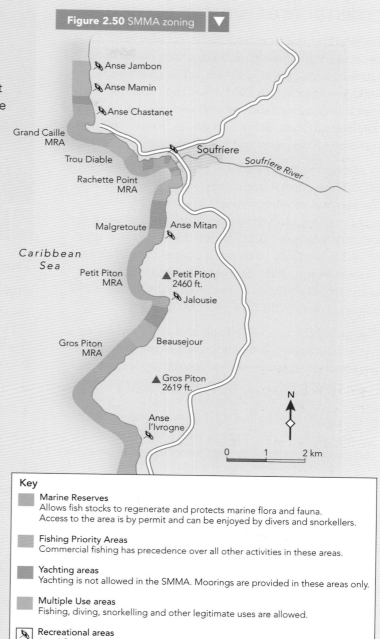

Figure 2.50 SMMA zoning ▼

Anse Jambon
Anse Mamin
Anse Chastanet
Grand Caille MRA
Trou Diable
Soufríere
Soufríere River
Rachette Point MRA
Malgretoute
Anse Mitan
Caribbean Sea
Petit Piton MRA
▲ Petit Piton 2460 ft.
Jalousie
Gros Piton MRA
Beausejour
▲ Gros Piton 2619 ft.
Anse l'Ivrogne
N
0   1   2 km

Key

Marine Reserves
Allows fish stocks to regenerate and protects marine flora and fauna. Access to the area is by permit and can be enjoyed by divers and snorkellers.

Fishing Priority Areas
Commercial fishing has precedence over all other activities in these areas.

Yachting areas
Yachting is not allowed in the SMMA. Moorings are provided in these areas only.

Multiple Use areas
Fishing, diving, snorkelling and other legitimate uses are allowed.

Recreational areas
Areas for public recreation – sunbathing, swimming.

Activities

1 Why are coral ecosystems seen as valuable economic and environmental resource?

2 Study the case study

  a Explain the need for planning and management in this area.

  b Describe and explain the methods used to manage the coral ecosystem in this area.

Discussion point

Why are coral ecosystems often described as 'fragile'?

⊙ 'Take it further' activity 2.10 on CD-ROM

The Mediterranean – a coastal area under pressure

The Mediterranean coastal zone includes over 20 countries from three continents and is home to over 160 million people (Figure 2.51). The coastline is 46000km long and in 2003 it was reported that the area had 584 coastal cities, 750 yachting harbours, 286 commercial ports, 68 oil and gas terminals, 180 power stations, 112 airports and 238 desalinisation plants.

In addition to the residential population, tourist flows double the coastal population in some areas during the summer months. Tourism is a major economic asset in many Mediterranean countries and since the industry is largely based on the 'sand and sea' type of visitor, pressure to develop coastal space has been intense (Figure 2.52). Estimates suggest that tourist demand will continue to rise from the estimated 190 million visitors in 2005 to over 300 million by 2025.

The growth of coastal tourism and other industries has encouraged the development of roads and linear urbanisation. Despite environmental protection in some coastal areas it is estimated that an additional 200km of coastline is being developed each year and that by 2025 over half of the Mediterranean coastline will end up being built upon, with some coastal conurbations extending over hundreds of kilometres.

Figure 2.52 A highly developed coastal area in the Mediterranean

The degradation of the coastal environment

The continual development of the coastal area has put increasing pressure on the physical environment. The major environmental pressures are related to the following.

◆ The growing population of the coastal area.

◆ The development of airports, holiday resorts and general urban sprawl leading to damage to or disappearance of fragile wetland ecosystems.

◆ Poor management of coastal areas leading to change in sediment flows. In addition, the removal of marine sediment for the construction industry has damaged seabed habitats.

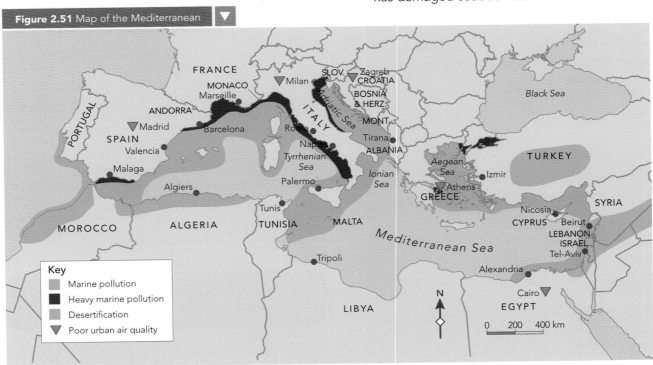

Figure 2.51 Map of the Mediterranean

Key
- Marine pollution
- Heavy marine pollution
- Desertification
- ▼ Poor urban air quality

- Oil and gas infrastructure development has seen the number of oil tankers increase significantly. Currently about 30 per cent of all marine freight and oil freight transits through the Mediterranean.

- The use of chemicals in agriculture has increased rates of river and sea pollution. There is evidence of increasing rates of **eutrophication**.

- Industrial development, which has increased the levels of chemical discharge into rivers and coastal waters.

- Uncontrolled waste management leading to serious problems of leaching from some coastal landfill sites.

- Untreated waste water being discharged to the sea. It is estimated that the majority of urban waste water is not properly treated.

> **Key term**
>
> **Eutrophication:** growth of algae as a result of fertiliser leaching from farmland.

Managing the Mediterranean region

In 1975, the Mediterranean Action Plan (MAP) was set up as part of the United Nation's Environmental Programme (UNEP). The aim of the MAP was to protect the marine environment of the Mediterranean sea. In 1995, the brief of the MAP was widened to include the whole coastal region and to consider the issue of sustainable development.

In the early part of this century, the UNEP/MAP commissioned a report of the current state of the Mediterranean region. Over 300 scientific experts contributed to the report, which was presented in 2006 and called 'A Sustainable Future for the Mediterranean – The Blue Plan'. The report will be used to monitor the European Union's (EU) 'Mediterranean Strategy for Sustainable Development', an initiative to clean up the Mediterranean sea by 2020. This strategy makes the following recommendations for the Mediterranean region:

- Ten per cent of all coastal and marine habitats should be protected, adding to the current 80 protected wetland areas.

- The development of 'green areas' between urban areas in order to reduce linear development.

- The reduction of linear road building.

- Inland tourism should be encouraged in order to increase the possibility of sustainable development in coastal areas.

- Future tourist development should show awareness for the environment in its planning and an economic responsibility for the environment once completed.

- Stricter implementation of rules to combat pollution from ships.

- Improved energy management in order to reduce the need for new coastal power stations.

- All waste water should be fully treated before being discharged into the sea.

U1
2

Coastal environments

Discussion point

Why is the Mediterranean coastal area considered to be 'under pressure'?

Activities

Write a paragraph explaining each of the following threats to the Mediterranean coastal area: population growth; tourism growth; industrial growth.

 'Take it further' activity 2.11 on CD-ROM

Knowledge check

1 Outline the features associated with coastal erosion and deposition.

2 Explore the factors that affect the development of these features and the processes responsible for them.

3 Using named examples, explain the need for, and methods of coastal protection in some areas.

4 Discuss the planning, management and environmental issues associated with coastal protection.

5 Compare two coastal areas in terms of the:
   a range of activities
   b conflicts that may arise.

6 Compare the planning and management issues in two coastal areas.

A list of useful websites accompanying this chapter can be found in the Exam Café section on the **CD-ROM**

ExamCafé
Relax, refresh, result!

Relax and prepare

What I wish I had known at the start of the year…

Hot tips

Kirsten

"I always try to get hold of past papers to find out how the examiner words the questions. I use these to practise reading and answering questions within the time allocated. Even better, ask your teacher for an examiner's report, as this contains helpful advice on how to improve performance."

Ahmed

"I find revising with a friend or in a small group helps. We can discuss questions and help clarify any misunderstandings. We learn from one another by realising what we do and don't know. Working together in a group encourages us to revise and help each other."

Common mistakes – James

▷ One of the commonest mistakes I've made is not fully reading the question. All too often I rush into answering the question and miss some vital wording. Other times I see it as very complex when really it is straightforward. I now use a highlighter pen to highlight the bits of the question that I should focus on, like this:

▷ With reference to **named** examples, **explain** how **sustainable development strategies** can reduce the **threats to communities** and the **environment** in **cold** environments.

▷ In the past I would have stopped reading at communities and missed the environment aspect – which actually I know more about. I can see at a glance the key aspects expected and this is a useful check, as I approach my conclusion, to see if I have missed anything.

Refresh your memory

2.1 What processes and factors are responsible for distinctive coastal landforms?	
Physical	Climate – wind speed and direction, precipitation
	Relief – slope, altitude, sea level, water depth, direction of coast
	Rock type – geology, structure: faults, beds, porosity, tilt of rocks
	Vegetation – type and percent cover
Human	Material supply – gravel abstraction
	Coastal management – coastal defences, planning, flood prevention
	Tourism and recreation, e.g. trampling
	Urbanisation
	Transport – ports, bridges, airports
Time	Features develop
	Climatic change
	Tectonic changes
Processes	Marine – erosion, transport, deposition (on-shore versus long-shore)
	Other – weathering, mass movement, river, wind, human activity
Landforms	Erosion – cliffs, caves, arches, stacks, bays, platforms
	Deposition – spits, bars, beaches, salt marshes, dunes, cusps

2.2 How can coasts be protected from the effects of natural processes?	
Coastal protection	To protect property, land, the transport infrastructure, to aid tourism and conserve the local environment (historic and biotic). It can have negative impacts in other areas so you need to weigh up costs versus benefits
Methods	Hard and soft engineering, planning restrictions, planned retreat, do nothing
Where	Beaches, cliff foot, cliff face, cliff top
Types	Beach replenishment, rock armour, groynes, sea walls, revetments, tetrapods, gabions, rip-rap, grading cliffs, planting vegetation, drainage, piling

Refresh your memory

2.3 In what ways can coastal areas be a valuable economic and environmental resource?

Residential development	Settlements are located near the coast because of the flat land, e.g. Thames Gateway, Canvey Island
Power source	Thermal, nuclear and tidal, e.g. oil fired power station at Tilbury
Industrial development	Fishing, tourism and heavy industry that requires flat land and a water supply
Minerals	Sediments, oils and gas, e.g. sand and gravel from the river bed
Services	Tourism, recreation, waste, e.g. London's waste
Agriculture	Fish farming and grazing on salt marshes, e.g. Kent
Transport	Ports, bulk cargo ports, e.g. ferry and container, Tilbury
Conservation	Nature reserves, e.g. Maplin sands
Other	Army ranges, e.g. Shoeburyness

2.4 What are the management challenges associated with the development of coastal areas?

Planning	Balancing environmental and socio-economic needs including costs (short term and long term); technology, political will, time, scale, knowledge of the issues and the wider impact
	Coastal areas are overseen by a range of planning authorities, e.g. local authorities, charities, (National Trust, National Park Authorities), private landowners (Duchy of Cornwall), conservation bodies (the RSPB) and other groups such as the military or industrial and mining companies

Top tips . . .

▷ At AS it suggests half the material will be asking you to regurgitate what you know from the course (AO1). These questions are usually the middle ones in short section questions whilst AO2 is often the starter based on unfamiliar data and the last part where you are expected to apply case studies (AO1) in an unfamiliar setting to discuss or evaluate. AO3 tends to be most important in extended answers especially essays. You can revise parrot fashion for AO1 but it is your fundamental understanding of the subject that is the key to AO2.

Get the result!

Sample question

Outline how rock type influences the rate of coastal erosion. [6 marks]

Student answer

Rock type is the basic framework which marine and other processes work on (1). Where the strata is end on to the sea (Atlantic coastline), differential erosion by the sea is easy (2). Soft rock, e.g. clay, is more easily eroded so forms bays, etc. (3). Where the strata is parallel to the sea, e.g. south Dorset, then differential erosion is difficult and coasts tend to be less eroded (4). Hard rocks such as granite in Cornwall resist erosion (5) but even here weaknesses such as faults, joints and beds of weaker rock allow the sea to erode to form caves, etc. (6)

Examiner says

This scores maximum marks because (1) basic concept is explained, (2) there is good use of geographical terminology, (3) the cause-effect linkage to a coastal landform is explored, (4) contrast is given with an example, (5) another factor in role of rock type is explained with an example, (6) and there is further elaboration on rock structure and its role linked to a landform.

Examiner's tips

Examinations are based on the need to test certain Attainment Objectives (AOs). The overall balance is set out in the specification. At AS:

- 50 percent of the marks come from AO1 — knowledge and understanding of the content, concepts and processes, so a lot is based on what you have been taught.

- 20 percent of the marks come from AO2 which is analyse, interpret and evaluate. This is tested by how you analyse and interpret information (data) and viewpoints and how you apply understanding in unfamiliar contexts, i.e. not straight from your notes. Evaluation is the most demanding aspect as it asks you to make a judgement based on the data. This is often considered more of an A2 task.

- The remaining 30 percent is AO3 which covers the use of a variety of methods, skills and techniques to investigate issues, reach conclusions and communicate findings. Hence the importance of using maps and diagrams where appropriate in your answer.

Chapter 3

Cold environments

You will have seen some of the media coverage of global warming; the plight of the polar bear, for example, is well known. However, there are many different types of cold environments and all of them are under threat of one kind or another. In this chapter you will examine three cold environments: glacial, periglacial and upland (mountain) including the climate, processes and major landforms associated with them. You will look at the ecosystems contained within these environments and examine how plants and animals adapt to the harsh conditions. You will also explore the opportunities and constraints of cold environments and consider how humans can manage cold environments to ensure sustainability.

Questions for investigation

- What processes and factors give cold environments their distinctive characteristics?

- Why are cold environments considered to be 'fragile'?

- What are the issues associated with the development of cold environments?

- How can cold environments be managed to ensure sustainability?

Consider this

Cold environments are a major resource for millions of people around the world.

What will happen to these people if cold environments begin to disappear?

3.1 What processes and factors give cold environments their distinctive characteristics?

The distinctive characteristics of cold environments are a result of climatic and geomorphological processes. These processes are influenced by a range of factors which vary from place to place.

Impact of climate and weathering on the landscape

There is a wide range of climates associated with cold environments. When we sub-divide cold environments into glacial, periglacial and mountain environments we create three types of climate. Yet there is great variety within each of these sub-divisions. For example, temperatures in Antarctica are very different from those on glaciers in southern Iceland. Similarly, a periglacial area near to the sea (maritime location) will be much milder than a periglacial region located within a continental interior (Figure 3.2). Likewise, temperatures in mountainous areas such as the English Lake District will be much warmer than in the Himalayas or Andes. The one factor that all these places have in common is that at some stage in the year (or in the past) they are cold – just how cold varies.

The concept of climatic geomorphology suggests that in a given climate certain processes predominate. For example, in cold environments, freeze-thaw weathering can be very common (see page 9).

Glacial (polar) climates

Glacial environments contain snow and ice all year round. Polar climates are very cold and relatively dry (Figure 3.1). Antarctic temperatures, for example, are low all year round, varying from −60°C on the ice sheet summit to −10°C on the coast. Winter temperatures vary from −70°C to −25°C, while summer temperatures vary from −40°C to −2°C. There is considerable variation in amount of precipitation that falls in Antarctica. Cold mountain winds blow out from the centre of Antarctica towards the edge of the continent. These katabatic winds form as dense cold air over the central plateau drains into depressions and valleys.

Figure 3.1 Antarctica's climate

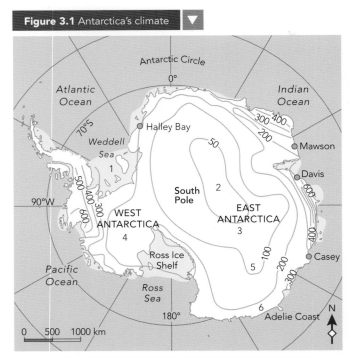

Key
— Isohyets of annual precipitation in terms of water equivalent

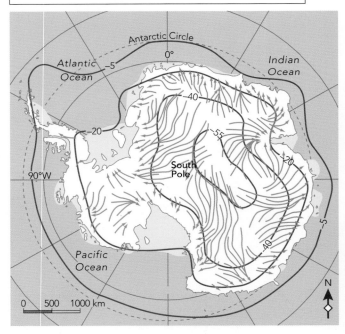

Key
— Isotherms of annual mean temperature
→ Katabatic wind direction

Key term

Katabatic winds: cold mountains winds – the dense air flows down from the mountains to the lowlands chilling the ground as it passes over.

U1

3

AS Geography for OCR

90

Periglacial climates

Periglacial zones have a significant cover of snow and ice but not all year round. They are found in areas of high altitude or high latitude. Periglacial areas are characterised by large seasonal temperature ranges. Temperatures in calm clear winter weather may fall to −50°C although spring temperatures rise above freezing and in summer temperatures may rise to over 20°C. These areas are cold because of:

◆ high latitude – they receive a relatively small amount of insolation due to the low angle of the overhead sun

◆ high altitude – temperature declines on average 1°C for every 100m climbed

◆ albedo – they reflect much solar radiation (average absorption is 40 per cent; on dark soils it is 90 per cent, on snow and ice 10–20 per cent).

> **Key term**
>
> **Insolation:** incoming solar radiation, that is, heat received from the sun.

Precipitation levels in periglacial environments are generally low. This is largely due to the low air temperatures, since cold air is only able to hold small amounts of moisture. Many periglacial areas are affected by high pressure conditions that also reduce the amount of rainfall. In the Arctic, rainfall declines away from oceans, as the westerly depressions progressively lose moisture as they travel.

Mountain (upland) climates

These are areas which were once covered in ice but are now free from snow and ice. However, the landforms that were formed during the cold periods continue to exert an influence over the lives of those who live there. Many upland areas are periglacial but this is not always the case

Some mountainous areas such as the Alps and the Himalayas are so high that they contain glaciers and ice caps. Others are not so dramatic. Nevertheless, there are certain things we can say about climates in mountainous areas.

◆ They are cool – they lose about 1°C for every 100m (or 10°C per km).

◆ They are wet – mountainous areas often cause relief or orographic rainfall.

> **Key term**
>
> **Orographic rainfall:** rain that is produced as air is forced to rise over high ground such as a mountain barrier, it subsequently cools, condensation occurs, and precipitation is produced.

Ambleside	Jan	Feb	Mar	Apr	May	Jun	Jul	Aug	Sep	Oct	Nov	Dec	Year
Temperature													
Daily max °C	6	7	9	12	16	19	20	19	17	13	9	7	13
Daily min °C	0	0	2	4	6	9	11	11	9	6	3	1	5
Average monthly °C	3	4	6	8	11	14	15	15	13	10	6	4	9
Rainfall mm	214	146	112	101	90	111	134	139	184	196	209	215	1851

Tromso, Norway	Jan	Feb	Mar	Apr	May	Jun	Jul	Aug	Sep	Oct	Nov	Dec	Year
Temperature													
Daily max °C	−2	−2	0	3	7	12	16	14	10	5	2	0	5
Daily min °C	−6	−6	−5	−2	1	6	9	8	5	1	−2	−4	0
Average monthly °C	−4	−4	−3	0	4	9	13	11	7	3	0	−2	3
Rainfall mm	96	79	91	65	61	59	56	80	109	115	88	95	994

Verkhoyansk, Siberia	Jan	Feb	Mar	Apr	May	Jun	Jul	Aug	Sep	Oct	Nov	Dec	Year
Temperature													
Daily max °C	−47	−40	−20	−1	11	21	24	21	12	−8	−33	−42	−8
Daily min °C	−51	−48	−40	−25	−7	4	6	1	−6	−20	−39	−50	−23
Average monthly °C	−49	−44	−30	−13	2	12	15	11	3	−14	−36	−46	−16
Rainfall mm	7	5	5	4	5	25	33	30	13	11	10	7	156

Figure 3.2 Climate data for Ambleside (English Lake District), Tromso (Norway) and Verkhoyansk (Russia)

1 Describe the pattern of temperature for Antarctica in Figure 3.1 and suggest reasons for the variations in temperatures.

2 Describe the main variations in precipitation shown in Figure 3.1 and decide if there is any link between the pattern for temperature and the pattern of precipitation.

3 Copy the data in Figure 3.2 into a spreadsheet. Draw three climate graphs to show variations in temperature and precipitation in each of the three locations.

4 Describe the main differences in the climate that each region experiences. (When describing the results, think about maximum and minimum temperatures, range of temperature, length of season below 0°C, rainfall total and seasonality.)

 'Take it further' activity 3.1 on CD-ROM

Discussion point

In what ways might global warming affect the climate of cold environments?

Processes in cold environments

Processes in glacial environments

Glaciers are often described as 'rivers of ice'. Like rivers they carry out erosion, transport and deposition; unlike rivers, they have the ability to carry huge materials. For example, the Madison Boulder in New Hampshire, USA, is a glacial erratic – a rock transported from its source to an area of differing rock type – estimated to weigh over 4660 tonnes! A glacier's load comes from materials falling onto the glacier as a result of mass movements and weathering, as well as through erosion by the glacier. The load helps the glacier to erode the land. Although glacial erosion is commonly associated with mountain areas, glacially eroded lowland areas are more extensive – if not as spectacular. However, it must be remembered that there is little direct evidence of the processes involved as they are happening under the glacier (Figure 3.3). In many cases, geographers must study the landforms and make assumptions about the processes involved.

 Figure 3.3 Glacial and meltwater erosion taking place under the Solheimajokull Glacier, Iceland

Glacial erosion consists of plucking, abrasion, freeze–thaw and meltwater on the valley sides. Although freeze–thaw and meltwater are not, strictly speaking, glacial, they are essential components of the glacial erosion system. Glaciers erode because they are mobile and they contain material at their base and sides. Little erosion occurs in polar glaciers because they are too static – however, in fast moving warm-based glaciers rates of erosion are high.

Plucking

Plucking is the ripping out of material from the bedrock. The more fractured and broken the bedrock, that is, the more freeze–thaw there has been previously, the more effective plucking becomes. It occurs mainly at the base of the glacier, but also at the sides. Plucking involves downward pressure caused by the weight of the ice and then downhill drag as the ice moves, slow enough for meltwater to freeze onto obstacles. Once the material has been prised out of the bedrock, it can be used for abrasion of the landscape (Figure 3.4).

Abrasion

Abrasion is often referred to as the sandpaper effect. It is the erosion of the bedrock by material carried by the glacier (just like abrasion in a river, see page 16). The larger and more angular the load, the greater the potential for erosion. The coarser material will scrape, scratch and groove the rock, leaving **striations** and **chatter marks**; the finer material will smooth and 'polish' the rock.

> **Key terms**
>
> **Striations:** scratch marks on a rock caused by abrasion.
>
> **Chatter marks:** discontinuous scratch marks on a rock caused by abrasion.

Observing abrasion is difficult; it involves digging tunnels through a glacier to gain access to basal cavities. Some of the best results are provided by Boulton's (1974) work on the Breidamerkurjokull glacier in Iceland (Figure 3.5). He monitored the movement of a basalt fragment over a large roche montonnée 20m below the surface. The basalt was removed and examined – it had been in contact with the bed in three places. The basalt contained 3m striations – varying from 3mm to 1mm. The decrease in depth was related to an increase in crushed debris – which spread the load over a wide area. This allowed the glacier ice to slide forward – when the sliding finished the build up of ice again abraded the basalt deeply. Boulton therefore observed

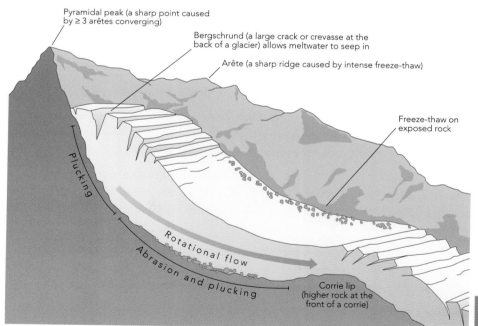

The Formation of a Cirque

Pyramidal peak (a sharp point caused by ≥ 3 arêtes converging)

Bergschrund (a large crack or crevasse at the back of a glacier) allows meltwater to seep in

Arête (a sharp ridge caused by intense freeze-thaw)

Freeze-thaw on exposed rock

Plucking

Rotational flow

Abrasion and plucking

Corrie lip (higher rock at the front of a corrie)

 Figure 3.4 Plucking and abrasion in a cirque

Locality	Average abrasion rate		Ice thickness	Ice velocity
	Marble plate	Basalt plate		
Breidamerkurjokull	3mm/yr	1mm/yr	40m	9.6m/yr
Breidamerkurjokull	3.4mm/yr	0.9mm/yr	15m	19.5m/yr
Breidamerkurjokull	3.75mm/yr		32m	15.4m/yr
Glacier d/Argentiere (French Alps)	36mm/yr		100m	250m/yr

Figure 3.5 Abrasion and plucking under Breidamerkurjokull

regular variations in the depth of striations related to the results of abrasive activity.

> **Key term**
>
> **Roche montonnée:** rock that has been plucked and made smooth by glaciers. Upstream side (stoss) is smooth due to abrasion, while the downstream slope (lee) is steeper and rougher due to plucking.

Freeze-thaw weathering

Rock fracture occurs due to pressure release, exfoliation and dilatation. Pressure release occurs during and after deglaciation. The removal of the overlying glacier leads to a massive decrease in the weight or pressure pushing down on the underlying rock. With the reduction in pressure, the rock expands upwards and outwards often leading to the formation of cracks and lines of weakness in the upper parts of the underlying rock. Following rapid erosion, exfoliation may occur (causing rocks to split parallel to the surface). Freeze-thaw will cause the expansion and contraction of joints – dilation – which leads to their

fracturing. The fractured rocks are then more likely to be attacked effectively by a glacier in a future glacial advance.

Chemical weathering by meltwater

Chemical weathering below glaciers is important, especially on carbonate rocks (limestones). CO_2 is more soluble at low temperatures, hence meltwater streams have the capacity to hold much CO_2. As the streams become more acidic, they are able to weather carbonate rocks more effectively.

The amount and rate of glacial erosion depends on a number of factors (Figure 3.6). These include:

- *local geology* Areas with well-fractured, jointed bedrocks are easily plucked; for example, the carboniferous limestone of the Burren in County Clare, Ireland.

- *velocity of the glacier* This is somewhat dependent on gradient – areas of fast-flowing ice lead to increased erosion.

Factors	Comments
Presence of debris in basal ice	Pure ice is unable to abrade solid rock. The rate of abrasion will increase with debris concentration up to the point where effective basal sliding is restricted.
Sliding of basal ice	Ice frozen to bedrock cannot erode unless it already contains rock debris. The faster the rate of basal sliding the more debris passes a given point per unit time and the faster the rate of abrasion.
Movement of debris towards glacier base	Unless particles at the base of a glacier are constantly renewed they become polished and less effective abrasive agents. Thinning of the basal ice by melting or **divergent flow** around obstacles brings fresh particles down to the rock–ice interface and increases abrasion.
Ice thickness	The greater the thickness of overlying ice, the greater the vertical pressure exerted on particles on the glacier bed and the more effective is abrasion. This is the case up to a depth where friction between particles and the bed becomes so high that movement is significantly restricted and abrasion decreases.
Basal water pressure	The presence of water at the glacier base, especially when at high pressure, can reduce the effective pressure on particles on the bed and therefore abrasion rates by buoying up the glacier. However, sliding velocities may tend to increase because of the reduced friction.
Relative hardness of debris particles and bedrock	The most effective abrasion occurs when resistant rock particles in the glacier base pass over a weaker bedrock. If the debris particles are soft in comparison with the bedrock, the debris particles are abraded and little bedrock erosion is accomplished.
Debris particles size and shape	Since particles embedded in ice exert a downward pressure proportional to their weight, large blocks should abrade more effectively than small particles. Angular debris will therefore be a more efficient agent of abrasion than rounded particles.
Efficient removal of fine debris	To sustain high rates of abrasion, fine particles need to be removed from the ice–rock interface since they abrade less effectively than larger particles (assuming larger particles are continually being supplied from above). Meltwater appears to be the main mechanism for the removal of fine (<0.2 mm) debris.

Figure 3.6 The main factors affecting the amount and rate of abrasion

◆ *amount and character of the load carried by the ice* If the load is coarse, resistant and angular, it will erode more than a load that is fine, weak and rounded.

However, some areas experience very little glacial erosion. The ice-protectionist theory is that ice protects the landscape underneath. This may be true for cold-based glaciers (which are immobile) but not for fast-moving temperate ones. In Antarctica, ice in valleys can be highly erosive, whereas that on plateaus is thin, slow moving, and protects the bed from erosion.

┌─ **Key term** ─────────────────────────────

 Divergent flow: ice flow which is around an object rather than over it.
└───

Activities

1   Distinguish between plucking and abrasion.

2   Describe and explain two types of weathering that are found in cold environments.

3   Study Figure 3.6, which shows rates of erosion under the Breidamerkurjokull glacier. How do rates of erosion vary with:

   a ice thickness

   b ice velocity?

 'Take it further' activity 3.2 on CD-ROM

Processes in periglacial environments

Ice action

Periglacial environments are dominated by freeze-thaw weathering. This occurs when temperatures fluctuate above and below freezing point (Figure 3.7). As water freezes it expands by 10 per cent, exerting pressures of up to 2100kg/cm^2. Most rocks can only withstand up to 210kg/cm2. It has most effect on well-jointed rocks, which allow water to seep into cracks and fissures. Congelifraction refers to the splitting of rocks by freeze-thaw action. Frost heave is the process where water freezes in the soil and pushes the surface upwards and churns it. Ice-lensing refers to the growth of ice crystals in soil (Figure 3.8). Geliturbation and congeliturbation are other terms for frost heave. Nivation refers to freeze-thaw weathering under a snow bank. The broken material is removed in spring and summer by the melted snow.

Solifluction

Solifluction literally means flowing soil and is an accelerated form of soil creep (see page 6). In winter water freezes in the soil causing expansion of the soil and segregation of individual soil particles. In spring the ice melts and water flows downhill. It cannot infiltrate into the soil because of the impermeable permafrost. As it moves over the permafrost, it carries segregated soil particles (peds) and deposits them further downslope as a solifluction lobe or terracette.

Chemical weathering

Chemical weathering is also effective in periglacial regions. Carbonation (see page 10) is an important process because of the low temperatures. CO_2 is more soluble at low temperature hence the water becomes quite acidic. It is aided by the slowly rotting vegetation which releases organic acids. Hydrolysis is also important because of the large presence of organic acids in the marshy soil. Hydration is the process where certain minerals absorb water, expand and change. For example, anhydrate is changed to gypsum during hydration. Although it is often classified as a type of chemical weathering, mechanical stresses occur as well.

┌─ **Key terms** ─────────────────────────────

 Carbonation: a form of weathering in which calcium carbonate reacts with an acid water to form calcium bicarbonate which is soluble and removed in solution.

 Hydrolysis: a chemical weathering process where water reacts with minerals such as feldspars to produce clay minerals.
└───

Other processes

Fluvial (river) activity is effective because of the regime of the rivers (snowmelt), the highly weathered nature of the bedrock and the nature of surface (unconsolidated sands and gravels). Wind action is also important because of the lack of trees and the disturbed nature of the ground.

Cambering is the process where segments of rock become dislodged from the main body of rock and begin to move downhill. It is aided by freeze-thaw. Avalanches are a type of mass movement commonly found on slopes steeper than 22°. Dry snow avalanches occur on north and east facing slopes, where the snow is unstable, whereas and wet snow flows generally result due to rapid snow melt.

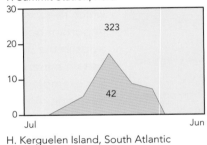

A. Yakutsk, Russia
(chart: No. of days vs Jan–Dec) — 126, 42, 197

B. Tuktoyaktuk, Canada
(chart) — 214, 108, 43

C. Spitsbegen
(chart) — 215, 91, 59

G. Sonnblick, Alps
(chart) — 215, 91, 59

D. Fenghuo Shan, Tibet Plateau
(chart) — 11, 354

E. Mont Blanc Station, Peru
(chart) — 27, 337

F. Summit Station, Peru
(chart) — 323, 42

H. Kerguelen Island, South Atlantic
(chart) — 22, 223, 120

Key
- Days >0°C
- Freeze-thaw days
- Days <0°C

A–C High latitude, low elevation
D–F Low latitude, high elevation
G Mid latitude, high elevation
H Subarctic oceanic, low elevation

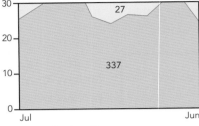

Figure 3.7 Freeze-thaw cycles

Activities

1. Study Figure 3.7. Under which conditions is freeze-thaw most effective as a type of weathering.

2. Explain briefly any two types of erosion in periglacial environments.

3. Describe the process of ice lensing as shown in Figure 3.8. How might it help weathering and erosion in periglacial areas?

Features of cold environments

Erosional features

Large-scale features formed by glacial erosion include cirques, glacial troughs (U-shaped valley) and roche moutonées, while small-scale features include erratics and striations (scratches on a rock). Many erosional features – especially those in lowland areas – have been protected and hidden by deposits of till laid down on top of them and are therefore not very visible.

— **Key term** —

Till: sediment deposited by a glacier – unsorted, angular and of variable sized material, till refers to unsorted deposits with a wide range of grain size, deposited directly by the ice – whether on land or below a floating glacier – and not subsequently changed. Sometimes called moraine or boulder clay.

Cirques

A cirque is an armchair-shaped hollow surrounded by knife-edged ridges called aretes. Cirques, along with troughs, are among

Figure 3.8 Ice lensing

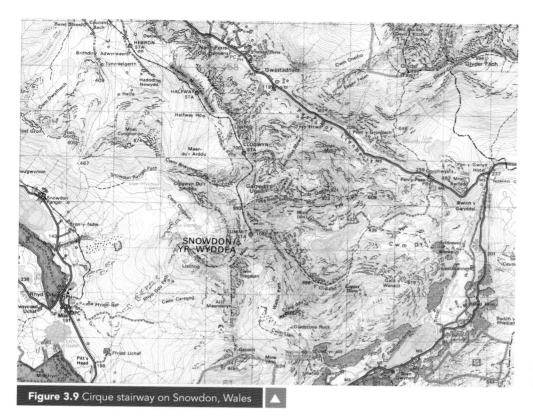

Figure 3.9 Cirque stairway on Snowdon, Wales ▲

different elevations. Cwm Llydau is a typical over deepened cirque with its own headwall. Within this headwall is the small cirque, Cwm Glaslyn, and above this a smaller cirque below the summit, Y Wyddfa (Figure 3.9). The stream that links them now forms waterfalls over the headwalls. In glacial times icefalls would have been here.

The formation of cirques (Figure 3.10) is complex and several processes are likely to be involved. In Britain, cirques develop best on north- and/or east-facing slopes where

the most characteristic features of glacial highlands. The length to height ratio from the lip to the top of the headwall ranges from 2.8:1 to 3.2:1. For example, Western Cwm of Everest is 3.2, whereas Blea Water Corrie in the English Lake District is 2.8.

Cirques are heavily influenced by the joint pattern of rocks. Rocks need to be resistant enough to withstand complete destruction, but weak enough to be heavily weathered and eroded.

Many cirques are compound features containing several stages of cirques (Figure 3.10). Small cirques were generally formed during periods of glacial re-advance. On Snowdon in Wales, for example, there is evidence of a cirque stairway – several cirques at

▼ **Figure 3.10** Formation of a cirque

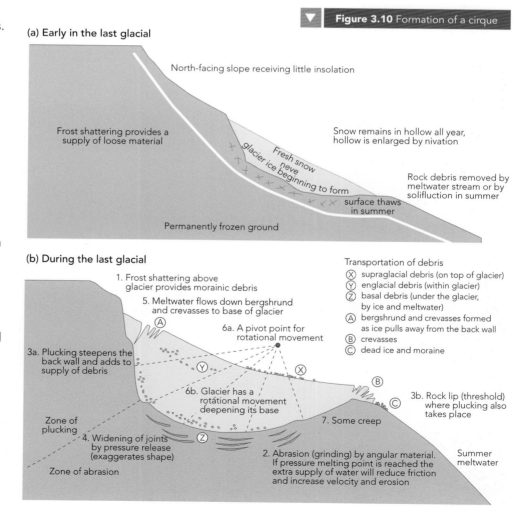

(a) Early in the last glacial

North-facing slope receiving little insolation

Frost shattering provides a supply of loose material

Fresh snow
névé
glacier ice beginning to form

Snow remains in hollow all year, hollow is enlarged by nivation

Rock debris removed by meltwater stream or by solifluction in summer

surface thaws in summer

Permanently frozen ground

(b) During the last glacial

1. Frost shattering above glacier provides morainic debris

5. Meltwater flows down bergshrund and crevasses to base of glacier

Ⓐ

6a. A pivot point for rotational movement

3a. Plucking steepens the back wall and adds to supply of debris

Ⓨ

Ⓧ

6b. Glacier has a rotational movement deepening its base

Ⓑ

Ⓒ

3b. Rock lip (threshold) where plucking also takes place

Zone of plucking

4. Widening of joints by pressure release (exaggerates shape)

Ⓩ

7. Some creep

Zone of abrasion

2. Abrasion (grinding) by angular material. If pressure melting point is reached the extra supply of water will reduce friction and increase velocity and erosion

Summer meltwater

Transportation of debris
Ⓧ supraglacial debris (on top of glacier)
Ⓨ englacial debris (within glacier)
Ⓩ basal debris (under the glacier, by ice and meltwater)
Ⓐ bergshrund and crevasses formed as ice pulls away from the back wall
Ⓑ crevasses
Ⓒ dead ice and moraine

insolation is lowest, allowing the rapid accumulation of snow. The most common theory of cirque formation is that a shallow, preglacial hollow is the original site of snow accumulation. The hollow is enlarged by freeze-thaw weathering at the edge of the snow patch (this freezing and thawing of the thin snow is called nivation). Continued nivation enlarges the hollow and so **neve** proper can form, and gradually, as the basin further develops, ice can accumulate. At a critical depth and weight of ice, the ice begins to move out of the hollow by **extrusion flow** in a rotational manner. This rotational movement of the ice helps to erode the hollow further by plucking and abrasion, and so a true cirque is formed. Meanwhile, meltwater especially that which makes its way down the **bergschrund** and **randkluft**, helps in continuing cirque growth; as the meltwater trickles down into the cirque it is involved in freeze-thaw weathering of the rock exposed at the back of the basin causing this backwall to retreat (headwall recession) and remain steep. The meltwater, once in the basin, is also involved in freeze-thaw weathering, preparing the rock for erosion by the moving ice.

Key terms

Neve (firn): a transitional stage between snow and ice. It has survived at least one summer's melting, and has been compressed by snowfall the following winter.

Extrusion flow: the movement of ice as a result of becoming too deep or heavy and therefore unstable. It 'collapses' and begins to flow outwards as a result of its own weight and pressure.

Bergschrund: a large crack or crevasse.

Randkluft: a gap between the rock face and ice in the hollow, caused when heat from the rock melts the ice.

When the ice finally disappears, an armchair-shaped hollow remains, often containing a small lake (llyn or tarn) (Figure 3.11) dammed back by the cirque lip left as a result of rotational movement of the ice. If several cirques develop in a highland region, they will jointly produce other erosional features such as when two cirques lie back-to-back, cirque enlargement by headwall recession will create a narrow, steep-sided ridge between the two, called an arête (Figure 3.12). If three or more cirques develop, the central mass that

Figure 3.11 Blea Tarn ▼

Figure 3.12 Arêtes ▲

will temporarily survive between them will become
a pyramidal peak, usually sharpened by freeze-thaw
weathering; for example, the Matterhorn. A horn is an
isolated upstanding mass of rock. There are few horns
in Britain, although Cir Mhor on the Isle of Arran and
Schiehallion in the Grampians are good examples.
Often these features are very much subdued, being
covered with scree which helps to protect them from
further weathering.

Activities

1 Using an annotated diagram, explain how cirques are
   formed.

2 Using a series of annotated sketches, show how arêtes
   and pyramidal peaks are formed. Why are there so
   few pyramidal peaks in Britain?

Glacial troughs and U-shaped valleys

Glacial troughs are characterised by steep sides
and a relatively flat bottom (Figure 3.13). Although
usually referred to as U-shaped, few are actually
shaped that way. Their present form depends on
subsequent activity (mass movements, deposition
and erosion by rivers) since the last glaciation. It is
more accurate to refer to glacial troughs as parabolic
rather than U-shaped. Glacial troughs – or U-shaped
valleys – are one of the most recognisable glacial
features. They result from the channelling of ice
through valleys, combining freeze-thaw, plucking and
abrasion.

Glacial troughs are formed as the result of many
processes. Before the actual onset of glaciation, active
freeze-thaw weathering under periglacial processes
will weaken the floor and sides, so preparing it for
rapid erosion. During interglacial phases, periglacial
periods will return, further weakening the already

Figure 3.13 Glacial trough – Seefeld, Austria

increased erosion associated with a thicker mass of ice. Some glacial troughs end abruptly at their head in a steep wall, known as the trough end, above which lie a number of cirques. Probably a whole series of cirque glaciers developed and joined at one point to feed the main glacial trunk; at the point of merger a sudden increase in the amount, weight and eroding power of the ice, may have formed the wall. In the preglacial valley, any river tributary would cut down to meet the level of the main river. With tributary glacial valleys this is not the case. The small tributary glacier has neither the weight nor the power to cut down to the depths of the main trough; the tributary ice will cut down so that its ice can slip onto the top of the main glacier.

eroded rock. In addition, at the end of glacial periods (deglaciation) there will be **pressure release** as the weight of the ice is reduced. During glaciation the eroding power of the moving ice will cause the valley to become U-shaped in cross section with a flat floor and steep sides. In plan view, the valley will become straight (as opposed to the winding nature of the former valley) as the interlocking **spurs** are bulldozed to leave truncated spurs. Extrusion in the ice can cause the ice to erode deep rock basins in the valley floor, later occupied by long, narrow **ribbon lakes** (Figure 3.14).

▼ Figure 3.14 Glacial trough

- Pyramidal peak
- Arête
- Cirque lake or tarn
- Truncated spurs
- Hanging valley with waterfall
- U-shaped valley
- Ribbon lake
- Deposits of moraine (boulder clay)

Key terms

Pressure release: a type of weathering in which a rock is able to expand outwards as a result of the 'unloading' of weight (pressure) and as a result fractures or cracks appear in the rock as it expands.

Spur: a projection of land from a ridge or mountain. Interlocking spurs occur in the upper course of river valleys, where streams tend to follow winding courses. The projecting spurs appear to overlap or interlock. Truncated spurs are those which have been eroded in their lower parts by glaciers.

Ribbon lakes: long, linear lakes which fill a glaciated trough such as Lake Windermere in the Lake District.

When the ice disappears, these tributary glacial valleys are left high, as hanging valleys, today often the site of waterfalls (Figure 3.15).

Key term

Hanging valleys: a small U-shaped valley formed by a small glacier that joins and hangs above a large U-shaped valley formed by a larger glacier.

As the ice moves along, it may also erode rock steps by scouring out lines of weaker rock, by opening up major lines of weakness (joints etc) or by experiencing periods of intense and potent extrusion flow. The addition of a tributary glacier can also develop rock steps due to the

Major modifications of these smooth, straight glacial troughs have occurred in post-glacial times. Frost shattering has produced **scree** which now often covers the sides of the troughs, masking the former U-shaped. **Moraines** (once covering the sides of the ice as lateral moraine, the middle as medial moraine and the base of the glacier as ground moraine) now lie dumped on the trough floor, damning back the water to form more ribbon lakes or being reworked on the current river valley.

Figure 3.15 Hanging valley, Co. Mayo, Ireland ▲

Key terms

Scree: angular sediment that collects at the foot of a mountain range or cliff. The rock fragments that form scree are usually broken off by the action of freeze–thaw weathering.

Moraine: glacial deposits consisting of poorly sorted, often angular, loose rock fragments. Many forms exist including englacial, lateral, medial, recessional and terminal.

Fjords

Fjords are glacial troughs below or partly below sea level (Figure 3.16). Up to one-quarter of fjords have active glaciers. Fjords are more likely to be U-shaped than land-based glacial troughs. This is because most fjords are cut into resistant rocks whereas land-based troughs are often covered with deposits of scree and soil.

Many fjords have a long profile which shows a greater degree of glacial erosion at the source of the fjord compared with the mouth. Fjords deepen quickly near their source, then decline very gently towards the seaward end. The mouth of a fjord is often marked with a sill or threshold, an area of shallow rocky ground or moraine. This is thought to be due to a reduction in glacial erosional power. The Sogne Fjord in Norway has a maximum depth of over 1300m but at its mouth is only 200m deep. In Scotland, at Loch Morar and Loch Maree, the sills have been

▼  Figure 3.16 Fjord, south-west Norway

raised above sea level by isostatic uplift following deglaciation.

> **Key term**
>
> **Isostatic uplift:** the rising of the land when the weight of a glacier has been removed as a result of deglaciation (melting of glaciers).

Activities

1   How are troughs formed? Why are few troughs U-shaped?

2   Make an annotated sketch diagram of Figure 3.14 to show the main characteristics of a hanging valley. How are hanging valleys formed?

Features resulting from meltwater (fluvioglacial) erosion

Meltwater channels

Meltwater from an area of melted ice is capable of intense erosion. Meltwater streams can carve deep meltwater channels in the land, which have in many cases, helped lead to modifications of drainage such as the drainage modification in the North York Moors (Figure 3.17).

During the last glaciation, ice sheets up to 300m thick in some places, spread south and south-west from north England and the North Sea, surrounding the North York Moors area, leaving the moors as an island of uncovered ice-free land. As the ice moved south, it first blocked the mouth of the River Esk and stopped its waters from reaching the North Sea. At the same time, ice moving from the western edge of the moors blocked the alternative outlet. Gradually water gathered in the low-lying dales, forming lakes that found temporary outlets across the ridges between one dale and another. Here the water cut cols or spillways. Eventually the lake waters joined and the water surface stood 215m above present sea level. At this height, the water spilt over the lowest point of the main divide and began to flow southwards in an ever-increasing torrent, which eroded the trench at Newtondale (Figure 3.18). Newtondale is thus known as an overflow channel. This feature is a well-defined trench being 75m deep with very steep sides cut into the surrounding plateau of the moors.

> **Key term**
>
> **Col or spillway:** a low point in a ridge or high ground, often formed as a result of back-to-back cirques. Some may be formed by the movement of glacier ice from one valley into another and some by glacial diversion of drainage.

Figure 3.18 Overflow channel, Newtondale, Yorkshire

Figure 3.17 Drainage modification in the North York Moors

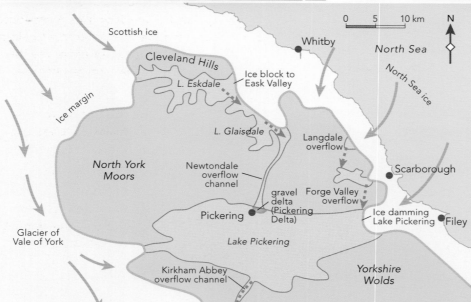

By this time both the east and west ends of the Vale of Pickering were also blocked by ice so that another vast lake was created. The Newtondale overflow water entering this lake (Lake Pickering) deposited the Pickering Delta.

This explanation of the North York Moors was worked out by Percy Kendall in 1902. Since that time further research has shown that some of the 'overflow channels', are in fact channels eroded by subglacial streams, that is, subglacial channels. Other types of channel that can be attributed to meltwater are ice marginal channels (formed between the ice and the highlands) and submarginal channels (formed near the edge of the ice but under it). Thus in some cases, lakes need not have existed.

Features resulting from glacial deposition

Ice is able to transport vast amounts of material, ranging in size from fine rock flour to huge boulders. This material is carried on the ice surface, within the ice itself and dragged along the base of the ice.

Often, this material is deposited directly by the ice – dropped from its lowest layers or redeposited at the ice front/margin. This direct deposition usually results in **unstratified material** called till or boulder clay. In contrast, meltwater, flowing from the ice, also transports and deposits material. This fluvioglacial deposition results in rounded, stratified deposits (Figure 3.19). The term 'drift' describes all glacial and fluvioglacial deposits left after ice retreat.

Figure 3.19 Glacial and fluvioglacial deposits ▼

(a) During glaciation

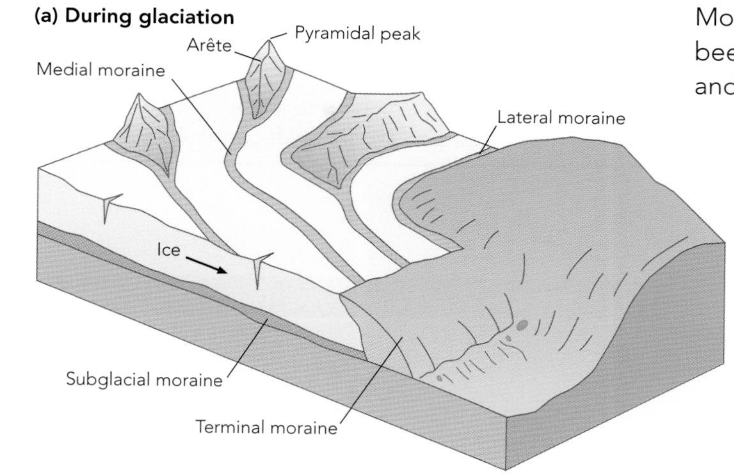

Labels: Arête, Pyramidal peak, Medial moraine, Lateral moraine, Ice, Subglacial moraine, Terminal moraine

(b) After glaciation

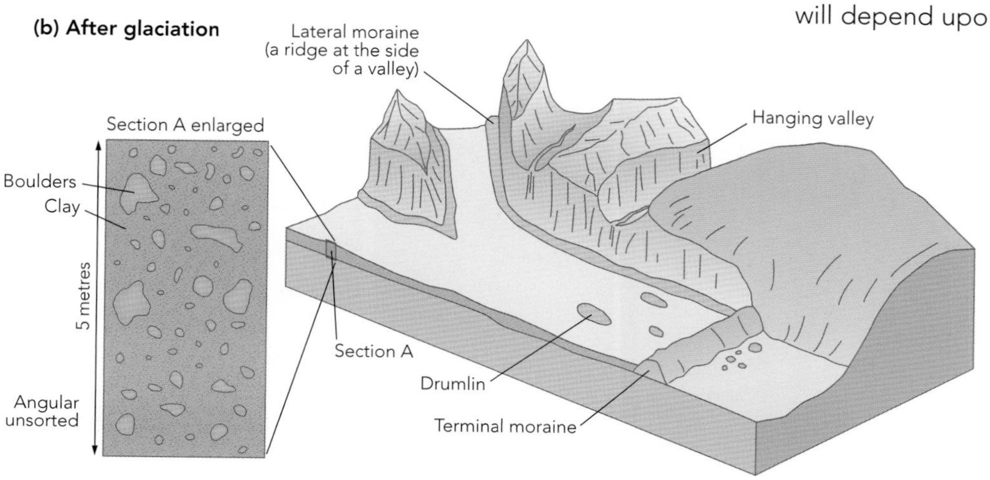

Labels: Lateral moraine (a ridge at the side of a valley), Section A enlarged, Boulders, Clay, 5 metres, Angular unsorted, Section A, Drumlin, Terminal moraine, Hanging valley

Key term

Unstratified material: usually glacially deposited material which is unsorted, variable in size and has no distinct layers.

Erratics

Erratics are large boulders foreign to the local geology that have been dumped by the ice, usually on flat areas (Figure 3.20). Some erratics have been left stranded in precarious positions as perched blocks such as the bluish micro-granite from an island off the Ayrshire coast has been found in the Merseyside area and in Fishguard in Dyfed.

Till

Till is a common, widespread and unstratified glacial deposit composed of finely grained rock flour (sands and clays) mixed together with rocks of different shapes and sizes. Its composition is variable, depending on the nature of the rocks over which the ice moved. Till may be divided into two types: lodgement till dropped by actively moving glaciers, and ablation till dropped by stagnant ice.

Moraines

Moraines are lines of loose rock fragments which have been weathered from the valley sides above the ice, and have fallen downslope onto the ice. The lines of material lying near the valley sides are lateral moraines. Where two glaciers meet, the lateral moraines of each will join to form a medial moraine down the centre of the glacier (Figure 3.21). At the snout of a stationary or slowly moving glacier, much material is deposited as a crescent-shaped mound or terminal moraine. Similar, but much smaller, recessional moraines mark the site where the snout halted for brief periods during the retreat of the ice front. The character of the terminal moraine will depend upon:

◆ the amount of material carried by the ice

◆ the rate of ice movement

◆ the rate of **ablation** and thus the amount of meltwater.

Figure 3.20 Glacial erratic, Sliabh na Caillagh, Meath, Ireland ▲

> **Key term**
>
> **Ablation:** the removal of material (ice, debris) from a glacier such as by melting, evaporation, sublimation.

> **Key term**
>
> **Moulins:** a circular sink-hole or portal into a glacier. They may be caused by meltwater entering a crevasse.

The angle of slope of the moraine facing the ice (ice contact slope) is always steeper than the lee slope. The largest terminal moraine in Britain is the Cromer Ridge in Norfolk: it forms a belt of hummocky hills, composed of sands and gravels, 8km wide and 90m high.

Debris is also carried within the ice (englacial moraine) having made its way down crevasses and moulins within the ice; the debris found at the base of the ice is called subglacial moraine; this is left after a steady retreat of the ice.

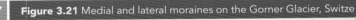

▼ Figure 3.21 Medial and lateral moraines on the Gorner Glacier, Switze

Drumlins

In some low-lying, undulating regions, particularly where a valley glacier flows into a lowland plain (causing a thinning of the ice), the till has been deposited as small, oval mounds a few metres long and high to considerable hillocks 1.5km long and 90m high such as in Cavan, in Ireland, the Central Lowlands of Scotland and the Solway Plain. These drumlins (Figure 3.22) usually consist of sandy rather than clayey till. The long axis of the drumlin is aligned with ice flow with the steeper, thicker end of the feature facing the direction from which the ice came.

It seems that the ice deposited each mass of till (possibly from an area of the ice where the base was more heavily loaded with material), where friction between the ice and the rock floor was stronger than the adhesion between the till and the ice. The mass of till was then streamlined by the moving ice. It is also possible that some drumlins were produced by the pressure of the active ice moulding sheets of older drift, while other drumlins were formed where a rock obstacle has been coated with till. This coating is often very thin. Drumlins formed in this way are called false drumlins.

Fluvioglacial deposits

The melting (wasting) of ice results in the formation of meltwater streams on the ice surface (supraglacial streams), within the ice (englacial streams) and beneath the ice (subglacial streams). These meltwater streams carry much debris and deposit this during active ablation or deglaciation, or during a prolonged period of ice stagnation, in stratified drift formations.

There are two major forms of stratified fluvioglacial drift: prolonged drift and ice contact stratified drift.

Prolonged drift

Prolonged drift consists of material dropped in glacial lakes (such as **varved clays**), along lake shorelines and as lake deltas such as Vale of Pickering. With the recession of continental ice sheets and large ice bodies, lakes frequently formed as temporary features. Into these lakes, material was deposited, brought down by seasonal meltwaters. Layered varve represents one season's melting. Summer deposition is indicated by a layer of coarse deposits (more vigorous meltwater is able to carry coarser, heavier material); winter deposition is indicated by a thinner layer of finer material (less meltwater, more sluggishly flowing water cannot carry coarse material).

> **Key term**
>
> **Varved clays:** A distinctive banded layer of silt and sand deposited in lakes near the margins of ice-sheets.

Figure 3.22 Drumlin, Carrick, West County Meath, Ireland ▼

Outwash plains (also called sandur) contain material that is well sorted and stratified by the meltwater streams (Figure 3.23). The coarsest materials (gravels) are deposited first near the ice margin, while the finer materials (sands) are carried further down the valley. The seasonal flow of meltwater also causes vertical layering.

Ice contact stratified drift

Ice contact stratified drift consists of stratified sand and gravel that is sorted by the action of meltwater streams and deposited next to the glacier. In addition, they are often modified as a result of ice retreat.

Eskers

Eskers are elongated ridges of coarse, stratified fluvioglacial material (sands and gravels). The ridges usually meander, such as in Scandinavia where they wind for over 100km between lakes and marshes. The popular theory is that material was deposited in subglacial meltwater tunnels during a period of lengthy ice stagnation. Another theory is that some eskers could be elongated deltas, deposited by streams flowing out from tunnels at the front of a continuously and rapidly retreating ice front. This theory is supported by the occurrence of some 'beaded eskers' – the 'bead' or 'bump' in the esker represents summer deposition when rapid ice wasting produced increased meltwater able to transport and deposit much material; in winter, deposition would be less but more regular. An excellent example in the British Isles is the Trim esker, in Ireland.

Kames

Kames are irregular mounds of bedded sands and gravels, arranged in a chaotic manner (Figure 3.24). Major types of kames may be distinguished.

The true kame (or kame delta), which probably represents a small delta, formed where a meltwater stream flowed out beneath an area of stagnant or slowly decaying ice, into a lake dammed between the ice front and drift material (probably the terminal moraine). Many kames are found along the northern sides of the Southern Uplands. One of the most characteristic features of a kame landscape is the small, shallow hollow (called a kettle hole) (Figure 3.25) amid the kame mounds, hence the term 'kame and kettle

▼ **Figure 3.23** Outwash plain, Solheimajokull, Iceland

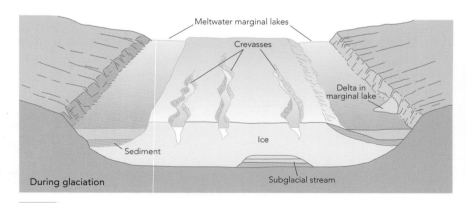

Meltwater marginal lakes

Crevasses

Delta in marginal lake

Sediment

Ice

During glaciation

Subglacial stream

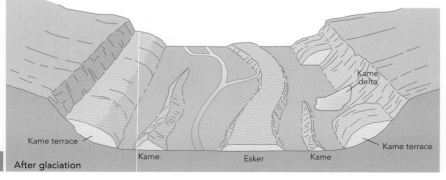

Kame delta

Kame terrace

After glaciation

Kame   Esker   Kame   Kame terrace

Figure 3.24 Fluvioglacial deposits ▶

106

country'. The holes are probably due to deposition of material around blocks of ice broken off from the front of the stagnant ice body. Each block of ice would finally melt leaving a hole. The holes are often filled with water as numerous small lakes. An excellent area for all kinds of fluvioglacial features is at the margins of the retreating Breidamerkurjokull glacier in south-east Iceland (Figure 3.26).

Kame terraces are formed along an ice edge, laid down by streams occupying the trough between the ice and valley side. They appear as narrow, flat topped, terrace-like ridges such as along the edges of valleys in the Lammermuir Hills in Eastern Scotland.

▲ **Figure 3.25** Kettle hole, near Solheimajokull, Iceland

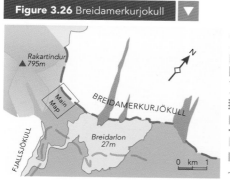

Figure 3.26 Breidamerkurjokull ▼

Ice margin (1965)
Ice margin (1945)
Medial moraine
Lakes and streams
Contours on land (m)
Contours on ice (m)
Kettle holes
Ground moraine
Moraine ridges
Sandur
Stagnant ice (partly buried)
Meltwater channel

Activities

1 Explain how you would tell the difference between each of the following pairs:

   a terminal moraine and a kame

   b lateral moraine and kame terraces

   c medial moraine and eskers.

2 Explain how drumlins, erratics, terminal moraine, kettle holes, eskers and kame terraces were formed.

⊙ 'Take it further' activity 3.3 on CD-ROM

U1

3

Cold environments

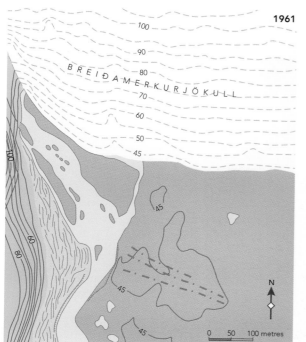

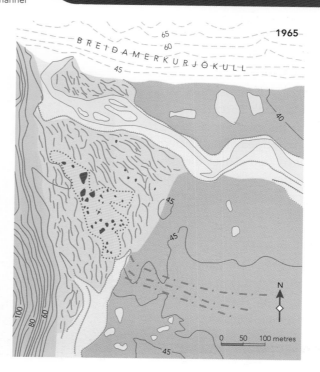

Discussion point

What are the links between glacial deposition and fluvioglacial deposition? Why might it be difficult to distinguish properly between the two?

Periglacial features

Patterned ground

Patterned ground varies from a few centimetres in size to over a 100 metres. It includes a variety of shapes such as garlands (elongated circles), polygons and stripes (Figure 3.27). Some patterned ground is sorted, some is not. Circles and polygons are more common on flat ground, while stripes are more common on slopes of between 5° and 30°. (On slopes over 30°, mass movement is rapid to allow patterned ground to form).

of the movement and freezing of water under pressure. Two types are generally identified: open-system and closed-system pingos (Figure 3.29). Where the source of the water is from a distant, elevated source, open system pingos are formed. These are largely found in areas of discontinuous permafrost. Groundwater, forcing its way to the surface, freezes. Very high levels of hydrostatic pressure are required for open system pingos to form.

By contrast, closed system pingos are isolated features on flat surfaces. They are associated with areas of continuous permafrost. They are formed when a lake in a permafrost area is in-filled with sediments. This causes an increase in the amount of insulation and the permafrost expands. This eventually traps a body

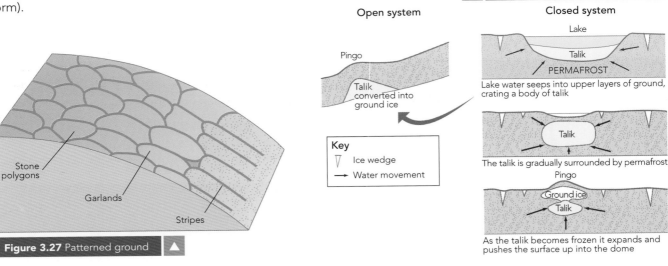

Figure 3.29 Formation of pingos

Open system

Closed system

Key
▽ Ice wedge
→ Water movement

Lake water seeps into upper layers of ground, crating a body of talik

The talik is gradually surrounded by permafrost

As the talik becomes frozen it expands and pushes the surface up into the dome

Figure 3.27 Patterned ground ▲

Frost heaving is a major cause of patterned ground. It helps move larger stones to the surface, fine grained material forms the raised core, while the stones form the edges of the pattern. In addition intense cold causes the ground to crack and stones may roll into these cracks. Surface wash is important for the formation of stripes (Figure 3.28).

Pingos

Pingos are isolated, conical hills up to 100m in height and may have a diametre of up to about 1km. They are found only in periglacial areas. They form as a result

Figure 3.28 Patterned ground ▶

of water (talik), which freezes, expands and creates a pingo. Alternatively, the diversion of a river or the draining of a lake may have the same effect. Nearly 1400 pingos are found in the Mackenzie Delta of Canada, 98 per cent of them closed system pingos. Examples of relict pingos can be found in the Vale of Llanberris in Wales. When a pingo collapses, ramparts and ponds are left, looking a bit like a volcanic crater.

Ice wedges and ice-wedge polygons

An ice wedge is a wedge-shaped mass of ice in the ground, which tapers downwards. When the ground freezes to very low temperatures (below −20°C), cracks develop in the ground, especially if the ground is formed of loose material. In the summer, water flows into these cracks, freezes as the temperatures drop and enlarges the crack. Over a long period of time these wedges may reach a depth of 10m. The annual rate of growth is about 1–20mm. Over hundreds of years they may reach sizes of between 10m and 30m in depth. As climate changes, sand and gravel may be washed into these wedges to form a fossil ice-wedge cast. A good example is the Long Hanborough Carrot near Oxford. Ice wedges commonly form polygonal patterns, hence the term 'ice-wedge polygons'.

Relict (past) features

Dry valleys

Dry valleys are river valleys without rivers (Figure 3.30). They are common on chalk and limestone such as The Manger at Uffington (Vale of the White Horse in Oxfordshire) and the Devil's Dyke near Brighton. During the periglacial period limestone and chalk became impermeable owing to permafrost, therefore rivers flowed over their surfaces. High rates of river erosion occurred because of springmelt, the highly weathered nature of the surface and high rates of carbonation. At the end of the periglacial period, normal permeability returns, water sink into the permeable rocks and the valleys are left dry.

> **Key term**
>
> **Springmelt:** the flooding that occurs in spring as a result of melting winter snow.

Misfit rivers

Misfit rivers (small rivers or streams that occupy large valleys) are also the result of periglacial activity. Some rivers, such as the River Windrush and the Evenlode in the Cotswolds, are too small and ineffective to create steep valleys with wide floodplains (Figure 3.31).

Figure 3.30 Bagg's Bottom – a dry valley feeding into the River Evenlode, Oxfordshire ▼

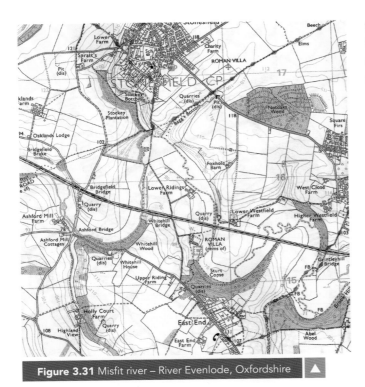

Figure 3.31 Misfit river – River Evenlode, Oxfordshire ▲

However, during the periglacial phase, rapid runoff from snowmelt would have allowed these rivers to carve steep, over deepened valleys, which bear no relation now to the small streams and rivers that flow in them.

Scree slopes

Scree slopes are slopes composed of large quantities of angular fragments of rock; for example, the slopes at Wastwater in the Lake District (Figure 3.32). Typically

they have an angle of rest, that is, the angle at which the slope is stable and no further mass movements occur, of about 35°. Large upland surfaces of angular rocks are known as blockfields.

Loess

Loess is a deposit laid down by the wind. It consists mostly of unstratified, structureless silt, but also includes angular and sub-angular pieces. It covers extensive areas in China, and northern Europe. In Britain it produces smaller deposits such as the Brickearth of East Anglia. It can be quite cohesive, as in the Chinese loess, and form steep, high cliffs when eroded by rivers. The source of the material is glacial abrasion, frost cracking, salt weathering and fine particles picked up from outwash deposits.

Activities

1. Explain how ice wedges are formed.
2. Describe the difference between open-system and closed-system pingos.
3. What is patterned ground? Explain the processes that are responsible for it.

Discussion point

How might periglacial processes and features affect glacial features in areas such as the British Isles?

Figure 3.32 Wastwater screes, Lake District ▼

3.2 Why are cold environments considered to be 'fragile'?

The links between ecosystems and climate in cold environments

As already shown, the climates associated with cold environments are varied and extreme. This includes extremes of temperature, precipitation and wind. Vegetation found in Arctic areas is adapted to the climate in a number of ways (Figure 3.33). The main characteristics – such as small size and predominance of perennials – are well adapted to a highly seasonal climate with a very cold winter and a short summer.

> **Key term**
>
> **Perennials:** plants living for several years.

Vegetation in cold environments

Towards the poles, the climate becomes drier and cooler, and summers are shorter. Plant cover is reduced, and plant height is decreased. Polar deserts have little vegetation. Less than 10 per cent of the area is vegetated, although locally, in favoured locations, moss and lichen may cover a high proportion of the area. In Antarctica there is a very limited cover of lichen and some mosses in crevices. Soils lack organic material.

Low Arctic tundra is found in Northern Alaska, Baffin Island and southern coastal Greenland. Vegetation includes dwarf shrubs, mosses, sedges, gross and lichen (Figures 3.34 and 3.35). Small-scale environmental changes are important – on well-drained sites dwarf shrubs such as willow and birch are

Figure 3.34 Dwarf trees are common in the Arctic – an adaptation to high wind speeds and low insolation

	Adaptation	How the adaptations help for life in the Arctic
1	Prostrate (low-lying) shrub	Insulation beneath snow, warmer microclimate.
2	Cushion plants	Low-lying plant close to the ground surface so that it reduces the impact of windspeed on water loss. Warm microclimate (cushion up to 25°C warmer than air).
3	Annuals rare	Growing season too short for full cycle.
4	Herbaceous perennials common	Large underground root structure, store food over winter.
5	Reproduction often by rhizomes, bulbs or layering	Avoids reliance on completing flower–seed production cycle.
6	Pre-formed flower buds	Maximises time for seed production.
7	Growth at low temperatures	Maximises length of growing season.
8	Optimum photosynthesis rate at lower temperature than most plants	Maximises length of growing season.
9	Frost resistance	Can survive at low temperatures – true of flowers, fruit and seed.
10	Longevity	Suitable for 'opportunist' life style; lichens may live for several thousand years.
11	Drought resistance	Suitable for rock surfaces or arid climates.

Figure 3.33 Adaptations of Arctic vegetation

Figure 3.35 Lichen colonising a lava flow in south-west Iceland

located, as well as heather and moss. In wetter parts grass, sedge and cotton grass dominate. By contrast, high Arctic tundra contains lower order species such as herbs and moss. Sedge moss bogs are found on poorly drained areas, and many areas contain bare rock rather than vegetation.

The productivity of cold environments is low, Figure 3.36 shows the typical lichen and green moss fauna. The more favourable areas are comparable with temperate grasslands, the least favoured areas are similar to deserts.

> **Key term**
>
> **Net primary productivity:** the rate of production of biomass that is available for consumption by herbivores – the next trophic level in the ecosystem.

Geomorphological processes and landforms also influence vegetation. Slope angle and aspect influences local climate and moisture availability. In East Greenland dwarf birch are located on the tops of hummocks while sedges are located in the wetter depressions. Frost

Figure 3.36 Net primary productivity (NPP) of the Arctic compared with other ecosystems.

	g/m²/yr
Polar desert	0–1
Arctic tundra	100–400
Temperate grassland	100–1500
Desert shrub	10–250
Tropical rainforest	1000–3500
Warm temperature	
mixed forest	600–2500
boreal forest	200–1500

heave makes it difficult for roots to establish, so areas of active soil movement are often devoid of vegetation.

Animal life in cold environments

Animal life in the tundra shows many characteristics (Figure 3.37). For example,

◆ very few species are involved (of 8600 bird species in the world only 70 breed in the Arctic; of 3200 mammals in the world only 23 are found in the Arctic)

◆ there are large numbers of a single species – for example, caribou and lemmings

◆ population numbers are very cyclical – for example, lemmings have a 3–7 year cycle.

Many species exist only in the Arctic. These are known as endemic species. Many of these are genetically unique, some are migratory, and many are located in limited ecological niches such as ice-margins. Plants and animals living in the Arctic are exposed to major

Figure 3.37 Adaptations to cold environments

Cold environment characteristic	Adaptation to cold environments
Severe climate	Low number of species
	Low mean densities
Low temperature	High-quality fur insulation
	Increased metabolic rates
Snow	Life below snow patch for smaller animals
	Large herbivores favour soft/thin snow
Short summer	Birds migrate
	Breeding cycle compressed
	Large clutch/litter size
	Simple food chain (Figure 3.38)

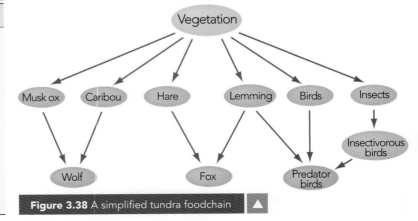

Figure 3.38 A simplified tundra foodchain

climatic variations even over very small distances. These variations lead to significant genetic variations within species.

Soils

Soils in periglacial environments are strongly affected by permafrost. This, and the low temperatures, are the dominant factors in soil formation. Bacterial activity is low and waterlogging sometimes leads to the formation of an acid humus. Just a few centimetres below the surface, blue-grey blotchy mud is found due to waterlogging (a process known as gleying). Ferric compounds are reduced to ferrous compounds as oxygen is lost and this produces the blue-grey blotches. Soils contain angular fragments of rock as a result of freeze-thaw action and frost heave. The most common soils to be found in tundra areas are known as tundra gleys (Figure 3.39). Not all periglacial soils are waterlogged. On the better drained sites podzols may develop (Figure 3.40). These are very distinct soils which show distinct soil horizons.

Nutrient cycle

The cycling of nutrients is essential for plant growth however in cold environments this can be prohibited by a number of factors. The amount of nutrients in the soils are limited because the rate of weathering is slow.

Figure 3.40 A podzol ▲

Figure 3.39 A gleyed (waterlogged) soil ▼

Similarly, as precipitation levels are also low there are few nutrients dissolved in rainfall. In addition frozen ground and/or snow cover can make it difficult for the plant roots to reach and absorb the limited supply of nutrients.

Key term

Podzol: light-coloured soil found under coniferous forests and on moorlands in cool regions where rainfall exceeds evaporation. The constant downward movement of water leaches nutrients from the upper layers, making podzols poor agricultural soils. The leaching of minerals such as iron, calcium, and aluminium leads to the formation of a bleached zone, often also depleted of clay. These minerals can accumulate lower down the soil profile to form a hard, impermeable layer which restricts the drainage of water down through the soil.

Activities

1 Explain the ways that Arctic vegetation is adapted to the environment.

2 Compare the productivity of tundra ecosystems with that of other ecosystems.

3 Explain how animals are adapted to cold environments.

Case study | Climate and ecosystems in the Alps

The Alpine climate

Many parts of Switzerland experience periglacial conditions. Switzerland has an alpine (mountainous) type of climate. In summer, maximum daytime temperatures reach up to 25°C. In winter, temperatures drop to between 2°C and 6°C in the valleys but at high altitudes temperatures are well below freezing. The coldest part of Switzerland is in Jura in the north-west, in particular the Brevine valley, which is a natural trap for cold air. Rainfall varies throughout the country with over 2500mm in Vaud in the south-west and less than 530mm in Valais less than 75km away in the south.

Alpine vegetation

The vegetation in the Swiss Alps is closely related to variations in climate. At high elevations, flowers bloom between April and July, whereas at lower elevations the growing season is longer. Typical plants include edelweiss and the alpine rhododendrons. Plants have developed the following adaptations for living in this environment.

- Bright pigments protect them from ultra-violet radiation.

- Colour attracts insects for pollination.

- Growing close to rocks to avoid trampling, for example orchids.

- Hairs on the leaves to reduce moisture loss.

- Waxy coatings to reduce water loss.

- Succulents store water.

- Growing close to the ground reduces moisture loss from lower wind speeds at ground level.

As temperature decreases by 1°C for every 100m, at an altitude of about 800m, coniferous trees replace deciduous trees (Figure 3.41). Coniferous trees are adapted to photosynthesise at lower temperatures whereas deciduous trees shed their leaves when the temperature becomes quite low. The red spruce is a typical tree found in the Swiss Alps. At around 2000m, tall trees are replaced by bushes and scrub which finally give way to Alpine meadows. Forests, however, are being affected by acid rain and pollution. As the forests die back and thin, the risk of avalanches increases since there is less vegetation to bind the soil and less interception of precipitation.

Figure 3.41 Swiss mountain environment ▼

Animal life

Despite environmental protection, animal life in Switzerland is under threat from increased human activity, especially tourism development. Typical species include the ibex (a type of mountain goat), the chamois (a type of antelope) and the marmot, a gregarious rodent. Characteristic birds include the chough and the golden eagle. Over 80 species are threatened with extinction.

Activity

Describe and suggest reasons for the changes in vegetation in the Swiss Alps with altitude.

Physical factors affecting cold environments

Ecosystems in cold environments are not, therefore, as fragile as commonly believed. It has tended to be assumed that the small number of species and the slow growth rates in cold environments, makes them fragile. Support for this view has traditionally come from the rapid changes in population numbers following disturbances. However, it is realised that population numbers fluctuate widely. Such large oscillations in fact provide polar and arctic ecosystems with the resilience and ability to survive dramatic changes in the environment, such as climate change. Another argument used to support the fragility claim was the disturbance of permafrost. This, it was claimed, led to irreversible changes. Recent research suggests that long-term changes are restricted to areas where there is a high proportion of ground ice, and even here, stability may be achieved in a number of years.

Another argument for the fragile tundra relies on the fact that there are relatively few species present in the tundra. Thus the decline of one species will have a significant impact on other species. Moreover, not only are there few species but the population density is low. Therefore, it is easy for indigenous people, armed with rifles, to exterminate species, such as caribou, over large areas.

Whilst all of the above are true, there is a key distinction between fragility and oscillation. Tundra ecosystems are characterised by large-scale oscillations. The lemming cycle is an excellent example. The fact that it recovers from low numbers illustrates its resilience rather than its fragility. However, under natural conditions, the tundra ecosystem has a great spatial range allowing smaller areas to recover from disturbance. If the ecosystem becomes fragmented, the opportunity for natural recovery becomes more limited.

Human impact on the environment

It is not just the disturbance to flora and fauna that has given rise to the idea that tundra areas are fragile. Disturbance to permafrost also causes concern. The problem is that ice-rich permafrost can be destroyed by heat from buildings, pipelines, changes in vegetation cover, and the impact of vehicles and machinery. If the ground contains excess ice, the heat will melt the ice and cause subsidence.

Thermokarst refers to the subsidence that occurs as permafrost melts. The most common cause of man-induced thermokarst on a landscape is the clearance of the surface vegetation for agriculture and/or construction purposes. Three factors control permafrost degradation.

◆ Ice content on underlying permafrost and, in particular, the presence or absence of excess ice.
◆ Thickness and insulating qualities of the surface vegetation.
◆ Duration and warmth of summer thaw period.

Man-induced thermokarst

The initial disturbance is often borrow pits, that is, where material has been removed for road, airstrip or other construction purposes. Thermokarst processes are rapid; typically hummocky relief forms through preferential subsidence along ice wedges. Stabilisation only begins 10–15 years after the initial disturbance and is not complete until >30 years. In environments of great summer thaw, the amplitude of the thermokarst mounds is greater, probably reflecting greater thaw depths.

The second cause of man-induced thermokarst is the movement of vehicles over permafrost terrain. If this occurs in summer when the surface has thawed and is soft and wet, surface vegetation can be destroyed and deep trenching and rutting can occur. This may be as much as 1m deep and 3–5m wide. Tracks favour continued thermokarst development by collecting water, and, if located upon a slope, promote gulleying by channelling snowmelt and surface runoff.

Changes to Europe's permafrost – how physical and human factors make the environment ecologically vulnerable

Permafrost throughout Europe is melting and threatening alpine villages and ski resorts with rockfalls and landslides (Figure 3.42). The main areas being monitored include the Murtel–Corvatsch mountain above St Moritz, and the Schilthorn, above the Muran and Gandeg resorts, near Zermatt.

A borehole sunk above St Moritz in 1986 showed that the ground temperature had risen between 0.5 and 1.0°C by 2000. If the temperature inside the mountain is only −2°C at the moment, then it will not take long to thaw. Already in Switzerland there have been rockfalls, landslides, mudflows, debris and slushflows as the ice has melted and weakened the mountains (Figure 3.43). These are likely to increase in the near future.

The Permafrost and Climate in Europe (Pace) organisation has been set up to monitor the creeping effects of climate change on the stability of mountains.

The combination of ground temperatures only slightly below zero, high ice contents and steep slopes, makes mountain permafrost in Europe particularly vulnerable to even small climate changes.

Research has shown that permafrost exists as far south as the Sierra Nevada mountains in Spain. Although ice was found only at the top of the Sierra Nevada mountains, in more northern parts of Europe, such as the Alps, it was found at 2500m. In Sweden it was found at 1500m. In Svalbard, in the high Arctic, ice was found at sea level. Britain's highest mountains are too low and too close to the warm westerly winds to have permafrost.

Other mountain ranges with permafrost being monitored include the Pyrenees, the Jotunheimen range in Norway and the Abisko range in Sweden. Boreholes have also been dug in Svalbard where coal is mined in the permafrost. The mine buildings have

Figure 3.42 Landslide near Zermatt – Switzerland ▼

Figure 3.43 Warm conditions in the Austrian ski resort of Meyerhofn ▲

their foundations on frozen soil but there are fears that the buildings will settle and fall if the frost melts. This could also be a serious problem in the higher European ski resorts where foundations of ski lifts and other buildings assume the ground will remain stable.

Activities

1 Suggest reasons why Europe's permafrost is melting.

2 Explain the environmental and economic implications of melting permafrost.

Discussion point

Why do some people claim that periglacial areas are fragile?

What is the evidence to suggest that periglacial areas are not as fragile as once thought?

'Take it further' activity 3.4 on CD-ROM

3.3 What are the issues associated with the development of cold environments?

Cold environments provide opportunities and challenges for development. Opportunities include resource exploitation, including agriculture, recreation and tourism. Challenges include environmental constraints, costs/remoteness, and conflicts with indigenous populations.

Resource exploitation

Agriculture

One of the main resources of cold environments to date has been its use as pastureland. Farming is nevertheless severely limited by extreme climate and a short growing season, which lasts just 90 days compared with 150–220 in mid-latitudes. The longer day length further north does not help much as most cultivated plants are not adapted to the long hours of daylight. One degree Celsius decline from the annual average temperature will reduce crop yields by 15 per cent and the carrying capacity of the land for livestock by 30 per cent. Soils are of a poor quality, with gleyed (waterlogged) soils widespread, although they respond well to fertilisers and being drained (Figure 3.44). However, in arctic environments, the large amounts of cheap geothermal energy has allowed for intensive cultivation in greenhouses, including tropical fruit (Figure 3.45).

Grazing land is important for the production of winter feed such as hay, grain and silage. The main crops grown include potatoes, barley, oats, spring wheat, rye and alfalfa. Most is grown for local consumption. Production of milk, vegetables, meat and eggs are barely competitive with similar products transported in from further south. Nevertheless, agriculture provides a livelihood for the people associated with it. However, much of the pasture has been affected by contamination. For example, military activity

Figure 3.44 Agricultural land in Iceland

Figure 3.45 Greenhouse cultivation from geothermal energy at Hvergerdi, Iceland ▲

and nuclear testing have been a major source of radioactive contamination in the Arctic. Radionuclides on moss and lichen become concentrated in caribou, which are in turn consumed by people.

There have been some conflicts with indigenous herders and other land users. For example, indigenous reindeer herders in Sweden – the Sami – say their traditional way of life is in jeopardy because the owners of private forests are using the law to exclude them from woodland. For the Sami, herding reindeer is a way of life. In the far north of Sweden, reindeer seek refuge from the bitter cold of winter in the forests. However, much of the forest exists on small private estates, whose owners are disputing the Sami's right to be there.

Energy and mineral extraction

Permafrost presents many problems to engineers. When the ground is frozen it has great strength, when thawed it is a jelly-like substance with no strength.

Three basic approaches to building a pipeline in permafrost terrain exist, they are to:

◆ bury the pipe in a trench

◆ suspend the pipe above ground on trestles (Figure 3.46)

◆ build a road along the proposed route with an adequate amount of fill to ensure stability and place the pipeline on the edge of the road surface, insulating and covering it with the appropriate amount of fill.

However, it is never quite so straightforward since the weight and heat (70–80°C) from an oil pipeline will cause thawing in permafrost. Moreover, if the soils near the pipeline contain large amounts of ground ice, they may become liquefied on thawing. Severe problems

▼ **Figure 3.46** Pipeline built on trestles in Nesjavellir, Iceland

may also arise when the pipe passes from a strong material to a liquefied region since the pipe would be put under considerable stress.

Therefore an alternative is to suspend it above the ground. Here the extreme temperatures make pumping difficult. The size and tremendous weight of the pipeline (a pipe of 12m diameter weighs about 230kg per 30cm) requires a dense piling network, which is extremely expensive. There are many places where gravel (which is a poor conductor of heat and is therefore used to build on) is in short supply, and where it is available, usually along streams, major disruption to salmon spawning may result if it is removed.

Mineral and energy resources – such as oil in Russia and Alaska, and iron ore at Kiruna, Sweden – causes local pollution and contamination. Pollution is a major problem in Arctic areas since oil breaks down more slowly under cold, dark conditions. For example, the Nyenski tribe in the Yamal Peninsula of Siberia have

suffered as a result of the exploitation of oil and gas. Oil leaks, subsidence of railway lines, destruction of vegetation, decreased fish stocks, pollution of breeding grounds, reduced caribou numbers etc have all happened directly or indirectly as a result of attempts to exploit this remote and inhospitable environment.

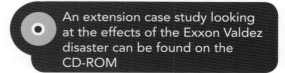

Activities

1 Describe the problems in piping oil across the Arctic.
2 Explain why permafrost is a problem for engineers.
3 Make a list of the techniques available to engineers to cope with permafrost.

An extension case study looking at the effects of the Exxon Valdez disaster can be found on the CD-ROM

Case study | Oil exploration in Siberia

Sibera has huge reserves of oil and gas. This provides Russia with the opportunity for great economic advance. Oil has been exploited in the area on a large scale since the 1970s. Nevertheless, there are serious implications for the region as a result of this industry.

The environmental impacts of development

The environmental impacts of petroleum development in western Siberia are well known: pollution of physical environment, disruption of natural processes through infrastructure development, destruction of wildlife habitat and resources. Soil pollution illustrates the scale of the problem. The two major sources of soil pollution are oil settling pits (for spent drilling fluids and production wastes) and broken pipelines.

Economic and socio-political impacts

Economic and socio-political impacts associated with oil development in western Siberia are less visible, but equally destructive. They include:

◆ population redistribution – there has been a significant redistribution of the Khanty population off their hunting territories

◆ economic dependency – where oil companies provide new technologies, such as outboard motors and snowmobiles, on which Khanty have become dependent

◆ deteriorating physical and mental health – depression and alcoholism abound due to loss of traditional way of life and lack of employment in new formal sector

◆ oil – responsible for the politicisation of the Khanty, some have become westernised and the traditional forms of authority, such as village headmen, are no longer the most important as they have been replaced by the well-off and rich. The headmen do not command the authority they once had.

Cultural impacts

Cultural impacts include:

- a reassessment by the Khanty of their relationship to and claims on the land of their family hunting territories

- the destruction of components of native religion including sacred places and culturally significant animal species.

Activities

1 Explain the reasons for the economic development of Siberia.

2 Outline the social costs of energy developments in Siberia.

3 'The people of Siberia should have full control of their resources'. To what extent do you agree with this assertion?

Case study | Ski tourism in Europe

While skiing brings great economic benefits – employment, investment and infrastructure – it also brings environmental costs to a region.

Europe is one of the most densely populated areas in the world and so therefore its mountains are increasingly popular tourist destinations. Most of the tourism in mountainous areas of Europe is concentrated in the Alps, which receives about 100 million tourists each year. Without tourists, many Alpine areas would not be economically viable. In some places over 80 per cent of jobs depend on them.

Environmental impact

The environmental impacts of skiing are far reaching. They include the construction of the ski pistes, related facilities such as access roads, parking, cafeterias, toilet facilities. Skiing removes habitats. It removes the natural protection against avalanches and it degrades the natural landscape. As most visitors travel by car, exhaust fumes lead to further forest damage and air pollution. Air travellers may be causing even more harm to the environment through the emission of fossil fuels at very high altitudes. Skiers can also damage trees by knocking off branches and killing young shoots underneath. Litter is another problem. For example, orange peel takes up to two years to break down.

One of the major impacts has been the development of ski runs and ski lifts (Figure 3.47). In Switzerland, the number of installations increased from about 250 in 1954 to near 2000 by 1990.

Figure 3.47 Ski run at Mayerhofn, Austrian Tyrol ▼

Figure 3.48 Forest is cleared for ski runs and chairlifts

Forest clearance (Figure 3.48) has led to an increased incidence of avalanches. Over 100km² of forest has been removed throughout the Alps and has led to higher rates of avalanches. New resort construction involves bulldozing, blasting and reshaping of slopes. This increases slope instability and leads to a higher incidence of avalanches.

Another increasing hazard is water pollution and sewage disposal. In the Alps chemicals used in preparing 36 glaciers for skiing have caused increases in nitrogen and phosphorus levels in drinking water.

The increasing popularity of skiing and the development of the ski industry has created a demand for larger accommodation blocks to be built in the popular resorts, such as Mayerhofn in the Austrian Tyrol. With limited space on the valley floor, this forces more development of the surrounding hillsides in order to fulfil accommodation demands.

Activities

1   Explain why is there so much pressure on Europe's mountains.
2   Describe the environmental impacts of skiing.
3   Outline the likely economic impacts of skiing.

Discussion point

What are the opportunities and constraints of living in, and exploiting, cold environments?

An additional case study on tourism in cold environments is available on the CD-ROM

3.4 How can cold environments be managed to ensure sustainability?

Managing tourism in the Alps

All eight Alpine states – Austria, France, Germany, Italy, Liechtenstein, Monaco, Slovenia and Switzerland – plus the European Union as a whole have signed the Alpine Convention, which aims to harmonise policies and to promote sustainable development.

The challenge of the convention lies in finding a balance between economic viability and sustainability. The implementation of the Convention may not be plain sailing. One problem is the lack of funding for promoting the environmental and economic policies agreed on. Another is that there are no sanctions that can be taken against any side who breaches the Convention.

The Alpine Convention is regarded as a model for other countries which share mountain ranges. As we have seen, the Alps are a popular destination for skiers. Skiing brings great economic benefits in terms of employment, investment and infrastructure. However, this development needs to be managed carefully. The following case study looks at the issues of tourism in the region more closely.

Case study | Sustainable development in the Alps: Part A

Environmental management

The Matterhorn ski region covers an area from the Rothorn and the Gornergrat to the Matterhorn glacier and Schwarzsee (Figure 3.49).

Figure 3.49 Map of the Matterhorn ski region ▼

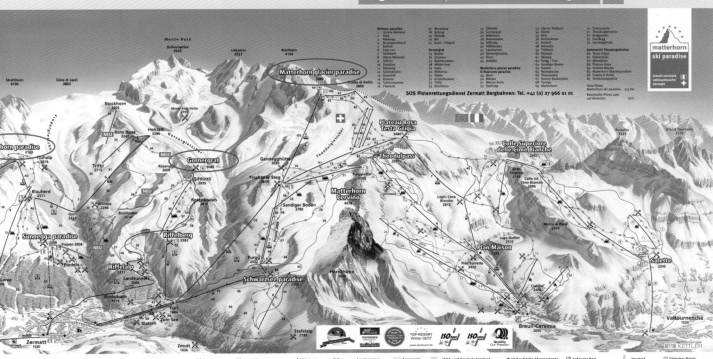

Electrically powered buses serve all three resort stations. There are 313km of marked pistes in Zermatt and Cervinia. To support the skiers there is one cog-railway, one underground funicular, nine cable-cars, five gondola-lifts, eight chair-lifts and nine drag-lifts (Figure 3.50).

Real environmental management is necessary to ensure that the natural beauty of the mountains remains unspoiled and that a modern railway business can still operate in spite of this. In 2002, therefore, Zermatt Mountain Cableways (ZMC) appointed a working party composed of environmental and planning firms to work out an overall plan for 'Sustainable skiing areas around Zermatt' which would highlight areas of conflict between construction projects and nature.

Figure 3.50 Ski lift infrastructure in Zermatt ▲

Damage inventory

The dramatic development of tourist infrastructure some decades ago left behind an accumulation of waste. In 2002, the Zermatt Mountain Cableways arranged for an inventory to be made of earlier damage (old ski lifts, pylons, huts etc) and of the legacy of this work that spoils the landscape. At the end of 2004, half of all prioritised damage had been made good. In 2005 the route of the Blattenlift (a lift for snowboarders) was restored by landscaping, reseeding and making the area look more natural.

Protection of forests and wild game

In 2003, the ZMC in collaboration with gamekeepers, forest wardens and biologists, drew up a forest and wild game protection programme with the aim of improving conditions for the forest and wild game in the winter. Segregated conservation areas have been fenced off and marked with notices. Specially constructed game observation points a short distance off the beaten track provide an opportunity to watch animals living in the wild without disturbing them. A wide-ranging information campaign makes

local people and winter visitors sensitive to wildlife protection – free-range skiing off-piste in particular often has fatal consequences for animals. Action at the Lower Schwarzsee involved the building of a dry-stone wall to protect sensitive marginal vegetation from being trodden on and contaminated.

Restoring nature

An extensive nature restoration initiative took place in the Gant region in summer 2005. This involved returning a large number of old ski routes to their natural state. The Damage Inventory programme included tidying up the natural landscape by the targeted removal of facilities that are no longer used.

Replanting

The replanting of damaged areas on high ground is exceptionally difficult. However, it is essential in many places to provide erosion protection and to reduce damage to the landscape. By sowing and – in especially tricky areas – bringing in plantlets, conspicuous gaps in the vegetation and areas at risk of erosion can be made green again. Jute netting has been used to stabilise the steepest parts of the slopes.

Environmental monitoring of building works

All building work, and also the work of returning areas to nature and replanting them, is monitored by a scientifically trained specialist. In addition to complying with environmental legislation, the aim is to work with consideration for conservation and long-term sustainability.

Environmental education

Zermatt Mountain Cableways is placing increasing emphasis on communicating ecological interrelationships. Information boards have been placed about the Schwarzsee Nature Conservation Area. In addition, in 2005, a glacier path was opened in the Gant-Findel glacier region near the mouth of the glacier (Figure 3.51).

Activities

1 Suggest reasons why Zermatt attracts many visitors.

2 Suggest ways in which visitors might damage the environment.

3 Outline ways in which it is possible to manage tourism sustainably in Zermatt.

The second part of this case study is on the CD-ROM and looks at how the Saaste region has balanced socio-economic and environmental needs

▼ **Figure 3.51** The start of a trek route on the Gorner glacier

Short term gains

Antarctica offers a very different example from the Alps. It has a very low population density and limited development but the pressures for change are immense. So far the continent has remained relatively unexploited and pressures for short term gains (such as mining the natural resource beneath Antarctica) have been resisted.

Antarctica is the world's last great wilderness. It is a continent almost entirely buried by snow and ice, so hostile and remote that it has no permanent inhabitants, apart from about 1000 winter scientists who stay in Antarctica all year round. Antarctica covers an area of 14 million km^3, which is almost as big as Europe. It is the highest continent, with an average height of 2300m. Over 99 per cent of the continent is covered by ice with an average thickness of 2450m. Antarctica is the coldest continent. However, on the Antarctic Peninsula temperatures have risen by around 2.5°C since the 1940s.

Sustainable management

According to the Antarctic Treaty (1961), Antarctica is to be used for peaceful purposes only and the environment is protected (Figure 3.52). Research is the major activity undertaken, although in summer, fishing and tourism also take place. Forty-six countries have now signed the Treaty, representing over 80 per cent of the world's population. Of the many countries that have taken part in the exploration of Antarctic, only seven have staked territorial claims in the region (Figure 3.53).

Mineral exploitation in Antarctica is very controversial. Although there are mineral occurrences in Antarctica, none are known in commercially viable quantities. Also, the technical, economic and environmental difficulties of extracting minerals are immense.

The Antarctic Treaty has been in operation for nearly 50 years and is regarded by many people as an outstanding example of international cooperation and an early example of a sustainable management policy. Some nations who are not part of the Antarctic Treaty believe that it is a 'rich man's club' and in the past have proposed that the continent should be managed by the United Nations as a global heritage for mankind.

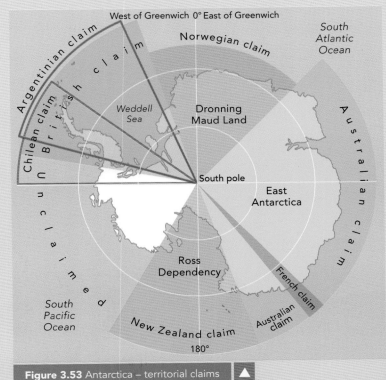

Figure 3.53 Antarctica – territorial claims

- Military activities in Antarctica are prohibited (e.g. military manoeuvres), although military personnel and equipment may be used for scientific research or other peaceful purposes.

- Freedom of scientific investigation and cooperation in Antarctica shall continue.

- Free exchange of information on scientific programmes and scientific data.

- Existing territorial sovereignty claims are set aside. No new territorial claims can be made whilst the Treaty is in force.

- Nuclear explosions and radioactive waste disposal are prohibited in Antarctica.

- The Treaty applies to all land and ice shelves south of latitude 60°S, but not to the high seas within the area.

- All Antarctica stations and all ships and aircraft operating in Antarctica have to be open to inspection by designated observers from any Treaty nation.

- Personnel working in Antarctica shall be under the jurisdiction only of their own country.

- Treaty nations will meet regularly to consider ways of furthering the principles and objectives of the Treaty.

- The Treaty may be modified at any time by unanimous agreement.

- The Treaty must be ratified by any nation wishing to join.

Figure 3.52 Summary of the Antarctic Treaty

Environmental Impact Assessment (EIA)

No activity, whether governmental or private, within Antarctica may proceed unless there is sufficient information available to determine that the environmental impact is acceptable. Strict regulations for waste disposal and waste management at stations and field camps are specified. Discharges into the sea of oil, chemicals and garbage, both from ships and stations, are all prohibited.

Tourism in Antarctica

There were more than 37 000 tourists to Antarctica in 2006, an increase of 14 per cent on 2005. In the early 1990s there were about 10 000 each year. Tourists first visited Antarctica in 1958. Antarctic tourism raises many issues. Of particular concern is how to protect the Antarctic wilderness while giving people the opportunity to experience its beauty for themselves. It has been suggested that tourism will inevitably lead to the degradation of the Antarctic environment and should be stopped. An alternative argument is that everyone has the legal right to visit the continent and that many of the tourists often act as advocates for the preservation of the environment.

Cruises normally last between 7–14 days, the majority in December to February. A hardy few sail yachts to the continent. Others wish to climb mountains or walk to the South Pole and they use the services of Antarctica's only land-based tour company – Adventure Network International. Overflying Antarctica without landing can also let people see the continent. Conservationists believe that no more than one tourist vessel at a time should be at a landing site, and that ships with over 500 passengers should be discouraged.

Whilst careful management and international co-operation has so far ensured that this fragile environment has remained relatively unexploited. Its future however, is increasingly uncertain. Countries such as the UK and Argentina have renewed their claims to parts of the continent; global warming is said to be having a measurable and negative impact on the fauna of the North peninsula and tourism looks set to continue growing. The challenge for the many stakeholders in Antarctica will be to balance their competing interests and ensure a sustainable future for the continent.

Activities

1. Describe and explain the effects of tourism on Antarctica.

2. How are pressures on Antarctica likely to change over the few decades. Justify your answer.

Discussion point

To what extent is it possible to manage cold environments sustainably?

 'Take it further' activity 3.5 on CD-ROM

Case Study: Managing Antarctica

U1

3

Cold environments

Knowledge check

1. What are the processes and factors that give cold environments their distinctive characteristics?

2. How are the distinctive characteristics of cold environments a result of climatic and geomorphological processes?

3. How are these processes influenced by factors which vary from place to place?

4. Why are cold environments considered to be 'fragile'?

5. How do climatic extremes lead to finely balanced ecosystems which can easily be damaged?

6. How can flora and fauna suffer as a result of change? Why is regeneration difficult in the harsh conditions?

7. What are the links between ecosystems and climate in cold environments?

8. What are the issues associated with the development of cold environments?

9. How do cold environments provide opportunities and challenges for development?

10. How can cold environments be managed to ensure sustainability?

A list of useful websites accompanying this chapter can be found in the Exam Café section on the **CD-ROM**

Exam Café
Relax, refresh, result!

Relax and prepare

What I wish I had known at the start of the year…

Mohammed

"I always read all of the questions first even though I can often do the first ones. I then put a cross by those I can't do and a double tick by those that I think are ideal. I then do what I think are my best questions first. This gives me confidence, gets the thought processes working and makes me feel less stressed."

Elisabeth

"I know that stress is quite normal and helps me to perform at my best, but if I'm too stressed my mind I just freezes. If this happens I try to breathe deeply and relax."

Hot tips

William

"I find it helpful to know what time of the day I work best. I find the mornings good for revising bits that I find difficult and the evenings good for the easier bits. I never revise after lunch because I'm not concentrating as well."

Common mistakes – Joyce

▷ "I never used to plan or structure my answer – I would start with an introduction but then I inserted paragraphs as I went along rather than with a clear idea of where they should go.

▷ I now do a rough plan with the headings I will use as the focus of each paragraph. The answer then shows a logical progression."

Refresh your memory

3.1 What processes and factors give cold environments their distinctive characteristics?

Weathering	Frost action, snow action (nivation), wet/dry, pressure release, hot/cold, vegetation, solution, acid action
Erosion	Glacial (bulldozing, plucking, abrasion), meltwater, wind action
Transport	Mass movement (solifluction, frost heave, avalanche, creep, flows, slides) meltwater, ice, wind
Deposition	Glacial (lodgement, ablation, englacial), meltwater, wind
Periglacial	Ice expansion, ice contraction, seasonal melt, mass movement
Factors	Physical – extreme climate, permafrost, glacial features, ecosystem
	Human – low population, little farming, long ignored, recreation

3.2 Why are cold environments considered to be 'fragile'?

Climate	Extreme with sudden events e.g. very cold snap
Permafrost	Delicate nature of supply (global warming)
Energy	Low energy environment (low inputs – cold)
Nutrients	Low stores, vulnerable flows
Species	Limited range of – limited gene pool and food chains
Population	Largely undisturbed until twentieth century

3.3 What are the issues associated with the development of cold environments?

Opportunities	Settlement – military and research bases
	Power production – wind
	Industry – tourism, furs, fish products
	Mining – oil, gas, ores
	Agriculture – grazing, use of heated greenhouses
	Transport – pipelines, coastal, air

3.4 How can cold environments be managed to ensure sustainability?	
Physical	Climate – Harsh and extreme
	Relief – glacial and periglacial deposits
	Vegetation – thin and tough
	Drainage – role of permafrost – poor drainage
	Ecosystem – hostile or low productivity
	Soils – leached, permafrost
Human	Role of indigenous groups – exploitation, cultural viability etc
	Pollution from mining
	Resource exploitation – e.g. over fishing
	Impact of recreation and tourism
	Transport improvement – out versus in migration

Top tips . . .

▷ Always plan to revise in short bursts of about an hour with a clear focus on a particular topic. It's even better to split this up into a 30 minute slot, 5 minute break and then a 20 or 25 minute slot. It takes time to get the mind online and focused but you need breaks as a reward for working hard. It is easiest to do it in blocks starting on the hour or half past (so have a clock handy) but then take a 15 minute break when you get up and move about and drink some water. Some people prefer longer sessions of revision. It is a good idea to divide your day into three sessions of three or four hours. Then, when you first start revising work, start with one of these sessions and later increase it to two. Try to avoid doing all three as you need to relax and socialise.

Get the result!

Sample question

What is the role of seasonal meltwater in a cold environment? [6 marks]

Examiner says

This scores maximum marks because – (1) the basic concept is explained, (2) a good understanding of duration and timescale, (3) and an understanding of the cause-effect linkage to process. (4) There is an understanding of the process using examples, (5) and an understanding of another process with cause-effect linkage with (6) good knowledge of appropriate landforms.

Student answer

Summer melt of ice and the active layer of the permafrost produce large quantities of meltwater (1) often in a short period of a few months (2). This water initially high in energy and low in load can be highly erosive (3), eroding channels e.g. spillways and forming seasonal lakes (4). The water quickly loses energy on the flat outwash plains and so deposits sorted material (5) as braids and deltas or as old river channels (eskers) and fans (Kames) (6) formed within or under the ice.

Examiner says

An equally effective approach would be to look at the role of meltwater in weathering and mass movement processes.

Examiner's tips

Plans are looked at and can gain credit if the candidate has run out of time to finish the answer.

Some students find it difficult to understand why ice features may be a result of contraction and expansion. Here it is in a nutshell:

This is very complex and is the result of there being different types of ice depending on the temperature and pressure. Initially as water freezes it expands but as temperatures continue to fall or pressure increases (with depth) it contracts into a denser ice. Frost wedges (contraction features) can show both processes and so indicate intense freezing. Expansion is more likely in moist environments (which are thus warmer) and contraction is an indicator of drier, colder environments. This could indicate seasons in an area. Such evidence is used to reconstruct past histories of periglacial environments.

Chapter 4

Hot arid and semi-arid environments

Images from Hergé's *Adventures of Tintin* might suggest that all deserts are covered with sand dunes and contain palm-fringed oases, but in reality most are just empty expanses of rock and gravel. Nevertheless, processes of weathering, erosion and deposition have, in some localities, created dramatic landscapes such as deep canyons, sculptured rocks and huge sand dunes. Plants and animals living in these harsh environments have evolved a remarkable number of ingenious ways to survive high temperatures and drought. Spectacular scenery, rich mineral reserves and sophisticated methods of irrigation have stimulated economic activities such as tourism, mining and agriculture. Careful management will, however, be needed to ensure economic, environmental and cultural sustainability.

Questions for investigation

- What processes and factors give hot arid and semi-arid environments their distinctive characteristics?

- Why are hot arid and semi-arid environments considered to be 'fragile'?

- What are the issues associated with the development of hot arid and semi-arid environments?

- How can hot arid and semi-arid environments be managed to ensure sustainability?

Consider this

Despite a 1.5km steep climb across bare rock in temperatures of 36°C, thousands of tourists visit Delicate Arch in Utah each year.

What threats and opportunities are posed by popularising tourist icons like this in desert areas?

4.1 What processes and factors give hot arid and semi-arid environments their distinctive characteristics?

Defining hot arid and semi-arid environments

Hot arid and semi-arid environments are generally characterised by high temperatures, low and unreliable precipitation and sparse vegetation. An arid area has less than 250mm of annual precipitation, while a semi-arid environment has between 250 and 500mm per annum. More precise definitions, however, also take into consideration evaporation and transpiration losses to create an aridity index (AI). Using the recent United Nations aridity index AI = P/PET, where P = average annual precipitation in mm, and Pet = potential evapotranspiration in mm, arid areas have an AI of <0.20, and semi-arid areas an AI = 0.20–0.50.

Key term

Potential evapotranspiration: is the maximum amount of evaporation and transpiration which could occur if water supply was unlimited.

Using the Köppen climatic classification these environments are interpreted to include:

◆ BWh: hot deserts with annual mean temperatures of >18°C such as the Sahara (Figure 4.1). Salah in Algeria for example averages 37°C in July and 13°C in January.

◆ BWn: hot deserts as above but with coastal fogs; for example, Atacama and Namib.

◆ BShw: Semi-arid, tropical deserts located between the Equator and the hot deserts; for example, sub-Saharan Africa. Timbuktu, for example, in Mali averages 22°C in January and 34°C June. These areas experience a rainy season during their hottest months.

◆ BShs: semi-arid, subtropical deserts located polewards of the hot deserts such as in North Africa and northern Iraq. Temperatures in Mosul in Iraq for example average 7°C in January and 32°C in June. These areas experience rainfall during their coolest months.

Figure 4.1 Location of the main hot arid and semi-arid environments

Key
Hot desert climate BWh and BWn
Semi-arid climate BShW and BShs

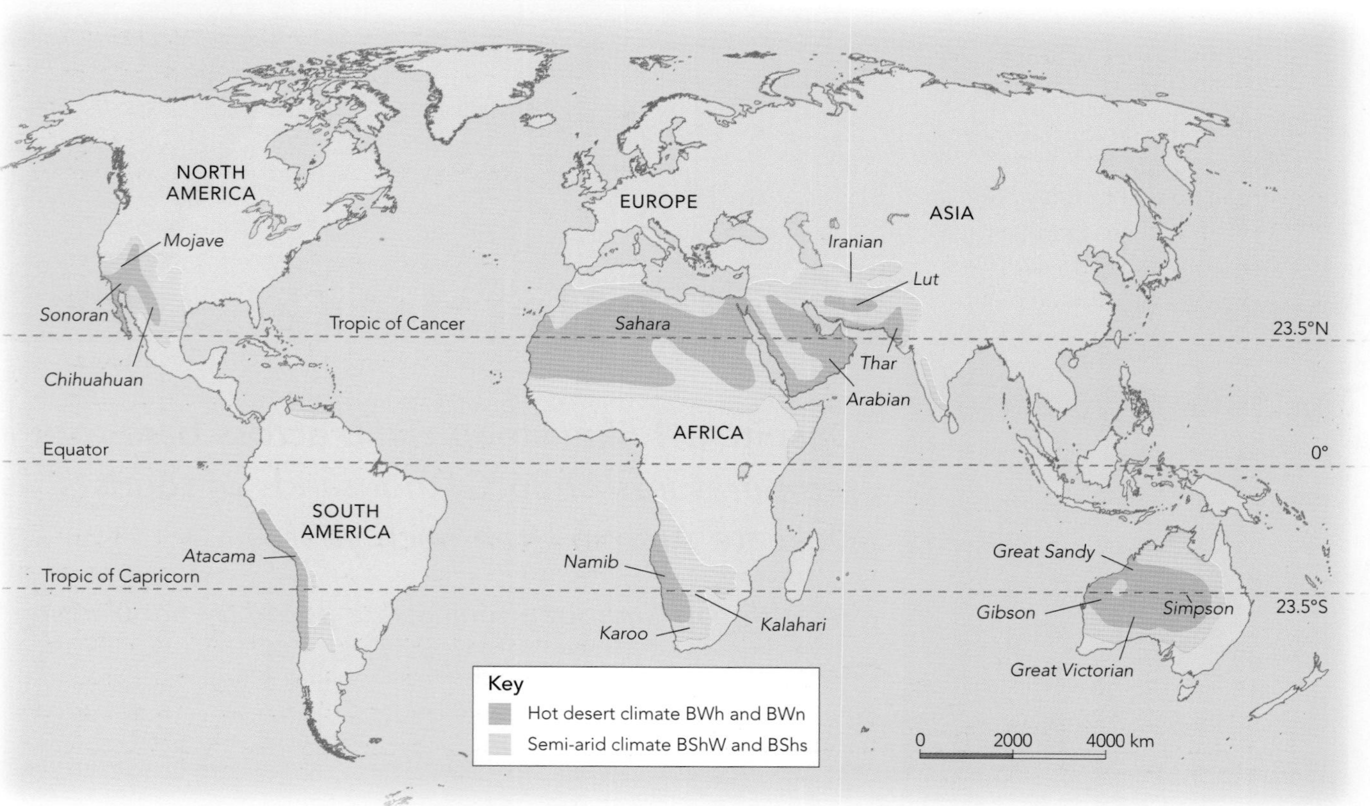

Cold, as opposed to *hot*, arid and semi-arid areas are found in polar regions and at high elevations and are mentioned in Chapter 3. Temperate arid and semi-arid areas within continental interiors, such as the Gobi, Turkestan and Takla Makan deserts in Asia, experience warm summers and cold winters and therefore are excluded from this chapter.

Processes and factors influencing climate

Temperature

Figure 4.2 shows the climate of Alice Springs, in the centre of Australia located at 23°S of the Equator, almost on the Tropic of Capricorn. In January (which is summer in Australia) the average maximum temperature is 36°C in the day, while the average minimum at night is 21°C. The large daily or diurnal range is explained by the lack of cloud. During the day, under cloudless skies, intense solar radiation heats the ground, which in turn warms the air above by conduction and convection. At night under clear skies, terrestrial radiation rapidly escapes and the ground becomes cool, which in turn chills the air directly above it by conduction.

Key terms

Solar radiation: radiation emitted from the sun.

Conduction: transfer of heat loss or gain through direct contact.

Convection: transfer of heat through convective movement of air.

Terrestrial radiation: radiation emitted from the earth.

The average annual temperature in Alice Springs is 20°C, much warmer than London at 11.5°C. The main reason hot arid and semi-arid areas experience high temperatures is because they are located at low latitudes. The high sun elevation here has two

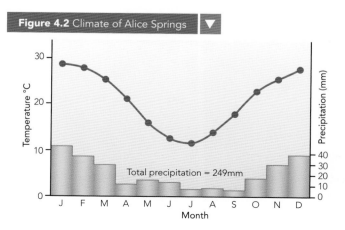

Figure 4.2 Climate of Alice Springs ▼

Total precipitation = 249mm

important consequences for temperature: first, the radiation per unit area is more concentrated than it is at higher latitudes (Figure 4.3): second, the solar beam has a shorter distance to travel through the atmosphere and therefore is less diluted by reflection, scattering and absorption than it is at higher latitudes. Given this information, you might expect Equatorial areas to be hotter than hot arid regions, but this is not the case because there is more cloud cover at the Equator, which reduces the amount of solar radiation reaching the ground.

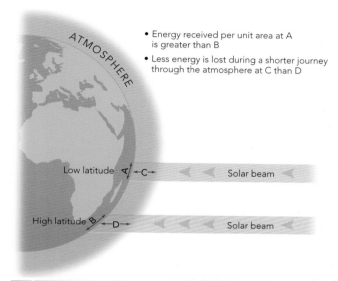

- Energy received per unit area at A is greater than B
- Less energy is lost during a shorter journey through the atmosphere at C than D

ATMOSPHERE

Low latitude A C Solar beam

High latitude B D Solar beam

▲ **Figure 4.3** The influence of latitude on the solar energy received on earth

In Alice Springs the hottest months of the year are December and January. By June the sun is overhead at the Tropic of Cancer in the northern hemisphere, which means that in Alice Springs the angle of the noon day sun is low and temperatures are therefore lower than in January.

Different hot arid and semi-arid areas experience different temperatures depending on their latitude, height above sea-level, distance from the sea, proximity to cool ocean currents and reflectivity of the ground surface or albedo. Tropical deserts are hotter than those located at higher latitudes because the sun is more overhead. Mountainous areas, such as the Tibesti Range in the Sahara, are cooler than surrounding areas. Deserts located in the centre of continents heat up and cool down faster than those by the sea and therefore have a greater range of annual temperature. Salt encrusted, dried-out lake surfaces reflect more solar radiation than areas of dark rock and are therefore cooler. Cold offshore ocean currents adjacent to coastal deserts depress temperatures.

U1

4

AS Geography for OCR

Key term

Albedo: the percentage of solar radiation reflected, as opposed to absorbed, by the earth's surface.

Activity

1 Describe in words the climate of Alice Springs in Figure 4.2.

2 Explain why Alice Springs is hotter in January than July.

3 Why might the average monthly temperature shown in the graph be misleading?

4 Find Alice Springs in an atlas and explain why:

    a it is hotter than Perth

    b it is slightly cooler than places further east in the Simpson Desert (contours may give you a clue).

Precipitation

Precipitation, or its absence, is a defining feature of arid and semi-arid areas. Amounts of precipitation vary enormously from about 10 to 500mm per annum. Antofagasta in the Atacama, which is very arid, receives 18mm of precipitation, much in the form of coastal fog. Hot deserts, such as the Sahara, get their rainfall from flash floods that are triggered by intense heating of the ground and convectional activity. Semi-arid areas such as the Kalahari experience seasonal rainfall brought about by global shifts in wind and pressure patterns. Tshane, for example, in the Kalahari receives 361mm of rainfall annually, most falling between December and March.

Factors causing aridity

There are five main factors responsible for aridity in hot (tropical/subtropical) arid and semi-arid environments.

◆ *Latitude* Hot arid and semi-arid areas are affected by subtropical high pressure cells which are found about 30°N and S of the Equator (see Figure 4.4). Air in these cells is subsiding and becomes compressed and consequently drier. Subsidence also prevents air from rising from the ground surface, cooling, condensing and forming cloud and rain (see

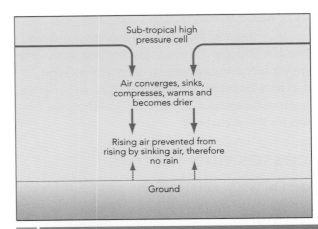

▲ **Figure 4.5** How subtropical high pressure cells prevent rainfall

Figure 4.5). Just occasionally, however, the subtropical air allows up-draughts to rise producing thunder clouds and flash floods. Subtropical, high pressure cells are best developed over the sea although their influence extends to land areas. In summer over the land masses, local heating encourages low pressure cells to form whereas adjacent seas are dominated by high pressure. In winter, both land and sea areas are affected by high pressure (Figure 4.4).

◆ *Offshore winds* Many hot arid areas are located where the prevailing winds blow from land to sea and therefore carry little moisture. The aridity of the Sahara for example is partly explained by the fact that North East Trade Winds blow across North Africa towards the Atlantic Ocean.

Key term

Trade Winds: surface prevailing winds which blow from subtropical high pressure cells towards the Equator.

◆ *Continentality* Areas in the centre of land masses, such as the Simpson Desert in central Australia, are dry because they are remote from rain-bearing winds which collect moisture from the sea.

Figure 4.4 A simplified summary of the world patterns of pressure and winds in June

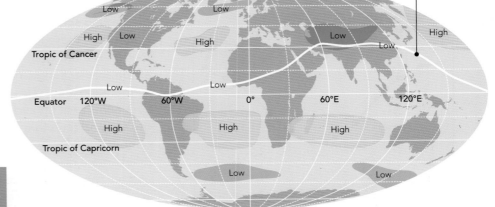

Note position of sub-tropical high cells at 30°N and 30°S of the equator

◆ *Relief* Mountain ranges block the passage of rain-bearing winds thus increasing the aridity on their leeward sides. Moist air forced to rise on the windward side of a mountain expands, cools and condenses to produce cloud and rain. On the leeward side a rain shadow develops because the air has already lost most of its moisture. Another reason the leeward side is dry is because as air starts to descend it becomes compressed and therefore warmer and drier. The Coastal Ranges in the SW of the USA prevent moisture spreading inland, which increases the aridity in the Sonoran Desert. In South America the moist SE trade winds that cross Argentina from the Atlantic are checked by the Andes, increasing dryness in the Atacama. In Australia the Great Dividing Range prevents SE Trades bringing rain into the central Australian deserts.

◆ *Cold offshore ocean currents* As you have seen, cold offshore ocean currents depress coastal desert air temperatures. Cold air holds less moisture than warm air which otherwise might have spread inland to produce rainfall. Instead, air above the cold current is chilled to its dew point to produce advection fog (Figure 4.6). The fog drifts inland carried forward on day-time sea breezes but eventually dissipates. Fog is often very persistent near the coast because it is trapped under an inversion layer by sinking air in a tropical high pressure cell. Where fog condenses it contributes to plant growth and weathering processes. Ocean currents are significant in increasing the aridity in the Namib, Atacama and Sonoran deserts.

Key terms

Advection fog: fog created when warm moist air comes into contact with a cold surface

Inversion layer: air layer where temperature rises rather than falls with height.

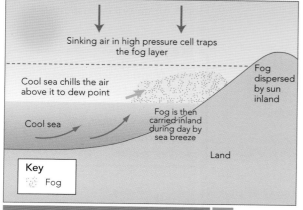

Figure 4.6 Formation of advection fog ▲

Activity

1 Use an atlas to name the mountain ranges and cold ocean currents which contribute to the aridity of the named hot arid deserts in Figure 4.7.

2 Figure 4.8 shows the Namib Desert.

 a Draw a diagram to explain the coastal fog.

 b Explain why the average temperature in the Namib is only 15–20°C, much lower than expected for latitude.

 c The average rainfall is just 125mm. Give reasons for the aridity.

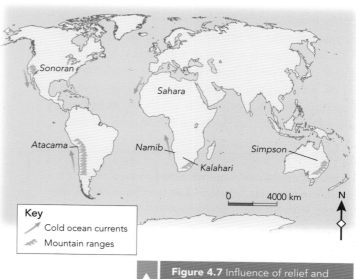

Figure 4.7 Influence of relief and ocean currents on hot arid deserts

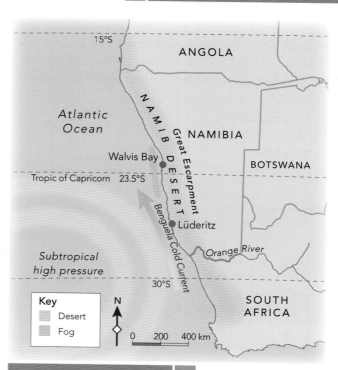

Figure 4.8 The Namib Desert ▲

U1

4

Hot arid and semi-arid environments

Seasonal rain in semi-arid areas

Seasonal rainfall occurs in tropical/subtropical semi-arid areas which fringe the hot arid deserts. Tropical semi-arid areas that lie between hot deserts and the Equator experience a short period of summer rainfall. This is because of the Hadley cell (Figure 4.9). At the Equator intense heating causes air to rise, expand, cool and condense producing heavy convectional rainfall. Air is drawn in at the surface and diverges at higher levels to create an area of low pressure. This area of heavy convection rain and converging winds is known as the inter-tropical convergence zone (ITCZ). At altitude the rising air spreads out and moves polewards. At about 30°N and S of the Equator, as it cools, the air begins to sink. As the air sinks, air aloft compresses the air below making it warmer and drier. Sinking air creates a zone of subtropical high pressure already mentioned. To complete the circulation, the air on reaching the ground surface moves towards the Equator. These surface winds which are deflected by the coriolis force produce the NE and SE Trades.

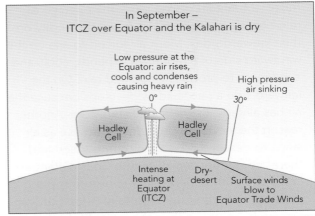

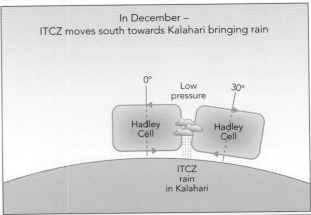

Figure 4.10 How seasonal rain is produced in the Kalahari Desert

Key terms

Inter-tropical convergence zone: an area of rising air, low pressure, converging surface Trade Winds and high temperatures coincident with the Equator.

Coriolis force: deflection of global winds by the rotation of the earth.

The Hadley cell shifts seasonally with the apparent migration of the overhead sun between the Tropic of Cancer and Capricorn, thus when the ITCZ moves seasonally from the Equator into the tropical semi-arid areas, it brings with it convectional rainfall (Figure 4.10). The rain occurs at the hottest time of the year and much is lost to evaporation, which makes it is less effective for plant growth.

Subtropical, semi-arid areas located poleward of the hot arid deserts experience winter rather than summer rain, and its origin is different. Poleward of the subtropical high pressure cells is an area where winds converge along the polar front. Depressions develop along this front producing rain. Global wind and pressure belts move with the apparent migration of the overhead sun and depressions on the polar front can track across North Africa bringing winter rainfall.

Key term

Polar front: boundary about 50°–60°N and S of the Equator where cool air from the poles meets warm air from the subtropical high pressure cells producing rain.

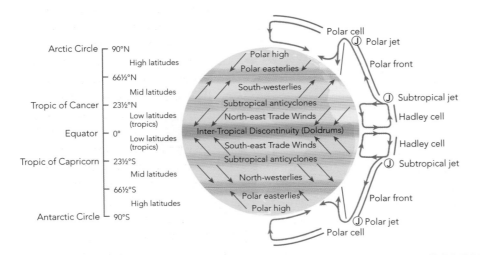

Note: the Trade Winds blow from subtropical anticyclones (high oressure) to the Equator (low pressure)

Figure 4.9 Global circulation of wind and pressure

(See Geographical Skills, pages 326–7.)

1 Draw a climate graph of Tshane in the Kalahari Desert located at 24°S of the Equator (Figure 4.11). Use the information in the climate graph of Alice Springs (Figure 4.2) to help you with the construction.

	J	F	M	A	M	J
Temperature 0°C	25	26	24	20	16	12
Precipitation mm	85	74	68	35	7	3

	Ju	A	S	O	N	D
Temperature 0°C	13	16	21	23	25	26
Precipitation mm	1	0	4	16	32	36

2 Why are November, December, January and February the hottest months?

3 What causes the seasonal rain in January and February?

4 Draw a sketch of the Kalahari and add labels highlighting the factors responsible for the aridity.

Figure 4.11 The Kalahari Desert ▼

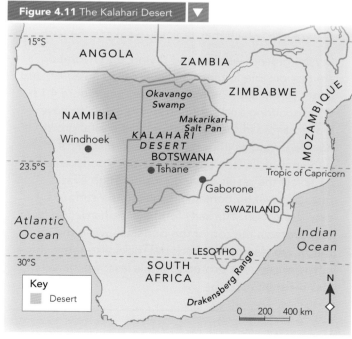

Winds

Winds in deserts can be strong and gusty and have a considerable influence on landforms in arid and semi-arid areas. Large-scale and local winds affect desert areas. The centre of a subtropical high pressure cell or anticyclone is calm, but circulating around the edge of this system are global winds such as the NE Trades. These winds shift seasonally with the apparent migration of the overhead sun. This means that for part of the year an arid area may be calm, but at other times it is very windy; for example, the Harmattan, a hot, strong, NE Trade Wind, blows across the southern Sahara in winter creating major dust storms and carrying sand which is eventually deposited in the Atlantic Ocean.

Another type of weather system to affect arid and semi-arid areas is monsoonal winds which are drawn towards continents in summer; for example, the Asian monsoon draws south-west winds across western India and southern Pakistan, and a north-west wind called the 'shamal' blows wind across Arabia and Iran causing major sandstorms.

The poleward margins of arid tropical and semi-arid areas are seasonally affected by westerly winds associated with depressions which track east. For example, a warm, dusty wind called the sirocco blows across North Africa towards southern Europe in spring.

Mountains, by obstructing wind flow, create local winds on their leeward sides. They can also funnel winds. Daily heating and cooling of sea and adjacent land areas produces local differences in atmospheric pressure leading to local sea and land breezes.

Climate and the physical landscape

The impacts of climate on the physical landscape are summarised in Figure 4.12.

Climate and weathering

The two main forms of weathering operating in hot arid and semi-arid areas are insolation weathering and crystal growth.

◆ *Insolation weathering* High temperatures during the day heat up surface layers of rock which expand, while at night they cool and contract. Repeated cycles of expansion and contraction cause the layers to peel or flake in a process referred to as 'onion-skin weathering' or 'thermal exfoliation'. Meanwhile, repeated heating and cooling of well-jointed rock causes 'block disintegration'. On a smaller scale, light and dark minerals within rocks such as granite heat and cool down at different rates and in time this leads to 'granular disintegration'. Collectively, the mechanical fracture of rocks by repeated heating and cooling is known as insolation weathering. Modern opinion is, however, divided as to whether it can occur without the presence of at least some moisture.

Figure 4.12 Impact of climate on the physical landscape of deserts ▼

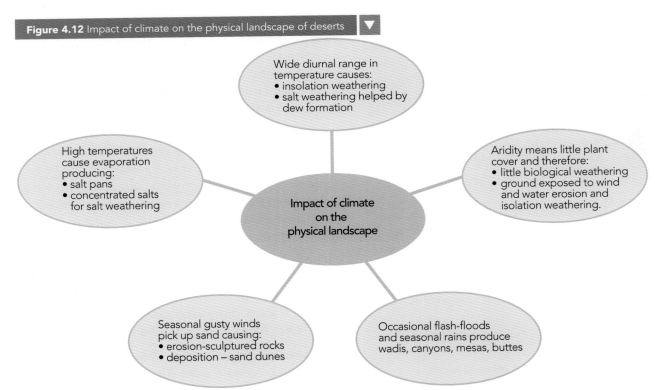

Wide diurnal range in temperature causes:
• insolation weathering
• salt weathering helped by dew formation

High temperatures cause evaporation producing:
• salt pans
• concentrated salts for salt weathering

Impact of climate on the physical landscape

Aridity means little plant cover and therefore:
• little biological weathering
• ground exposed to wind and water erosion and isolation weathering.

Seasonal gusty winds pick up sand causing:
• erosion-sculptured rocks
• deposition – sand dunes

Occasional flash-floods and seasonal rains produce wadis, canyons, mesas, buttes

◆ *Crystal growth* This is a major cause of weathering in deserts especially in porous, sedimentary rocks such as sandstone. High temperatures draw saline groundwater to the surface and the water evaporates. Salt crystals grow between pores and joints and leads to granular and block disintegration. Crystal growth also works in conjunction with insolation weathering, causing flaking or exfoliation. Coastal, fog-bound deserts are particularly weathered by crystal growth.

Other forms of weathering which occur in desert areas include:

◆ *Wetting and drying* This occurs when flash floods, seasonal rains, coastal fog and dew encourages clay minerals in rocks to swell. Repeated expansion on wetting and contraction on drying causes the rock to disintegrate.

◆ *Frost-shattering* In mountains, repeated cycles of freezing and thawing of water between joints and pores in rock causes it to shatter, producing scree.

◆ *Hydration* Minerals such as anhydrite absorb water and expand, which causes stress and granular disintegration. When water is added to anhydrite, it forms gypsum, which is susceptible to other weathering processes such as carbonation.

◆ *Solution* Water dissolves minerals such as rock salt, which is then removed in solution.

◆ *Oxidation* Oxygen dissolved in water reacts with minerals such as iron and manganese to create oxides and hydroxides. The result is often seen as a red staining on rocks such as sandstone and basalt. Minerals that are oxidised increase in volume, which mechanically weakens the rock.

Crystal growth, wetting and drying, and hydration are responsible for small hollows called alveoles, and larger tafoni seen on rock surfaces (see Figure 4.13). Wind scouring and case hardening may also have played a role in their formation. Salt crystal growth is now regarded as important in the formation of many pedestal rocks (see Figure 4.14), once considered to be solely the result of wind erosion. High evaporation rates draw water containing salts to the surface by capillary action. The salts form a case-hardened layer that protects the top of the pedestal, while the rocks below are weathered. Pedestal rocks also form where a harder cap rock protects, and is less weathered than, less resistant rocks below.

─ **Key terms** ─

Alveoles: small hollows, 5–50cm in size, occurring in clusters or honeycombs with thin partitions strengthened by case hardening.

Tafoni: hollows a few metres across with arch-shaped entrances, often developed along lines of weaknesses such as joints and bedding planes.

Case hardening: a hard layer of salt-encrusted rock formed where salts have been drawn to the surface by capillary action.

Pedestal rock: isolated pillar of rock with an indented profile.

Figure 4.13 Alveoles developed in sandstone, Utah ▲

With little vegetation, biological weathering is limited, but lichen and algae can cause micro-morphological changes to rocks. Respiration releases carbon dioxide for chemical processes such as carbonation, and the excretion of oxalic acid also causes weathering.

Although not strictly a weathering process, the gradual removal of the surface layers of rock can cause those underneath to expand. This process, known as 'pressure release', 'unloading', or 'sheeting', produces slightly curved, horizontal lines of weakness or pseudo-bedding planes along which other weathering processes can operate. Unloading contributes to the formation of inselbergs, such as in the tropics (Figure 4.15 and case study of Ulu_r_u).

> **Key term**
>
> **Inselbergs:** upstanding masses of crystalline rock that project above plains in deserts and semi-arid areas.

Finally, it is worth remembering that several landforms seen today in semi-arid areas were formed at a time when the climate was different from now. Some screes were formed when the climate was colder and wetter than today. Inselbergs formed when the climate was wet and warm, which encouraged deep chemical weathering, particularly hydrolysis.

> **Key term**
>
> **Hydrolysis:** a weathering process where water reacts with minerals such as feldspars to produce clay minerals.

Figure 4.14 Pedestal rock ▼

Case study | Weathering processes operating on Uluru, Australia

Figure 4.15 Uluru, Australia ▲

Uluru is an example of an inselberg and stands 340m above the surrounding plain in central Australia. It is 2.5km long, 0.5km wide and is made of arkose sandstone (sandstone with more than 25 per cent feldspar). The strata dip almost vertically and harder bands within the sandstone stand proud of the surface. Flaking of the rock surface is the result of insolation weathering and crystal growth. The rock layers are also etched with tafoni. Unweathered arkose is grey in colour but the rock appears red because water has dissolved the iron that has been re-deposited as a thin skin of iron oxide. Unloading, or pressure release, is thought to be responsible for sheet fractures, which are up to 2m thick and best seen in a feature called the 'Kangaroo Tail'. Curved, or flared slopes, which occur at the base of the rock, are believed to have formed by deep chemical weathering, particularly hydrolysis, at a time when the climate was warmer and wetter than today.

Discussion point

Look at the photograph of Delicate Arch at the beginning of this chapter. What weathering processes might be operating on this landform? Can you suggest how the arch might have been formed?

The role of wind and water in shaping landforms

Wind erosion

The three main processes of wind or aeolian erosion are: deflation, corrasion and attrition. Deflation occurs when wind picks up and removes unconsolidated material creating deflation hollows, salt pans and desert pavements. Corrasion happens when sand carried in the wind abrades rock surfaces, producing sculptured rocks and ventifacts. Attrition describes the way sand grains collide with each other and in doing so become smaller. The amount of erosion depends on the strength and duration of the wind, the composition and structure of the rocks, the vegetation cover and moisture content of the soil.

> **Key terms**

> **Deflation:** the entrainment and removal by wind of unconsolidated material.

> **Corrasion:** when wind-blown sand abrades rock surfaces.

> **Attrition:** the action of sand grains colliding with each other and in doing so becoming smaller.

> **Deflation hollows:** large enclosed depressions partly created by deflation; for example, the Qattara Depression in the Sahara.

> **Salt pan:** flat areas encrusted with salt sometimes filled with shallow, saline lakes.

> **Desert pavement:** surface of stones resting on a finer material such as sand, silt or clay. Known as gibber plains in Australia, and stone pavements in USA. Formed where wind or water selectively removes finer material to leave behind larger stones. Alternative theories suggest that coarse material is brought to the surface by cycles of heating and cooling, freezing and thawing, and wetting and drying.

> **Ventifact:** wind-polished stones which look like Brazil nuts and are a few centimetres in size.

Wind erosion creates sculptured rocks such as yardangs, a Turkish term for steep-crested, linear ridges of rock orientated parallel to the prevailing wind (Figure 4.16a and b). Yardangs occur in groups, or 'fleets' and the ridges are separated by wind-scoured grooves. They develop in easily eroded soft, but cohesive, sediments such as silt and clay as well as more resistant rocks such as limestone and sandstone. In appearance they resemble the shape of an upturned boat with smooth sides and a sharp keel. Ratio of length to width is typically 4:1 and the end facing the wind is often the highest and broadest. They vary in height from a few centimetres to over 100m and can be several kilometres in length. Good examples of

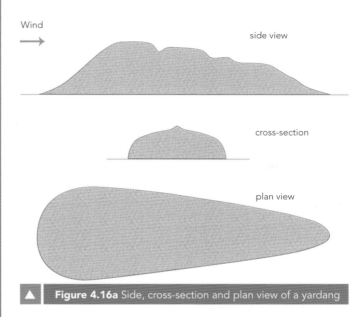

Wind

side view

cross-section

plan view

▲ **Figure 4.16a** Side, cross-section and plan view of a yardang

▼ **Figure 4.16b** A yardang

medium-sized 'meso-yardangs' occur in the Kharga yardang field in Egypt. Mega-yardangs up to 200m high and many kilometres long are found near the Tibesti Mountains in the Sahara. Abrasion smoothes pits and grooves on the sides of the yardang, while deflation helps form the aerodynamic shape. Some yardangs may have initially formed in wetter climates where water created gullies that were later exploited by wind (see case study of Namib Desert for other examples of yardangs).

> **Key term**
>
> **Yardang:** linear ridge of clay, silt or rock sculptured by abrasion and deflation.

Many pedestal rocks and zeugens (Figure 4.14) are now attributed to weathering rather than wind abrasion. Where wind abrasion occurs, it is most effective within 1–2m of the ground. Abrasion, particularly if the rock is soft and protected by a harder cap rock, helps to produce an indented profile.

> **Key term**
>
> **Zeugen:** a type of pedestal rock with a resistant cap rock.

Wind deposition

Sand covers only 20 per cent of the world's arid and semi-arid areas, but varies proportionally from less than 1 per cent in south-west USA, to 26 per cent in the Sahara and 38 per cent in Australia. Wind transports material in three ways (Figure 4.17). Very small particles <60 micrometres (μm) are carried in suspension and may remain aloft for long periods. Sand-sized particles between 60 and 500 μm move in a series of hops along the ground by saltation and is the main way sand is moved in desert. Heavier particles >500 μm are rolled along the surface by surface creep.

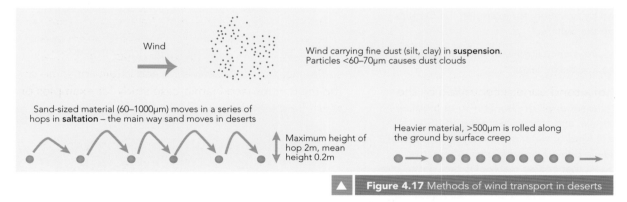

Figure 4.17 Methods of wind transport in deserts

Figure 4.18 Sand sea

Key terms

Suspension: fine sediment carried within the air.

Saltation: movement of sand grains in a series of hops along the ground.

Surface creep: rolling or pushing of sediment along the ground.

Sand is largely composed of quartz and feldspar minerals originally weathered from quartz-rich rocks such as granite and sandstone. Strong winds in non-aeolian areas on the fringes of subtropical high pressure cells pick up sand from near rivers, lakes and seas and can carry it huge distances into deserts where it is deposited. About 60 per cent of the world's sandy arid and semi-arid areas are covered by extensive sand seas or 'ergs', which contain a variety of different types of sand dune (Figure 4.18). Major sand seas occur in North Africa, Asia and Australia. Dunes develop around obstacles such as rocks and plants, while others form where a rough surface, such as a patch of sand, causes frictional drag and deposition. Yet others form when wind flows converge from different directions. The shape and speed of movement of a sand dune is controlled by the strength and direction of the wind, the volume and grain size of the sand supply, the shape of the land and the presence of vegetation. Sandy areas without significant dunes are called sand sheets, the largest example being the Selima Sand Sheet in north-west Sudan. The dune type found in sand sheets are 'zibars', low forms rising a few metres made of coarse sand and orientated normal to wind flow.

Key terms

Sand sea: extensive area of sand containing a variety of dune types.

Sand dune: mound of loose, sand-sized material created by grain on grain movement.

Sand sheet: extensive flat or gently undulating area of sand with no significant dunes and sparse vegetation.

Sand dunes can be classified into two types: fixed or anchor dunes, which form around an obstacle such as a rock or plant, and free dunes where no such obstruction is present. Free dunes are further sub-divided into transverse dunes where sand movement is normal to wind direction and linear dunes where sand movement is parallel to the wind. Figure 4.19 summarises the characteristics and origins of some of the main types (see Namib case study for examples of sand dune types).

Case study | Features produced by wind erosion and deposition in the Namib Desert

The Namib is 1600km long, 50–160km wide and bounded to the west by the Atlantic Ocean and to the east by the Great Escarpment (Figure 4.8). The desert is comprised of gravel plains, isolated inselbergs and five large dune fields, the largest of which is the Namib Sand Sea located between Lüderitz and Walvis Bay. Straight, linear dunes, which are orientated north to south and rise to between 50m and 150m high, occupy 75 per cent of the Namib Sand Sea. Other types of dune include barchans and barchanoid ridges, which occur along the coastal strip. Star dunes rising to 145m are found on the eastern margin of the desert. Features created by wind erosion include yardangs, which occur in the southern part of the desert and are made of dolomite, a type of magnesium rich limestone. They are orientated parallel to south-east winds. Ventifacts, developed in dolerite, an igneous rock, are widely found inland from Walvis Bay.

Discussion point

You might be surprised to learn that barchans and pedestal rocks are relatively rare desert landforms and arid areas are not typically full of sand dunes and palm trees. What influences our perception of deserts?

Water erosion

Seasonal and convective rainfall in hot arid and semi-arid areas is often heavy and therefore water runs across the surface rather than infiltrates. Moreover, there is little vegetation to intercept the rainfall and high temperatures bake the soil to a hard crust. Water initially spreads across the surface as sheets and then collects into channels. Landforms created by water erosion include wadis, canyons and canyon-lands.

A wadi is an Arabic term for a dry river valley with steep sides and a wide floor covered with channel deposits. It forms when flash-floods and seasonal rain creates an ephemeral river. High discharge,

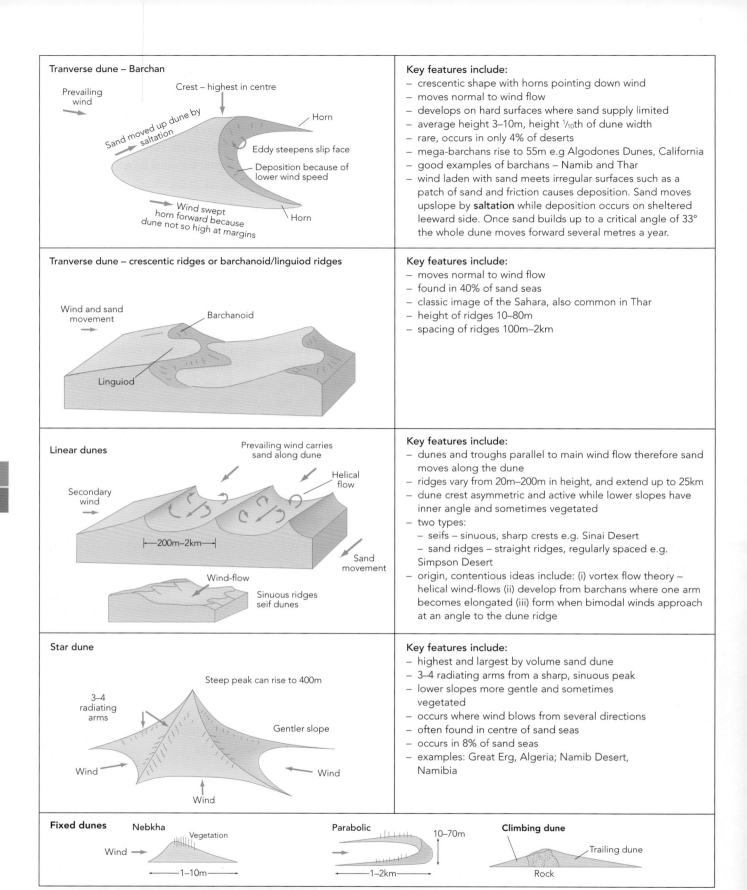

Tranverse dune – Barchan

Prevailing wind

Crest – highest in centre

Sand moved up dune by saltation

Horn

Eddy steepens slip face

Deposition because of lower wind speed

Wind swept horn forward because dune not so high at margins

Horn

Key features include:
- crescentic shape with horns pointing down wind
- moves normal to wind flow
- develops on hard surfaces where sand supply limited
- average height 3–10m, height $\frac{1}{10}$th of dune width
- rare, occurs in only 4% of deserts
- mega-barchans rise to 55m e.g Algodones Dunes, California
- good examples of barchans – Namib and Thar
- wind laden with sand meets irregular surfaces such as a patch of sand and friction causes deposition. Sand moves upslope by **saltation** while deposition occurs on sheltered leeward side. Once sand builds up to a critical angle of 33° the whole dune moves forward several metres a year.

Tranverse dune – crescentic ridges or barchanoid/linguiod ridges

Wind and sand movement

Barchanoid

Linguiod

Key features include:
- moves normal to wind flow
- found in 40% of sand seas
- classic image of the Sahara, also common in Thar
- height of ridges 10–80m
- spacing of ridges 100m–2km

Linear dunes

Prevailing wind carries sand along dune

Helical flow

Secondary wind

200m–2km

Sand movement

Wind-flow

Sinuous ridges seif dunes

Key features include:
- dunes and troughs parallel to main wind flow therefore sand moves along the dune
- ridges vary from 20m–200m in height, and extend up to 25km
- dune crest asymmetric and active while lower slopes have inner angle and sometimes vegetated
- two types:
  - seifs – sinuous, sharp crests e.g. Sinai Desert
  - sand ridges – straight ridges, regularly spaced e.g. Simpson Desert
- origin, contentious ideas include: (i) vortex flow theory – helical wind-flows (ii) develop from barchans where one arm becomes elongated (iii) form when bimodal winds approach at an angle to the dune ridge

Star dune

Steep peak can rise to 400m

3–4 radiating arms

Gentler slope

Wind

Wind

Wind

Key features include:
- highest and largest by volume sand dune
- 3–4 radiating arms from a sharp, sinuous peak
- lower slopes more gentle and sometimes vegetated
- occurs where wind blows from several directions
- often found in centre of sand seas
- occurs in 8% of sand seas
- examples: Great Erg, Algeria; Namib Desert, Namibia

Fixed dunes

Nebkha

Vegetation

Wind

1–10m

Parabolic

10–70m

1–2km

Climbing dune

Trailing dune

Rock

Figure 4.19 Characteristics and formation of the main types of sand dune

together with the availability of loose, dry sediment in the channel bed encourages the transport of large amounts of sediment, albeit for a short period or distance. Boulders are abraded and the underlying bedrock is scoured. Discharge, however, declines downstream because the storms are localised and some water evaporates, or infiltrates the river bed. As the discharge declines, the stream loses its competence and braiding occurs. Wadis vary in size from a single channel a few metres long to dense networks many kilometres in length. They are known as arroyos and washes in south-west USA. Arroyos tend to have steeper sides than washes. Large wadis may have been formed during wetter periods in the past.

> **Key terms**
>
> **Ephemeral river:** temporary river which flows intermittently or seasonally.
>
> **Braiding:** the tendency for a river to spit into smaller channels and deposit material when attempting to carry a heavy load.

A canyon is a gorge with a deep, narrow channel bounded by resistant rocks (Figure 4.20). The floor of the valley is often occupied by an exogenous river; for example, the Colorado. Aridity limits weathering and surface erosion, which preserves the steepness of the valley sides. Very deep canyons form where an area experiences tectonic uplift while the river incises vertically into its bed; for example, the Grand Canyon. Seasonal rainfall and flash-flooding causes the river discharge to vary and when low it cannot carry its load and braiding occurs. Tributary streams swollen by flash-flooding transport boulders to the main river creating rapids.

> **Key term**
>
> **Exogenous river:** a permanent river which derives its water from beyond the desert margin.

The south-west USA contains a variety of 'canyon-land'-type landforms. Wadis and canyons dissect the desert surface. Steep-sided plateaus of rock called 'mesas' and smaller 'buttes' are also common features (Figure 4.21). Many of these are composed of horizontally bedded layers of rock capped by more resistant bands that protect the softer rocks below. At the base of a mesa is often a pediment sometimes partly covered with scree from rock falls. Mesas and buttes are believed to represent the remains of what were more extensive plateau surfaces that have since been dissected by river erosion.

Figure 4.20 Grand Canyon ▼

Figure 4.21 Mesa and butte ▲

┌─ **Key term** ─────────────────────────────┐

Pediment: gently sloping bedrock surface extending from a mountain front to the alluvial plain below. Origin controversial, could be formed by wind or water erosion.

└──┘

Water deposition

Salt pans, playas and sabkhas are features created by water deposition. All are flat, low-lying areas with little vegetation and partly occupied by shallow, ephemeral saline lakes. Precipitation, groundwater seepage and surface runoff fill the lake, which usually has no surface outlet. When the lake dries out, the clay floor cracks and curls up. Water evaporates leaving behind deposits of sodium chloride, sodium sulphate, gypsum and calcium carbonate. Plants such as prickleweed and saltbush, which can tolerate salt, occupy the edges of the pan, while mesquite bushes grow where the water-table is near the surface (Figure 4.22).

Playas and pans vary in size and have a variety of origins. Large playas, such as those in Death Valley in California, stretch for several kilometres and

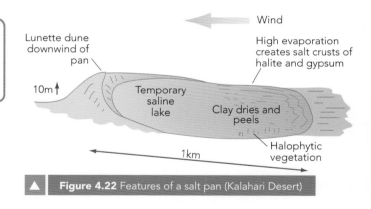

▲ **Figure 4.22** Features of a salt pan (Kalahari Desert)

occupy structural basins bounded by faults. Others, for example in the Kalahari and Australia in semi-arid areas, are only a few 100 metres wide and are formed by wind deflation. When the lake dries out, salt weathering loosens the material on the floor of the pan, which is then eroded by wind. The material is deposited as a crescentic, lunette dune on the downwind side of the pan. Yet other pans form where animals overgraze around waterholes, or develop in depressions between linear dunes. The term 'playa' and 'pan' are interchangeable, but sometimes playa is used to refer to areas within structural basins and reserve pans to describe smaller areas formed by wind deflation such as those in the Kalahari. The Arabic word 'sabkha' refers to a type of salt pan fringing coastal areas such as the Arabian Peninsula.

 'Take it further' activity 4.1 on CD-ROM

Activity

1 Draw a spider diagram summarising the main features of wind and water erosion and deposition.
2 Draw an annotated sketch of the mesa and butte in Figure 4.21.
3 Find other images of desert landforms such as yardangs, sand dunes, wadis and canyons and add labels to show features and how they formed.

4.2 Why are hot arid and semi-arid environments considered to be fragile?

Impact of climate on ecosystems

Climatic extremes lead to finely balanced ecosystems that can easily be damaged. As you have learned, temperatures can soar to 50°C in the shade, while at night temperatures plunge by 15 to 20 degrees and rainfall is low and unreliable. Plants and animals have evolved mechanisms to cope with heat and dryness, but suffer when exposed to long periods of drought. Strong winds and shifting sands create unstable surfaces that make it difficult for vegetation to become established. Soils are thin and infertile because lack of precipitation limits chemical weathering processes. High evaporation from rain, runoff and groundwater makes soils saline. Hard, salty crusts impede root growth. Without vegetation little organic matter is created that would normally decompose to release plant nutrients. High temperatures and low moisture also slow biological activity that would break down organic matter. Lack of vegetation exposes the soil to wind and water erosion.

Key term

Ecosystems: groups of organisms (plants, animals and bacteria) which interact with one other and the environment so that material is exchanged between the living and non-living (air, soil, water) parts of the system.

Plant adaptations to climate

Harsh conditions mean that plants, particularly those in hot arid areas, are sparse and small in size. Variations in temperature, rainfall, relief, geology and soils can nevertheless lead to wide differences in species diversity. In the Atacama, vegetation is quite rare, whereas in the Sonoran Desert it is quite diverse. There are a number of ways in which the vegetation in hot arid and semi-arid areas combat drought.

◆ Ephemerals escape drought by remaining as seeds until rain falls when they germinate, flower and seed within a matter of weeks. Species are small and have shallow roots. When ephemerals flower, they transform the desert into a wonderful display of colour.

Key term

Ephemerals: plants with a short life cycle which flower after rain and then seed and die.

◆ Succulents, such as cacti (Figure 4.23), cope with drought by storing water in their fleshy leaves or stems. They collect and store water during seasonal rains to draw upon during dry periods. To reduce transpiration they also close their stomata during the day and open them at night when it is cooler. Stomata are also sunk into grooves or recesses on leaf and stem surfaces. Many also have thickened, waxy leaves or cuticles which act like waterproofing to reduce water loss. Plants with adaptive mechanisms to survive drought such as these are called xerophytes. Remarkably, many succulents also have a CAM (crassulacean acid metabolism) which allows them to carry out photosynthesis when their stomata are closed during the day.

Figure 4.23 The prickly pear cactus ▲

Key terms

Succulents: plants which store water in fleshy stems and leaves.

Transpiration: loss of water vapour to the atmosphere mainly via plant stomata.

Stomata: pores in leaves and sometimes stems from which water vapour escapes and through which oxygen and carbon dioxide is exchanged.

Cuticle: waxy layer covering the outer plant wall designed to reduce water loss.

Xerophyte: plants living in dry areas which have special mechanisms to survive drought such as swollen stems, thick cuticles, sunken and sometimes closed stomata, small leaves and spines. They include cacti and shrubs such as the creosote bush.

CAM (crassulacean acid metabolism): a process by which carbon dioxide taken in at night is stored until the day when photosynthesis can occur without the plant opening its stomata.

◆ Phreatophytes obtain their water from long roots which extend to groundwater below the water-

table. An example is the mesquite bush which grows beside streams and on the edges of salt pans.

◆ Many shrubs and trees have small leaves and spines to reduce transpiration losses; for example, ironwood. Some plants loose their leaves in the dry season, but continue to photosynthesise on their green stem; for example, Palo Verde. Stomata are often more numerous on the underside of the leaf. Acacia trees in semi-arid areas have spreading canopies to reduce evaporation from the ground surface. Grasses have narrow leaves which often roll in to reduce transpiration. Leaves also die off in the dry season.

Key term

Phreatophytes: plants living in dry areas which have roots systems to groundwater supplies.

◆ High evaporation can create salt pans. Plants growing in these saline soils are called halophytes. Many of these plants are similar in appearance to plants growing in salt marshes found in the UK. The saltbush excretes salt onto its leaves when conditions become too saline. Succulents such as pickleweed tolerate a high salt content by compensating with a high water uptake.

> **Key term**
>
> **Halophyte:** plants adapted to growing in saline conditions such as salt marshes and salt pans.

Harsh climatic conditions make food supply scarce. Many plants, such as the prickly pear cactus and the acacia tree have evolved spines and thorns to protect themselves from overgrazing. Others, such as the sodium apple, contain bitter poisonous latex which animals avoid. Water scarcity encourages some plants, such as the creosote bush, to give off toxic substances to deter other plants from growing nearby. Some plants take water and nutrients directly from others, that is, they are parasitic. One example is orobanca, which lives off small succulent bushes growing in sand dunes. With few species scattered over large distances, seed dispersal is difficult. To solve this problem some plants have specially designed wind dispersal mechanisms. Others such as burr grass have barbs which can attach themselves to the coats of passing animals.

Different hot arid and semi-arid areas contain different types of vegetation. The Australian deserts are dominated by spinifex grass, eucalyptus and acacia trees. The Namib Desert has euphorbia succulents and a curious plant called the welwitschia which has two large, strap-like leaves onto which fog condenses.

Animal adaptations

Animals in hot arid and semi-arid areas tend to be small and light in colour. Some are highly mobile so that they can travel large distances to find water; for example, gazelles and kangaroos. To avoid the heat of the day, some animals, such as the golden mole, are nocturnal or live in burrows. Lizards and snakes seek shade in the heat of the day under rocks. Many animals store water in their tissues, often as fat and may concentrate their urine to reduce water loss (Figure 4.24). Desert foxes have large ears with veins close to surface to increase heat loss (see Figure 4.24). Reptiles and insects have hard skins to reduce moisture loss. Some snakes and chameleons change skin colour to reflect more heat. Many insects have long legs to keep their bodies away from the hot sand. The sidewinder snake moves in such as way to reduce contact with the ground. Some insects have mechanisms which enable fog to condense on their bodies. Many desert birds are carnivores and rely partly on the moisture from the animals and insects they eat.

Camels store food as fat in their hump. They have a large water capacity, and their thick lips and mouth enable them to eat thorny plants. Their long eyelashes protect their eyes in sand storms and their padded feet enable them to walk in hot, shifting sands.

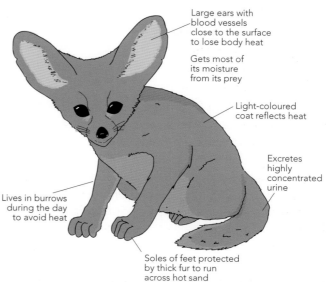

Fennec Fox – lives in Sahara
(weighs 1.5kg and is 200mm tall)

Large ears with blood vessels close to the surface to lose body heat

Gets most of its moisture from its prey

Light-coloured coat reflects heat

Excretes highly concentrated urine

Lives in burrows during the day to avoid heat

Soles of feet protected by thick fur to run across hot sand

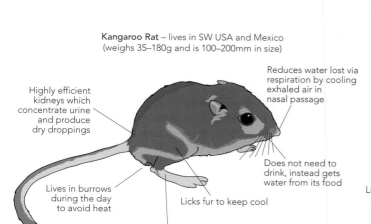

Kangaroo Rat – lives in SW USA and Mexico
(weighs 35–180g and is 100–200mm in size)

Reduces water lost via respiration by cooling exhaled air in nasal passage

Highly efficient kidneys which concentrate urine and produce dry droppings

Does not need to drink, instead gets water from its food

Lives in burrows during the day to avoid heat

Licks fur to keep cool

Does not perspire

Figure 4.24 Animal adaptations to drought

Case study | The impact of climate on the Sonoran Desert ecosystem

The Sonoran Desert extends over southern California, southern Arizona and northern Mexico. Its bi-seasonal rainfall encourages succulents such as the Saguaro cactus, which you may be familiar with from films about the Wild West (Figure 4.25). It can reach 15m in height and grows on well-drained slopes and lives for up to 175 years! It has an accordion-like stem which expands and fills with water during the winter wet season. Its ribbed stem reduces wind currents which would otherwise lead to high evaporation losses. Shallow roots catch water which falls during storms before it evaporates, or is lost down through the soil. Stomata are sunk into the stem. Many other types of succulent, such as the barrel, hedgehog and prickly pear cacti have similar mechanisms to the Saguaro to combat the effects of drought. The cholla cactus is covered with dense spines which help to reflect solar rays.

A common drought-enduring small tree growing on the upper slopes is the Palo Verde. This plant loses its leaves in the dry season, while its green bark enables it to carry out photosynthesis. Lower, gentler slopes are covered with creosote bushes, which have small, dark resinous leaves to reduce transpiration. Plant spacing is controlled by water availability, that is, creosote bushes grow further apart when supplies are scarce. Ephemerals such as the brown-eyed primrose commonly grow among the creosote bushes.

Phreatophytes, such as the mesquite bush, grow beside streams and on the edges of salt pans. Halophytes, such as inkweed, saltgrass and pickleweed occupy saline soils on salt flats. Where water reaches the surface, such as along the San Andean fault, groves of Californian fan palm grow. Their very large leaves indicate water supply is plentiful and there is no need to conserve supplies.

Soils in the Sonoran Desert are thin, relatively infertile and alkaline. Seasonal rains leach soluble salts down through the soil which are then drawn up again under high evaporation in the dry season. Flash-flooding can compact soil creating impermeable surfaces. Salt flats in the Salton Trough are covered with thick crusts of sodium chloride, gypsum and calcium carbonate.

▲ **Figure 4.25** The Sonoran ecosystem with Saguaro cacti

Activity

Each arid and semi-arid environment contains assemblages of plants and animals which match the regional climate. Use the information contained in this section and the case study of the Sonoran Desert together with web searches to complete the following activities. A further useful source is the article 'Plant distribution in the Sonoran Desert' by J. Dove (*Geography Review*, 15 (2) November 2001).

1 Find images of some of the plants found in the Sonoran Desert and add labels to show their climatic adaptations.

2 Find out about animals living in the Sonoran and the ways in which they are adapted to drought. Good examples to choose are: a gila monster, a rattlesnake and a roadrunner.

3 Bring your ideas together in a spider diagram which summarises plant and animal adaptations to living in the Sonoran Desert.

'Take it further' activity 4.2 on CD-ROM

Physical and human factors which make ecosystems vulnerable

Arid and semi-arid ecosystems are easily damaged by natural or human disturbance. Once the sparse vegetation cover is lost, the soil which has taken thousands of years to form is quickly eroded by wind and flash-floods. Slow growing plants, such as the desert agave, which only flowers after 25 years of growing, take a long time to recover after disturbance. Harsh climatic conditions mean that nutrients stores and flows are small and easily broken (Figure 4.26).

Physical factors which make ecosystems vulnerable include drought, frost and lightning strikes. Dust storms can bury vegetation, while intense flash flooding breaks foliage and stems and removes plants from the sides of swollen rivers. Human factors which disturb ecosystems include: urbanisation, water abstraction, road, rail and airport construction, grazing, mining, wildlife trading and the introduction of non-native species.

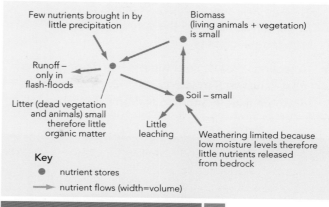

Figure 4.26 The desert nutrient cycle

Case study | Physical and human disturbances to the Mojave and Sonoran ecosystems

- Cities such as Phoenix have replaced deserts covered with cacti and creosote bushes.

- Taking water from rivers together with groundwater abstraction to supply a growing population has adversely affected mesquite bushes and cottonwood trees which grow along stream banks.

- Road construction and gas and water pipelines have displaced natural habitats. Fenced highways restrict the movement of animals such as the pronghorn (antelope) and the desert tortoise (Figure 4.27). Pronghorn normally roam over long distances in search of food and water, but in 2001–02 a drought caused their numbers to fall. The situation was made worse because fences prevented the pronghorn reaching water supplies.

- The growing popularity of using off-road vehicles for recreation and the establishment of military bases in the Mojave Desert have damaged soil and vegetation. Tank tracks and car tyres crush stems, leaves and seeds and compact the soil, reducing the soil's ability to hold water and air.

- Overgrazing by cattle has left only unpalatable species and trampling has compacted soils. Delicate 'cyptobiotic soil crusts' composed of cyanobacteria, lichen, algae and fungi, which prevent erosion, absorb water and fix nitrogen, have also been destroyed. Livestock have also crushed lizard and desert tortoise dwellings and eaten their food.

- Domesticated animals, such as the burro or donkey, have escaped into the wild and multiplied, which has caused a decline in the native desert bighorn sheep population.

- The introduction of non-native plant species such as tamarisk, first introduced as an ornamental or as a windbreak, has since displaced native cottonwoods and desert willows growing along river banks. Buffle grass, another non-native species, introduced into Mexico and Texas in the 1960s to improve pasture, has since spread into Arizona. By occupying gaps in the sparse vegetation cover, buffle grass has increased the fuel load and fire risk. In 1994, fire engulfed 1150 acres of the Saguaro National Park, destroying 340 acres of desert scrub and killing many desert tortoises. The Saguaro cactus with its fleshy stem was not fire resistant and 20 per cent subsequently died following the burn. The Palo Verde, which provides shade for Saguaro seedlings to become established, was also lost. Removal of native species has hastened the spread of the non-indigenous grasses.

◆ A recent series of cold winters has damaged the frost-sensitive Saguaro cactus, which grows at its northern limit in Saguaro National Park. This tall

cactus can also be struck by lightning. The Saguaro cactus is also sometimes illegally removed and used as a garden ornamental. The chuckwalla, gila monster, desert tortoise and some snakes are also targeted by collectors.

Figure 4.27 Desert tortoise ▲

Restoring natural vegetation after human impact is expensive and also difficult because climatic conditions are harsh. The Saguaro cactus and the desert tortoise are now protected by law and areas such as the Saguaro National Park have been established to protect desert ecosystems. Off-road, recreational driving is now banned on the Kelso Dunes because the area supports a number of rare, endemic species. Some tamarisk has been removed from river banks and replaced by native cottonwood, willow and mesquite.

Key term

Cyptobiotic soil crust: grey-brown soil crust composed of cyanobacteria, lichen moss and microfungi. Filaments of cyanobacteria bind soil particles.

Discussion point

Why should hot arid and semi-arid ecosystems be protected?

4.3 What are the issues associated with the development of hot arid and semi-arid environments?

4.4 How can hot arid and semi-arid environments be managed to ensure sustainablity?

Hot arid and semi-arid areas provide opportunities and challenges for development in tourism and recreation, agriculture and the exploitation of resources such as minerals.

Exploitation brings about short-term gains, but careful management is required to ensure sustainable development of fragile environments. Contrasting hot arid and semi-arid environments are identified which illustrate these issues (see Activities page 164). In addition the ways in which tourism and agriculture can be managed sustainably are examined.

Tourism opportunities and challenges

Hot arid and semi-arid areas contain a spectacular variety of natural and human attractions. High temperatures, lots of sunshine and warm, clear seas provide opportunities for resort development along desert-fringed coasts such as the Mediterranean and Red Sea. Spectacular scenery provided by canyons, rock arches, mesas and buttes can be found in south-west USA, while huge sand dunes cover the Namib and Sahara deserts. Wildlife attractions include black rhino and elephant in Namibia. Ancient temples and pyramids line the Nile Valley in Egypt. Rich cultural traditions, art and rituals are associated with native Indian groups in south-west USA (Figure 4.28) and aborigines in Australia. More active sporting activities include: scuba diving in the Red Sea; white-water rafting on the Colorado River in USA; and cycling across sand dunes in the Simpson Desert in Australia.

Figure 4.28 Petroglyphs carved in desert varnish (hard thin coating on rocks of iron maganese oxide) at Newspaper Rock, Utah

Attractions can, however, only become tourist destinations, once the infrastructure is in place. This means building roads, airports, railways, hotels, and installing gas, electricity and water supplies. These are formidable challenges in hot arid areas, and made even more difficult in politically unstable places such as Sudan and parts of the Middle East.

The economic, socio-cultural and environmental opportunities offered by tourism are that it:

- generates foreign currency earnings. This is particularly beneficial in hot countries with few other resources. Egypt derives 45 per cent of its foreign earnings from tourism. The income can be used to improve infrastructure

- creates direct employment – for example, a waiter in Tunisia – and indirect work – for example, a fishermen supplying seafood to a hotel. Multiplier effects occur because the income from hotel bills allow the proprietor to buy food from local farmers, who in turn can purchase goods such as clothes, which create more employment

- encourages local people to acquire new skills. In Tunisia, for example, the government, in recognising the value of tourism to its economy (6.3 million international arrivals in 2005) has set up hotel training schools

- assists, through tourist spending, the preservation or revitalisation of traditional crafts, customs and rituals. Money is also used to restore heritage, for example, ancient tombs in Egypt

- develops infrastructure such as roads and water supply, which benefit locals and other sections of the economy

- preserves natural landscapes from more damaging forms of development or activity, for example, Kunene National Park in Namibia protects the threatened black rhino and elephant from poaching.

The challenges posed by tourism are that:

- jobs are often low paid and low skilled. Hotels may close in the wet season in tropical semi-arid areas

- tourist revenue is transferred overseas, rather than benefiting the host country. This is known as external leakage and occurs particularly with 'all-inclusive packages' where profits made by foreign-owned hotels and airlines are sent overseas. Importing food also causes external leakage

- over-reliance on tourism is risky, particularly if the country is politically unstable; for example, the terrorist attack on the Red Sea resort of Sharm el-Sheikh caused a temporary decline in Egyptian tourism

- regional disparity can occur where tourism is focused in one part of the country, for example, the Mediterranean coast of Tunisia

- tourists who behave badly, or dress inappropriately cause offence to local people in the host country. Many hot arid countries in the Middle East are Muslim and some western tourists cause offence by the way they dress, for example, female tourists who wear shorts are forbidden from entering St Katherine's Monastery on Mt Sinai

- debases local culture where customs and rituals, such as native aboriginal dances, are simplified for a tourist audience. Such 'pseudo events' can over time be seen as the norm and local performers lose their understanding of traditional practices. This 'commodification' of the culture is also displayed in the mass production of tourist souvenirs which leads to a loss of traditional craft skills

- large visitor numbers can overwhelm small villages; disturb religious services and disrespect spiritual places, for example, tourists are asked not to climb Uluru because it is regarded as sacred by the aborigines, but some still do

- airport and hotel construction destroys natural landscapes and habitats, for example, tourist developments disturb loggerhead turtles which breed on sandy beaches on the Egyptian Mediterranean coast

- tourist attractions and facilities physically displace local people, for example, villagers on the west bank of the Nile near Luxor have been forced to move to prevent household waste water leaking into ancient tombs

- discharge of untreated sewerage from hotels into the sea, or local water supplies causes illness such as gastroenteritis. Untreated sewerage contaminates fish. Hotel and restaurant demand for local seafood deplete local fish stocks

- high demand for water in hotels for laundry, washing and swimming pools, and other tourist facilities depletes local supplies, for example, the watering of golf courses in Palm Springs, California

- tourist facilities designed, or constructed in unsympathetic materials which do not blend in with the landscape (Figure 4.29)

- overcrowding at popular attractions detracts from the quality of the visitor experience, for example, high visitor numbers at the Great Pyramids, Giza

- rubbish discarded during desert safaris creates an eyesore and endangers wildlife
- off-road vehicles used for recreational purposes damage fragile soils. Helicopter pleasure rides over the Grand Canyon disturb wildlife. Air pollution from tourist buses increases sulphate weathering on the Sphinx at Giza
- people damage ancient sites (see Valley of Kings case study), or steal mementos, for example, olive trees in the Garden of Gethsemane in Jerusalem are now fenced off to avoid people cutting branches for souvenirs.

THE GRAND CANYON SKYWALK

Yesterday saw the opening of the 'Grand Canyon Skywalk', a horseshoe-shaped concrete structure with a glass floor which extends 70 feet from the rim of the canyon and one mile above the Colorado River. The construction of the Skywalk has been controversial. Hualapai Indians, on whose land the Skywalk has been built, claim the tourist attraction will reduce overcrowding in the National Park further east. About 4 million people visit the park every year, but only 300 000 come to the Hualapai Indian Reserve.

Revenue from ticket sales will be divided equally between Mr Jin, the local businessman, who put up the £21 million to build the Skywalk

and the Hualapai Indians. The 2200 strong tribe need the money to offset high unemployment and poverty. A visitor centre and restaurant is planned to open in 2008, and a reconstructed Indian village should further benefit the Reservation. The current unpaved road to the Skywalk is also likely to be upgraded.

Not all of the local Indians are happy with the development and regard the Canyon as sacred ground, while others fear that not all members of the tribe will benefit equally from the development. Conservationists regard the Skywalk as an eyesore. They also oppose the expansion of the small airport at Grand Canyon West to accommodate the extra tourists.

Figure 4.29 Flagstaff Herald newspaper report

Case study | The Valley of the Kings

Ancient Egyptian tombs on the west bank of the Nile near Luxor are under pressure from tourism. Large numbers of visitors visit the tombs each year and their breath raises the humidity in the tombs, softening the

plaster so it no longer adheres to the wall. People touching or brushing against the paintings also cause damage. Irrigation of sugar cane in the Nile valley is also causing problems in that as the water-table rises, saline water evaporates inside the tombs and salt crystals grow, damaging the paintings. In the 1990s, cracks appeared in three tombs and in a fourth the ceiling collapsed. Rising water-tables and visitor pressure has resulted in the closure of the finest tomb in the valley, that of Seti I.

Efforts to reduce the tourist impacts have included installing dehumidifiers, placing glass screens in front of the paintings and closing selected tombs on a temporary basis to allow humidity levels to fall (Figure 4.30). Numbers entering the tombs at any one time are also controlled and popular tombs, such as that of Tutankhamen, attract a higher entry fee. A new

visitor centre opened in 2007 should help to spread visitor flows. Paths between the tombs have also been widened to prevent overcrowding. Other plans to improve the site include moving car parks and replacing existing tarmac roads with those of sand and gravel sprayed with polymer so they are less visually intrusive.

Extending opening hours, limiting the number of tombs tourists can visit and constructing full-size replicas of tombs such as that of Tutankhamen (which receives 300 000 visitors a year) are possible future options to reduce overcrowding.

Figure 4.30 Tomb closed for restoration, Valley of the Kings

Discussion point

In 2005 a new bridge across the Nile 10km south of Luxor was constructed. Given that this will further increase visitor pressure, is the construction of replica tombs a way forward?

Activity

Using the information in this section, the case study and the Flagstaff Herald article, outline the opportunities for and problems caused by the development of tourism in hot arid and semi-arid environments.

Sustainable tourism

Sustainable tourism means encouraging continual investment and maintenance of local jobs, protecting and enhancing local customs, livelihoods and cultures and conserving natural landscapes and ecosystems. Examples of sustainable tourism in hot arid and semi-arid areas include:

◆ involving local people in managing tourist facilities and attractions, for example, aborigines at Uluru work as rangers

◆ creating national parks such as Arches in Utah to protect landforms, plants and desert animals

◆ controlling visitor numbers at popular sites. In the Valley of the Kings in Egypt the number of people entering the tombs at any one time is limited (see case study). In the popular pilgrimage site of Nazareth, overcrowding has been reduced by pedestrianisation, better signage and improving tourist routes between holy sites. At Arches National Park in Utah, parking is permitted only at designated car parks and the number of spaces at each of these is restricted to reduce visitor pressure on nearby trails

◆ establishing tourist codes, for example, visitors entering Arches National Park are reminded that the removal of rocks, plants, fossils and artefacts is prohibited and not they should not feed wild animals or drop litter. Climbing the Great Pyramids at Giza is now forbidden

◆ restoring historical sites using entrance fees, for example, Egyptian tombs

◆ building hotels and tourist facilities using sympathetic materials and designs and using renewal energy sources, for example, Uluru Rock Resort was designed to have a low visual impact on the surrounding desert and uses solar energy.

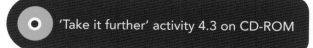
'Take it further' activity 4.3 on CD-ROM

Agricultural opportunities and challenges

Hunting and gathering is still practised by some indigenous people such as the Australian aborigines, but the main form of agriculture in the Middle East and North and sub-Saharan Africa, where water is scarce, is nomadism. True nomads wander with their herds of camels, goats, cattle and sheep in search of water and good grazing, while semi-nomads move only in the dry season. Some indigenous people have abandoned nomadism and instead have adopted a more sedentary lifestyle and live in villages and towns.

Key term

Nomadism: a wandering form of existence in search of good pasture and water for livestock and the collection of fruit and roots.

Oases, together with rivers which flow through deserts such the Nile, provide opportunities for irrigated agriculture, for example, the oasis city of Douz in Tunisia, produces dates, while olives, figs, oranges and pomegranates are cultivated in the shade of the palm trees. The date palm is a very useful desert plant in that the trunk is used to make roofs and fences, leaf fibres are made into mats and ropes, and dates provide nutritious food.

Barley, wheat, beans, sugar cane, rice, cotton, onions and tobacco are all grown on the fertile banks of the Nile (Figure 4.31). Methods of transferring the water to the fields vary from simple dykes to advanced sprinkler systems. The construction of the Aswan Dam has meant that the once seasonal river flow has been regulated to provide water for growing crops all year round.

A continuous growing season, lots of sunshine to ripen the crops, advanced irrigation methods, modern, fast transport systems and refrigeration has meant that a variety of crops are now grown for the export market. Hydroponics is practiced in countries such as the USA, Mexico, Kuwait, Abu Dhabi and Saudi Arabia. In Middle East countries such as Kuwait, desalinisation plants provide water for the cultivation of crops such as tomatoes, lettuce, peppers and cucumbers in greenhouses.

Figure 4.31 Agriculture in the Nile Valley

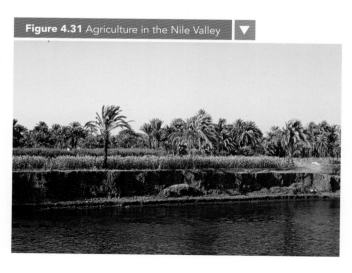

─ Key term ─

Hydroponics: the growing of crops without soil.

Scant and unreliable rainfall, strong winds, thin, infertile soil and remoteness from markets all pose serious challenges for agricultural development in areas such as the Sahel in sub-Saharan Africa. Periods of drought, together with human mismanagement of the land, has converted marginal areas into desert by a process called desertification. Human factors which have contributed to desertification include monoculture, overgrazing and deforestation. The underlying cause of many of these problems is poverty and high population growth, which has placed too many demands on the soil to produce food.

─ Key terms ─

Sahel: a semi-arid area in sub-Saharan Africa stretching from Senegal to Ethiopia where rainfall is unreliable and droughts occur.

Desertification: the conversion of marginal land to a desert brought about by naturally occurring periods of drought together with human mismanagement of land.

Another problem facing agricultural development is salinisation. This occurs when high temperatures encourage saline water to be drawn up through the soil and salt crystallises in the upper layers. Hard salt crusts impede root growth, reduce infiltration and increase surface run off (Figure 4.32). Salt may also become concentrated in an impervious layer causing water-logging. Salty soils can be managed by growing salt-tolerate crops such as date palm, cotton, sorghum, barley, rye and alfalfa. Irrigation of crops, however, without proper drainage encourages secondary salinisation because, as the water-table rises, high evaporation draws salts to the surface. In contrast, in the Nile Delta, excessive well-pumping has

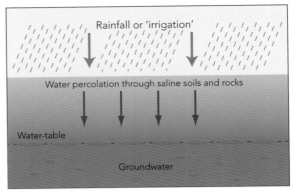

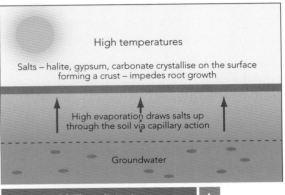

Figure 4.32 The salinisation process ▲

lowered the water-table, which has encouraged salt water to seep into coastal soils also causing salinisation. The challenges posed by agricultural development are discussed in the case study of the Thar desert.

─ Key term ─

Salinisation: the accumulation of salt in the soil.

⊙ 'Take it further' activity 4.4 on CD-ROM

Case study | Agriculture in the Thar Desert

The Thar Desert is located in the Indian states of Rajasthan and northern Gujarat and also south-east Pakistan (Figure 4.33). It is densely populated by humans and livestock and covers an area of 200000km². Precipitation varies from 100mm in the west to 50mm in the east, much falling in the summer monsoon season between July and September. In July temperatures can reach 50°C. Dust storms are common in May and June and cause dunes to migrate

blocking roads and railways. Composed of sand dunes and plains, low hills and saline lakes, the Thar Desert is vegetated by scrub and scattered trees. Wildlife, some of which are rare, include, the Great Indian bustard, the Indian gazelle and the Indian wild ass. The summer monsoon water is collected for water supply; the groundwater is too deep and saline for human consumption. Canals cross the area and supply irrigated water for the cultivation of wheat and cotton.

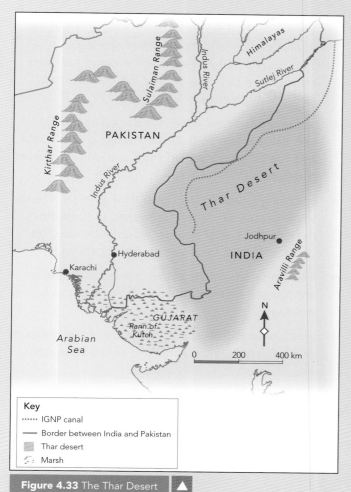

Figure 4.33 The Thar Desert ▲

Key
- ••••• IGNP canal
- ──── Border between India and Pakistan
- ▨ Thar desert
- ⋯ Marsh

has also declined as fodder has become scarce. Loss of plant cover has exposed the soil to wind erosion. Heavy trampling, especially around waterholes, has compacted the soil, limiting seed development, damaging plant roots and encouraging runoff rather than infiltration.

Under population pressure, trees and shrubs have been removed for building materials, fuel, hand-tools and medicine. Deforestation has caused a decline in organic matter which normally supplies nutrients and helps to hold soil particles together. Once the vegetation is removed the sun bakes the soil until it cracks and blows away. The loss of native trees such as the Khejari, a nitrogen fixer, deprives the soil of valuable nitrate fertiliser. Fallow periods which normally allow the soil to recover have been shortened, and the total area under crops increased in a effort to feed the rising population. The continual cultivation of millet, rather than intercropping with beans, which increases fertility through nitrogen fixation, also exhausts the soils.

In an effort to stop desertification and extend the area under irrigation, the Indira Gandhi Nahar Pariyojana Canal (IGNP), formerly known as the Rajasthan Canal, was constructed to bring water from the Himalayas to the desert. At 649km long it was one of the worlds biggest canal projects. Irrigated water has been used to grow crops such as rice, groundnut, mustard, wheat and sugar cane, but it has also created problems. Seepage from irrigation channels and over-watering of crops has resulted in water-logging and salinisation. A rise in the incidence of malaria is also blamed on the canalisation. Damper soils have also adversely affected desert fauna that prefer drier conditions, and numbers of desert fox and Indian gazelle have declined.

Hot dry conditions, strong winds and human activities such as deforestation, monoculture and overgrazing are causing soil erosion. The Thar contains huge numbers of livestock, and overgrazing by sheep, goats and cattle has removed palatable grasses leaving behind inedible plants. The quality of the animals

Sustainable agriculture

Desertification and salinisation are encouraging farmers to try more sustainable methods of agriculture. To reduce overgrazing, pastoralists in the Thar Desert are being urged to reduce their herd sizes to allow grass to recover and improve the quality of the livestock. They are also encouraged to use solar devises rather than dung for fuel, and instead to use the manure for fertiliser. The planting of trees such as *prosopis cineraria* is providing timber and fuel-wood and green leaves which can be fed to camels, goats, sheep and cattle. Another introduced species such as the ber tree produce large fruits and jojoba plants yield oil. Meanwhile, sand dunes are being stabilised with trees and grasses such as the exotic *acacia tortilis*.

The underlying issues are, however, how to control population growth so that less pressure is placed on the land. See the Niger case study (page 162) for other examples of agricultural sustainability.

Efforts to reduce salinisation have been tried in the Thar and more economically developed areas such as Arizona in the USA. Drip and sprinkler irrigation is an efficient method of watering because evaporation losses are reduced. Lining canals with concrete reduces seepage. Limiting water use, encouraging water recycling, building dams across wadis, digging ponds to catch rainwater and growing salt-tolerant crops are all methods of conserving water use.

 'Take it further' activity 4.5 on CD-ROM

Case study | Niger sustainable agriculture

Niger is a poor, semi-arid country in West Africa within the Sahel (Figure 4.34). Life expectancy is only 44 years and GNP is just $900 per capita. Soils cannot produce enough food to feed the population, which is increasing at 2.9 per cent per annum. Desertification has been caused by periods of drought and human mismanagement of the land. Sustainable practices which have been tried include:

◆ mulching – incorporating vegetation into the soil to increase organic matter, improve infiltration and reduce soil erosion. Mulches also stimulate termite activity which breaks up the soil. Millet stalks are useful as mulch, but these are often also needed for building material, firewood, medicine and fodder, so farmers use these sparingly. Other useful mulches include the native tall grass *andropogon gayanus* and the indigenous shrub *guiera senegalensis*.

◆ planting pits or 'za' – these are small holes dug with a hoe which collect runoff water and rainwater and into which are added small amounts of manure and seed

◆ manuring – pastoralists corral their animals in fields to be fertilised in return for money, or food, from the farmer. Mineral fertiliser would increase crop yields, but most local people cannot afford it.

◆ intercropping – trees, native grasses and shrubs are planted in strips between millet to act as a windbreak

◆ bunds –low stone walls are built along the contours of the fields to retain earth and hold back water.

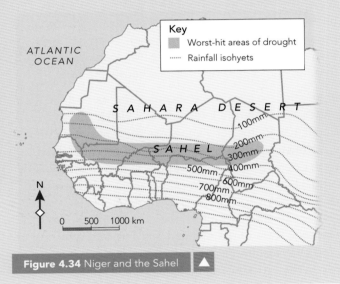

Figure 4.34 Niger and the Sahel ▲

Discussion point

1 What are the obstacles to adopting more sustainable agricultural methods and conserving the soil in Less Economically Developed Countries and how might these be overcome?

2 What are the pros and cons of importing food produced in hot arid countries using irrigated methods?

Resource exploitation opportunities and challenges

Mining

Some hot arid and semi-arid areas contain rich mineral resources. Oil reserves have radically altered the economies of countries in the Middle East such as Saudi Arabia and Kuwait. The Atacama in Chile contains the largest open-cast copper mine in the world, while Namibia has rich reserves of diamonds and rare minerals.

Exploitation, however, poses huge challenges because the reserves are often located in remote areas.

Considerable investment in roads and railways is required before mining can take place. Housing, water and electricity supplies also need to be provided for the miners. Even if capital is available, mining is risky because the world demand for minerals fluctuates. In the boom times mining towns flourish, but when the market collapses they become ghost towns. Some in time, however, become heritage tourist attractions. A case study of mining in the Australian desert illustrates some of the opportunities, challenges and sustainable management options.

Case study | Mining in the Australian deserts

Australia, once rich in gold, is now a leading producer of uranium, titanium, manganese, nickel and opals (Figure 4.35). Mining provides jobs in desert areas where few other opportunities exist, for example, the large copper and uranium mine at Olympic Dam in central Australia employed 3000 people in 2005. Moreover, the company operating the mine has contributed $45 million to the South Australian Government in royalty payments which has benefited the area regionally. Residents living in nearby Roxby Downs have also benefited from improvements in local services and the multiplier effects the mine has created. House prices have, however, increased and local people complain they cannot afford to buy.

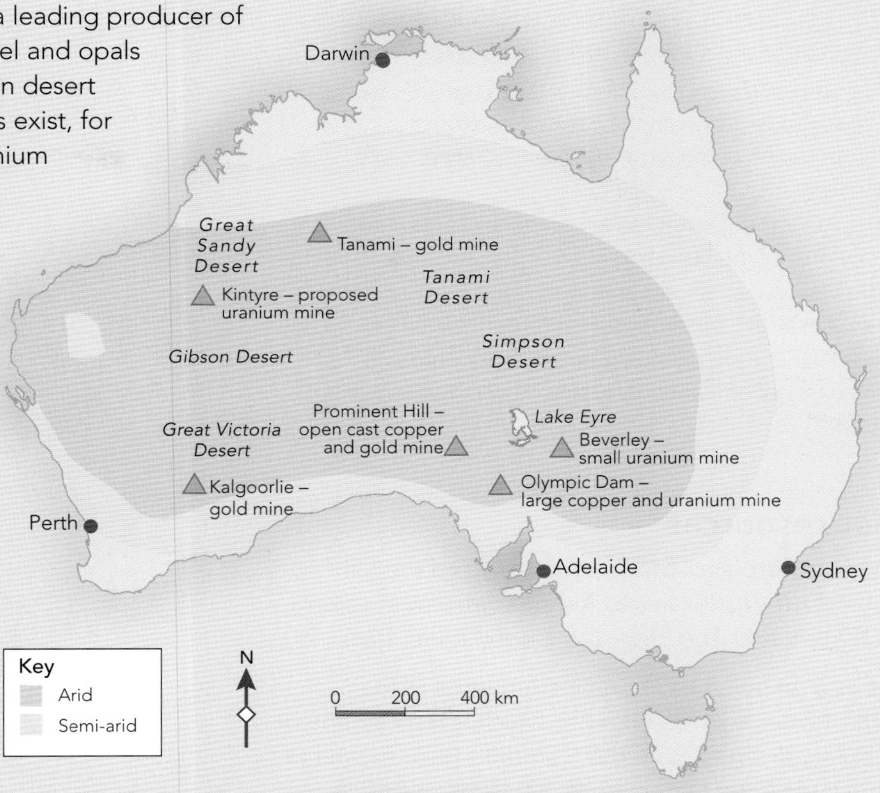

Figure 4.35 Mining in the Australian deserts

Aboriginal groups benefit directly if mining takes place on their land by receiving an annual rent. The tracks and roads built by mining companies also enable aboriginal groups to reach remote areas for hunting and gathering. Disputes over legal ownership of mineral rights do, however, sometimes occur and some indigenous groups oppose the opening of new mines.

From Figure 4.35 you can see that mining companies face big challenges moving the ore from the centre of Australia to the coast, for example, copper ore from the Olympic Dam mine has to be trucked 560km to the port at Adelaide. Modern mining methods also use large quantities of water, for example, the Olympic Dam mine is dependent on supplies brought by pipeline from 12 desalination installations on the coast together with groundwater sources. Opponents of mining at the Olympic Dam mine fear that water extracted from the Great Artesian Basin for refining and processing ore will endanger rare desert spring mounds in Lake Eyre. These springs rise from water stored in the underground aquifers and are an important source of surface water in central Australia and are of cultural significance to the aborigines.

Opponents claim that mining has a number of other negative environmental impacts. The construction of roads, pipelines and processing facilities removes desert vegetation and soils which are fragile. The low relief of the desert means that mining operations are visible for long distances, creating an eyesore. Open-cast mining, such as that at Prominent Hill where the hole is nearly 1km wide and 50m deep, creates a large visual impact. Uranium and copper ores, once mined, are crushed and treated with chemicals to concentrate the mineral content. Opponents argue that liquid waste from treatment plants and seepage from dumps and ponds contaminate water and soils. There are particular concerns over the mine at Beverley which will use an in situ leaching method to extract the ore, that is, acids will be forced into the groundwater to dissolve and remove the uranium. Mining also creates spoil heaps which scar the landscape and are a source of dust, while the movement of people and vehicles encourages the spread of weeds which colonise disturbed sites.

Mining companies, aware of their responsibilities, undertake environmental assessments before mining takes place. To reduce visual impacts, the height of waste dumps is limited and re-landscaping occurs when mining ceases, for example, the gold mines at Tanami will be contoured and laid with rock drains to prevent erosion once operations finish. Mining companies recycle water from treatment ponds and spray weeds to control their spread. Controlled burning of vegetation also encourages the spread of native eucalypts to obscure tracks when mining ceases.

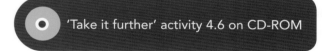
Chose two contrasting arid and semi-arid environments and describe and explain the opportunities and challenges each presents for economic development. Possibilities are:

◆ the Australian Desert, an arid/semi-arid area with valuable mineral reserves

◆ the Nile Valley, an arid area with significant ancient remains which present opportunities for tourism

◆ the Thar Desert, an arid/semi-arid area where an increasing population threatens to damage fragile soils.

Other resources

Dramatic desert landscapes have attracted the film tourist industry, for example Tunisia formed the setting for *Star Wars* and *The English Patient* (Figure 4.36).

Water as a resource has been instrumental in the growth of cities such as Phoenix, which is dependent on supplies diverted from the Colorado River as well as groundwater. Over-extraction is, however, raising the issue of whether such growth is sustainable.

Discussion point

What are short- and long-term benefits and problems of setting films in desert areas?

◉ 'Take it further' activity 4.6 on CD-ROM

Figure 4.36 Film locations in Tunisia ▲

Choose two contrasting arid and semi-arid environments, and describe and explain how fragile environments can be exploited for short-term gain and how careful management can help to ensure sustainability. Possibilities include:

◆ agricultural development in the Sahel region in Niger, a semi-arid area located within a poor country

◆ economic development in the south-west USA, a wealthy region where water extracted from the Colorado River has been used to develop agriculture and industry (alternatively research the Nile in Egypt)

◆ tourist development in the Nile Valley, Egypt, or at Uluru, Australia, or in the Saharan deserts.

Suggested resources to enable you to research the Colorado, the Nile and Uluru are included in the 'Take it Further' section on the CD-ROM.

U1
4

AS Geography for OCR

1   What and where are the main hot arid and semi-arid areas?

2   Why do hot arid areas experience a wide range of diurnal temperature?

3   Why are hot arid areas hot?

4   Why do different hot arid and semi-arid areas experience different temperatures?

5   What factors contribute to the aridity of hot arid and semi-arid areas?

6   Why do different hot arid and semi-arid areas experience different amounts of rainfall?

7   What causes seasonal rain in tropical semi-arid areas?

8   How does climate influence the physical landscape?

9   What weathering processes operate in hot arid and semi-arid areas?

10   How does wind and water shape landforms?

11   In what ways are plants and animals adapted to drought?

12   Why are desert ecosystems fragile?

13   What human and physical factors make ecosystems vulnerable?

14   What opportunities and challenges are posed by economic developments in:

   a   tourism and recreation

   b   agriculture

   c   resource exploitation?

15   How can careful management help to ensure sustainable development in hot arid and semi-arid areas?

A list of useful websites accompanying this chapter can be found in the Exam Café section on the **CD-ROM**

U1

4

Hot arid and semi-arid environments

Exam Café
Relax, refresh, result!

Relax and prepare

What I wish I had known at the start of the year…

Sanjay

"I use memory pegs (usually called mnemonics) to learn my work. These reduce your notes to just a few words that help me recall my notes in the examination. They're ideal for lists or for helping to remember connections. Some people choose a key word and make the letters stand for something. I prefer to pick out key words and use their first letters to spell something – usually that has some meaning to me.

For example, I use SPAMIST (you can imagine the image this gives me) to list human activities/land uses:

Settlement **P**ower **A**griculture **M**ining **I**ndustry **S**ervices **T**ransport

I then use these to organise my answer – they might form the basis of paragraphs in an essay. I look at these lists the night before the examination. I can usually remember at least 20 of these pegs by linking them to fingers and toes!"

Hot tips

Cery

"If I've got a memory peg with two words starting with the same letter, I add a number. For example,

Types of transport = **SCR2APE** = Sea, Canal, Road, Rail, Air, Pipe, Electronic

And if I can't make a word, I may have to add a vowel – I usually write that in lower case so I don't think it's a factor, such as

Rock structure = **F2aB JaM** = Fault, Folds, Beds, Joints, Minerals

Well it works for me!"

Vijay

"I struggled with the practice exam question '*How do we know that desert areas were wetter in the past?*', so I asked my teacher to summarise the key points for me:

◆ Cave paintings in the Sahara desert show animals that now live in the wetter savannah areas further south.

◆ Pollen analysis shows that Mediterranean types of vegetation once grew in the Sahara.

◆ Many desert areas show complex patterns of now dry river channels. Some ancient remains show weathering and water damage from a wetter time, such as the Sphinx in Egypt.

◆ Even some documentary evidence shows that North Africa was wetter in Roman times when it was the 'bread basket' for Rome.

◆ Ancient water deposits that probably date from post glacial times are common in desert areas.

This really helped!"

Common mistakes – Jerry

▷ "I've been known to answer a question that I expected rather than the one being asked. Some topics I like and know well, and we've been over questions based on them in class. The danger is that the examiner sees my answer as irrelevant.

▷ For example, the question 'Explain why water is of importance in hot arid areas' I saw as being about how water creates desert landforms but it isn't, is it?

▷ I've now learn't to slow down and re-read a question before answering it to make sure I'm covering everything"

Refresh your memory

4.1 What processes and factors give hot arid and semi-arid environments their distinctive characteristics?

Weathering	Exfoliation, granular disintegration, block disintegration, wet/dry, pressure release, hot/cold, vegetation, solution, acid action, salt crystalisation
Erosion	Wind action (deflation hollows, yardangs, zeugens) flash-floods (or historical – wadi, gullies, pediment) and exotic rivers (canyon)
Transport	Mass movement (creep, slides), flash-floods, wind
Deposition	Wind (ripples, ridges, barchan dune, seif dune), flash-floods (salt pan, sabkha, playa)
Factors	Climate – arid, windy, extreme temperatures, sudden storms Relief and landforms e.g. dunes Drought resistant ecosystem Traditional way of life – oasis versus nomadic Resource development – minerals, tourism

4.2 Why are hot arid and semi-arid environments considered to be 'fragile'?

Climate	Extreme with 'sudden' events, e.g. sand storm
Water	Delicate nature of supply (surface and sub-surface)
Energy	Low-energy environment (low inputs – dry but high sun energy)
Nutrients	Low nutrient stores, vulnerable flows – fierce competition
Species	Limited range – limited gene pool and food chains
Population	Traditionally low and self-sufficient

4.3 What are the issues associated with the development of hot arid and semi-arid environments?

Opportunities	Settlement – military bases, resorts, e.g. Las Vegas Power production – wind, solar Industry – tourism, hi-tech, e.g. USA Mining – oil, gas, ores Agriculture – grazing, irrigated farming Transport – railway, e.g. Southern Australia
Issues	Conflicts between activities Need to protect and conserve fragile environment Indigenous cultures – often nomadic. Desire to protect cultures Unclear international boundaries e.g. so called 'neutral territories' Cost of offsetting conditions

4.4 How can hot arid and semi-arid environments be managed to ensure sustainability?	
Physical	Climate – harsh and extreme (dust storms)
	Relief – shifting sands, steep slopes
	Vegetation – thin and tough
	Drainage – lack of water
	Ecosystem – hostile or low productivity
	Soils – extreme alkaline (salts)
Human	High-cost environment (water, cost of keeping cool)
	Remote and land transport difficult
	Pollution threats, e.g. oil
	Waste disposal
	Existing indigenous populations
	Need for conservation

Top tips . . .

▷ Introductions are vital as they create an immediate impact in the examiner's mind.

▷ For example:

    Q: Assess the impact on the environment and local communities of developments in hot arid areas.

▷ An effective introduction would be:

    Developments in hot arid areas can bring a variety of impacts – some positive others negative, some planned others accidental. **(1)** Hot arid areas vary in scale **(2)** from local such as Death Valley in Arizona to the almost continental scale of the Sahara **(3)** in North Africa. Impacts may vary over time and may be viewed differently by various groups within the community.**(4)**

This introduction is effective as:

1 It recognises that impacts are varied – it seeks to suggest a classification.

2 Much clearer idea of how geography may affect the issue.

3 Good use of geographical terminology – sense of place clear.

4 Even the introduction recognises that the discussion is not one sided.

Get the result!

Outline the ways in which vegetation adapts to survive drought in hot arid environments. [6 marks]

Examiner says

1 & 2) Student has provided two explanations of how vegetation adapts with an example.

Examiner says

The final sentence shows good knowledge of other adaptations. This answer would gain maximum marks.

Student answer

Some plants store water in bulbous roots (1) or in thick, protected stems e.g. Cacti. Others have long roots to reach underground water. Most have few leaves or very narrow, spikey ones to reduce transpiration (2). Some flower and seed very quickly when the rare rains do come — the desert blooms. Most are small to reduce the need for much water and some defend what water they have with thorns or poison e.g. creosote bush.

Examiner says

This is a good answer. The student has given an explanation of cause-effect linkage to adaptation (2), and supplied another method of adaptation with cause-effect.

Examiner's tips

There are two broad types of maps used in the exam. The first type is a map of an area with the instruction to identify the pattern or distribution of something such as population. You may be required to look where something is but equally where it isn't. Pattern usually suggests high, medium and low values with some aspect of shape e.g. concentric zones, linear etc. To get the higher marks reference should be made to the map in terms of direction and stating some of the values of the high and low areas.

The second type of map is one where you use a grid reference to locate a feature. Ordnance Survey maps are commonly used to do this. Many candidates seem reluctant to deal with such maps but they have a key supplied and are often easier than you think. Remember you must give evidence from the map – this means locate features using grid references (in most cases four figures will do). Do remember it is the numbers along the bottom of the map you use first, then those up the side. Again quote directions (the top of the map is always north) and distances in kilometres using the scale shown on the map.

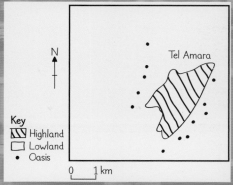

Distribution of highland and Oasis in the Tel Amara desert

Managing change in human environments

To develop an understanding of:
- the processes that produce a variety of human environments
- the principal changes, and the causes involved in these changes
- how humans are trying to manage these changes.

Chapter 5

Managing urban change

The majority of you live in urban areas, it is important, therefore, that you understand this particular 'half' of the world. You need to be aware of its distinguishing characteristics and to have some 'feel' for the scale and pace of the process of urbanisation. You should be alive to the diverse challenges that face those who live in the urban world, particularly those who are responsible for managing it. You should also be concerned about the future. Towns and cities are greedy consumers of resources and major polluters of the planet. For how much longer can this continue?

Questions for investigation

- What are the characteristics of urban areas?

- What are the social and economic issues associated with urban change?

- What are the environmental issues associated with urban change?

- How can urban areas be managed to ensure sustainability?

Consider this

Today most urban areas are widely recognised as places of pollution, poverty and personal insecurity.

So why are more people than ever before living in them? Are they doing so for good reasons or because they have no other choice?

5.1 What are the characteristics of urban areas?

What is urban?

Before we begin to explore the content of this part of Unit 5, we need to be clear about the meaning of the word that recurs throughout this chapter – urban. Most national censuses distinguish between urban and rural settlements on the basis of population size. The former are always taken to be larger than the latter, but the size threshold varies from country to country – mainly between 2000 and 10000. The term 'urban' is also frequently used to describe both places and people. Urban places show a number of features that distinguish them from rural places (Figure 5.1). Similarly, urban people are considered to be different from rural people. This is not just in terms of the environment in which they live, but also in what they do for a living and their general lifestyles. So in trying to identify what exactly is 'urban', it is helpful to work under three main headings environment, economy and people.

You should be able to pick out in Figure 5.1a some of the general features of urban areas indicated in Figure 5.2.

The geographical characteristics of urban areas including their functions, processes and patterns are explored on pages 178–180.

The geographical characteristics of urban areas including their functions, processes and patterns are explored on pages 178–180.

> **Key terms**
>
> **Urban:** the distinctive characteristics of towns and cities – their size and density, environments and ways of life.
>
> **Rural:** the character of country areas and the activities and lifestyles encountered in such areas.

Figure 5.1a Urban landscape

Figure 5.1b Rural landscape

Environment
- Exlusion of the natual world
- Dominance of buildings and transport networks
- High levels of environmental pollution, not just of water and air, but also sound, light, and visual pollution. Congestion, particularly of traffic, due to high building densities
- Pace of living

URBAN CHARACTER

Economy
- Employment mainly in secondary, tertiary and quaternary sectors
- The provision of commercial and social services for local residents and those living in tributary area

People
- High population densities
- Distinctive lifestyles, values and behaviour
- Diversity in terms of wealth, age and ethnicity
- Raised stress levels

Figure 5.2 The general features of urban areas

Activity

Are you able to add any other diagnostic features of your own to Figure 5.2?

Theory into practice

With others, come up with some examples that illustrate the 'distinctive lifestyles, values and behaviour' of urban people.

U2
5
AS Geography for OCR

What is urbanisation?

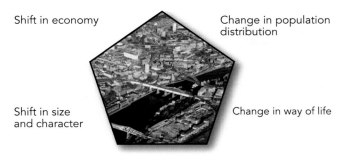

Shift in economy

Change in population distribution

Shift in size and character

Change in way of life

Spread of built-up area

Figure 5.3 The urbanisation pentagon

Urbanisation is the process by which places and people become more urban. Its most obvious outcome is a rise in the percentage of the population living in towns and cities. This figure (usually referred to as percentage urban) is the most widely used measure of the 'urban-ness' of a country or region. There is tendency among students to confuse rural–urban migration with urbanisation. Rural–urban migration is just one element of the urbanisation process. In reality, it has at least five different elements or facets (Figure 5.3).

◆ A shift in the economy of a country or region. The emphasis moves from farming and the primary sector to manufacturing and the provision of a wide range of services.

◆ A change in the distribution of population. People become more and more concentrated in the growing towns and cities. Rural–urban migration is a major contributor here.

◆ A change in the way of life of those people moving into towns and cities. It is not just a change in occupation. As indicated in Figure 5.2, there is a change in lifestyles, values, codes of behaviour and social institutions.

◆ Changes in the size and character of settlements. Some settlements, particularly those at favoured locations or with access to particular resources, grow more quickly than others. Differential growth in the settlement network undergoing urbanisation sees some villages grow into towns, some towns into cities and so on.

◆ The spread of the built-up area. The natural environment is progressively 'eroded' or lost beneath what is essentially an artificial, man-made environment.

Key terms

Urbanisation: the rise in the proportion of a population living in urban areas.

Percentage urban: the proportion of a population living in urban areas.

Rural–urban migration: the movement of people from the countryside into towns and cities.

It is important to understand that there is a close connection between urbanisation and another global process – development. In fact, it is development that generates urbanisation. This link is illustrated by Figure 5.4. As countries become more developed – LDCs become LEDCs, LEDCs become RICs and so on – they become more urbanised. They move along what is best described as an urbanisation pathway. The rise in the level of urbanisation along the pathway continues until a country becomes an MEDC.

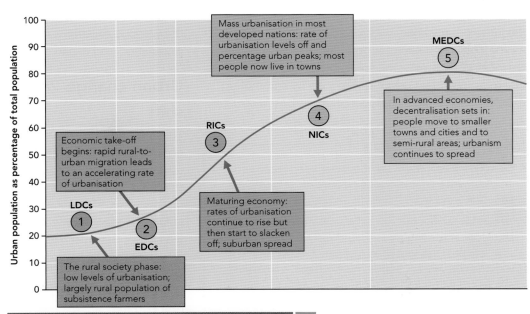

Figure 5.4 Development and the urbanisation pathway

Key terms

Development: the increased use of resources and technology that leads to a rise in the general standard of living of a country. It is usually measured in terms of material wealth.

LDC: least economically developed country.

LEDC: less economically developed country.

RIC: recently industrialising country.

NIC: newly industrialised country.

Urbanisation pathway: the course followed by all countries as they develop and become more urbanised.

MEDC: more economically developed country.

 'Take it further' activity 5.1 on CD-ROM

At present, there is some uncertainty about what will happen to the urbanisation pathway in the future. Some suggest that the percentage urban figure will begin to decline slightly. In support of their idea, they draw attention to a process of change that is becoming more evident. This process has been given the name – counterurbanisation. This term is particularly unfortunate for it gives the wrong impression that the new change is somehow 'anti-urban'. People are *not* abandoning towns and cities in their droves.

So what exactly is happening? The short answer is that two different changes are going on. The first is that some businesses and people are opting to move out of the largest cities and into smaller cities and towns. This move might be seen as 'anti-big city'. It is not reducing the level of urbanisation. It is simply diffusing urban growth down the urban hierarchy. The second element of counterurbanisation involves a relatively small number of urban businesses and people that are moving out into what are seen as rural areas. Many of these urban–rural migrants retain urban-based jobs by commuting, whilst others remain customers of urban-based services. The result is rural dilution, not ruralisation.

Key terms

Counterurbanisation: the movement of people and employment from major cities to smaller settlements and rural areas.

Urban hierarchy: the vertical classification of towns and cities according to a variable such as population size. It is best thought of as a pyramidal structure with many towns at its base. Above these is a smaller number of cities, and above these, an even smaller number of regional centres. At the top of the pyramid, there is a capital city.

Activity

Explain the difference between 'rural dilution' and 'ruralisation' by writing a definition of each term.

Discussion point

Why should counterurbanisation be seen simply as part of the urbanisation pathway?

What is urban change?

Having now some 'feel' for the meaning of urban and the nature of urbanisation, it is now time to clarify a third term, 'urban change'. The most obvious form of urban change is the growth of urban areas. Fortunately the converse, urban decay, rarely effects whole towns and cities but is found in localised pockets. Urban growth and decay are only one mode of urban change In fact, the term 'urban change' covers many changes which are happening today at a range of spatial scales, from the global to the local. They are also occurring at a variety of different speeds. Of these changes, the most important are those occurring globally. They are the direct outcomes of urbanisation. You need to highlight three of them.

Urban change 1 – the rural–urban shift

The first is the shift in where the majority of the world's people live and work. In 1800, only 5 per cent of the world's population lived and worked in towns and cities. Today, roughly 50 per cent of the global population is reckoned to be urban, and by 2030 the figure is expected to rise to 60 per cent (Figure 5.5). So the world is experiencing a massive growth in urban population and urban areas.

Urban change 2 – the MEDC–LEDC shift

This second urban change is a geographical one. One hundred years ago, the fastest rates of urbanisation occurred in what is often described as the Developed World – i.e. the MEDCs. Today, the fastest rates of urbanisation are taking place in the LEDCs of the so-called Developing World. Figure 5.6 shows that the great majority of the world's urban population already lives in the Developing World. The shift is forecast to continue well past 2030. Figure 5.6 also illustrates two other important points. First, since 1950, just over half the population of the Developing World has been classified as

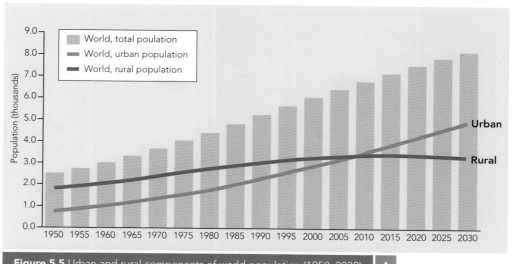

Figure 5.5 Urban and rural components of world population (1950–2030) ▲

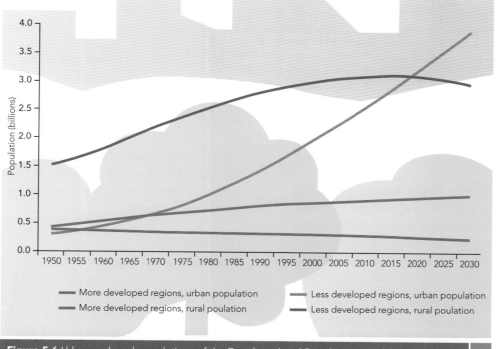

More developed regions, urban population — Less developed regions, urban population
More developed regions, rural poulation — Less developed regions, rural poulation

Figure 5.6 Urban and rural populations of the Developed and Developing Worlds (1950–2030) ▲

urban. Secondly, the percentage urban figure for the Developed World is slowly levelling off.

Urban change 3 – the shift in size

Not only is the world more urban, but also the rapidly increasing urban population is giving rise to ever larger urban areas – our third major urban change (Figure 5.7). Many of these cities are engulfing other towns and cities to form vast urban agglomerations. The term mega-city is now popularly used to describe the very largest of these. There are 17 with populations greater than 10 million. Below them, there are over 20 cities with populations between 5 and 10 million. But these large cities are not only growing in size – they are

changing their distribution. The majority of them are to be found in the Developing World, particularly south and south-east Asia.

Key term

Mega-city: a city or urban agglomeration with a population of 10 million or more inhabitants.

Discussion point

Does it not seem strange that many of the world's largest cities are located in the less urbanised Developing World?

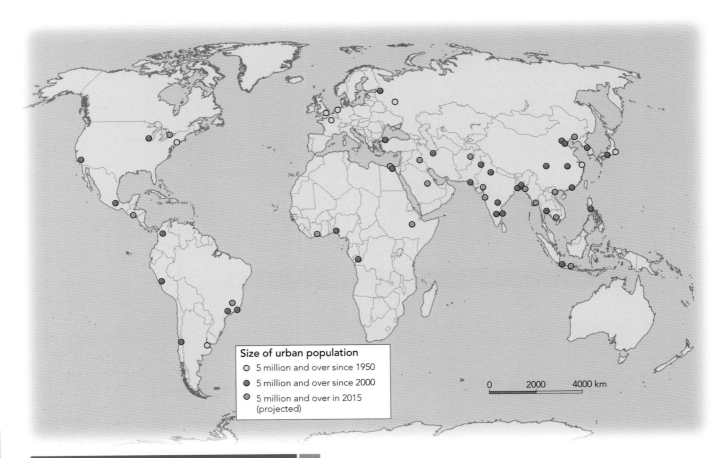

Figure 5.7 World's largest urban areas (1950–2015) ▲

Among these 'urban giants' is a very select band known as the **world cities**. They include London, New York and Tokyo and, because of their status and influence, they are power points of the global economy. They are the places where vital decisions about the global economy are made.

These three changes provide the important background or context for the more specific urban changes that make up the content of this unit.

> **Key term**
>
> **World city:** one of the world's leading cities – a major node in the complex networks being produced by economic globalisation. The influence of world cities is linked primarily to their provision of financial and producer services.

The characteristics of urban areas

Three major points need to be made that build on the list of urban characteristics set out in Figure 5.2. First, it is the functions or economic activities of urban areas that contribute much to the distinctiveness of urban

areas. Second, urban areas show recurring *patterns* produced by the spatial sorting of both land uses and people. Third, these urban patterns are the outcome of distinct *processes*.

Functions

For some people, the most fundamental of all the urban characteristics listed in Figure 5.2 are those to do with the urban economy and its **functions** (see also Figure 6.1 on page 216 which highlights the link between population size and the number of services provided). They argue that the activities located in urban areas are the root source of all the other characteristics. Whilst there are those who might contest this importance, it is vital that you have a clear idea of the functions and activities that are typical of urban areas. Of these functions, it often goes unrecognised that residence is the major land use of all urban areas, in many cases accounting for well over half the built-up area. Manufacturing used to be a core function of most MEDC cities. However, **deindustrialisation** and the **global shift** in manufacturing to NICs and RICs have changed all that.

Today, the most lucrative economic activities are those that fall within the tertiary and quaternary sectors. The provision of a range of services, not just for the urban inhabitants themselves, but also for those living in tributary areas, generates many jobs and much capital. Within the quaternary function, it is the handling of information, research and development, administration and financial management that can bring great prosperity to an urban area.

It is important to remember that not all functions or land uses are strictly 'economic'. Recreation is one example. While sports stadiums, swimming pools, golf courses and leisure centres may represent a commercial side to recreation, most urban areas contain patches of open space – parks, gardens, playing-fields – that are free to use. As such, they play an important part in day-to-day life of urban residents.

Key terms

Functions: the main activities of a town or city that are evident as land uses within the urban area.

Deindustrialisation: the relative or absolute decline in the importance of manufacturing in the economy, particularly in cities.

Global shift: the changing distribution of economic activities, particular the movement of manufacturing from MEDC to LEDC cities.

Activity

Make a list of the different land uses found within urban areas.

Patterns

One feature of all urban areas is the spatial sorting of functions and people. Similar activities and similar types of people tend to cluster together to create a sort of mosaic within the built-up area. The Central Business District, industrial estate, edge city and retail park are evidence of this sorting of land uses. Similarly, the ghetto, areas of fashionable housing, the sink estate and the gated community attest to the sorting of people. Clearly, the precise detail of the urban mosaic varies from settlement to settlement, but it is possible to make a few broad generalisations about the spatial pattern of urban areas.

◆ The general age of the urban area decreases from the centre, because towns and cities grow outwards from a historical nucleus.

◆ The overall density of development tends to decrease, because as a town or city grows its fringe expands. This allows access to still more rural space for the further spread of the built-up

area. More space for expansion more often leads to lower densities of development.

◆ As a consequence of the above, no matter where you are in the world, the urban area will show the same broad spatial components, namely a core, a suburban ring and a rural fringe. The transition from the core to the suburban ring often shows up as a fairly distinctive zone. It is referred to as the inner ring.

A major factor guiding the sorting of different land uses and people is the urban land value surface. Activities and groups of people differ in terms of what they can afford to pay for the urban space they occupy. The fact that urban land values generally decline outwards from the core also helps to reinforce the ring-like pattern. But there are other factors that make an impact and the more important of these are shown in Figure 5.8 on the next page. Clearly the physical geography e.g. steep slopes, flood plains, rivers and waterfronts is a strong influence. So too are the historical factors of heritage and the fact that the past makes an indelible mark on the present (inertia). Economic and social factors play a significant part, but the influence of urban managers has become particularly significant in recent years.

Key terms

Central Business District (CBD): the commercial centre of a town or city.

Edge city: a feature of suburban intensification where parts of the suburbs become more city-like through the agglomeration of offices, factories and large shopping complexes.

Ghetto: a residential area that is largely occupied by one ethnic or cultural minority group.

Sink estate: an area of poor housing occupied by disadvantaged households.

Gated community: an area of wealthy private housing with a secure perimeter wall or fence with controlled entrances for residents, visitors and their cars.

Inner ring: a transitional zone between the commercial core and the suburban ring. Usually involves a mix of non-residential land uses and old housing. A part of the urban area characterised by brownfield sites and regeneration.

Urban land value surface: the spatial variations in land values or rents within the built-up area created by the bidding process.

Urban manager: professionals who distribute and control resources in an urban area.

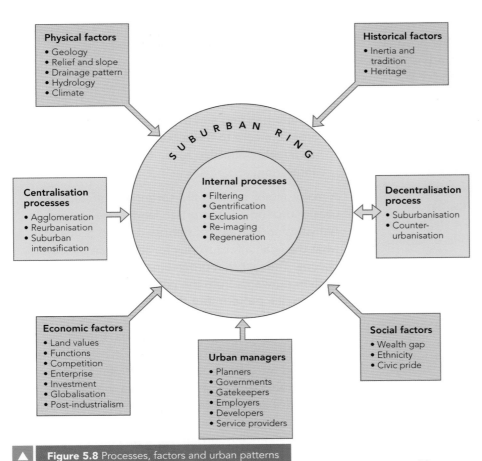

Activity

What sort of segregation is going on in i) the ghetto, ii) the sink estate and iii) the gated community?

'Take it further' activity 5.2 on CD-ROM

Figure 5.8 Processes, factors and urban patterns

Processes

You have now looked at two major processes – urbanisation and counterurbanisation. It would be sensible to see the latter as part of the former.

There are four more processes shown in Figure 5.8 that are undoubtedly part of urbanisation. They fall into a time sequence (Figure 5.9). The sequence starts with agglomeration as people are drawn into the newly emerging towns and cities and as new businesses are set up. But agglomeration soon leads to suburbanisation as people move out of the congested core into areas of lower density development. Also contributing to the process is the continuing accumulation of rural–urban migrants and new businesses at the margins of the urban area. Eventually, the process of urban regeneration kicks in as planners decide to redevelop brownfield sites vacated by closed factories and by people and businesses moving out to the suburban ring. Regeneration, and the re-imaging that frequently goes with it, might both be seen as forms of reurbanisation. So too when the outward expansion of the suburban ring and the rapid consumption of rural space create a situation in which it becomes necessary to intensify development within it. Suburban intensification involves raising housing densities and building on open spaces. Non-residential activities, such as retailing, offices and service industry become more evident.

Key terms

Agglomeration: the clustering of people and their activities around a particular location.

Suburbanisation: the outward spread of the urban area, often at lower densities compared with the older parts of the town or city. The decentralisation – of people first and then services and employment – is encouraged by transport improvements. The process is also encouraged by in-migrants settling around the urban fringe.

Urban regeneration: the revival and modernisation of old and rundown urban areas.

Brownfield: land that has been used, abandoned and is now awaiting a new use.

Re-imaging: a process where the unfavourable image of an urban area is changed. This is achieved by minimising its negative aspects and promoting its positives.

Reurbanisation: the movement of people and activities back into the central and inner parts of a town or city.

Suburban intensification: the process where suburbs become more urban as a result of increasing building densities, introducing non-residential activities and developing remaining open spaces.

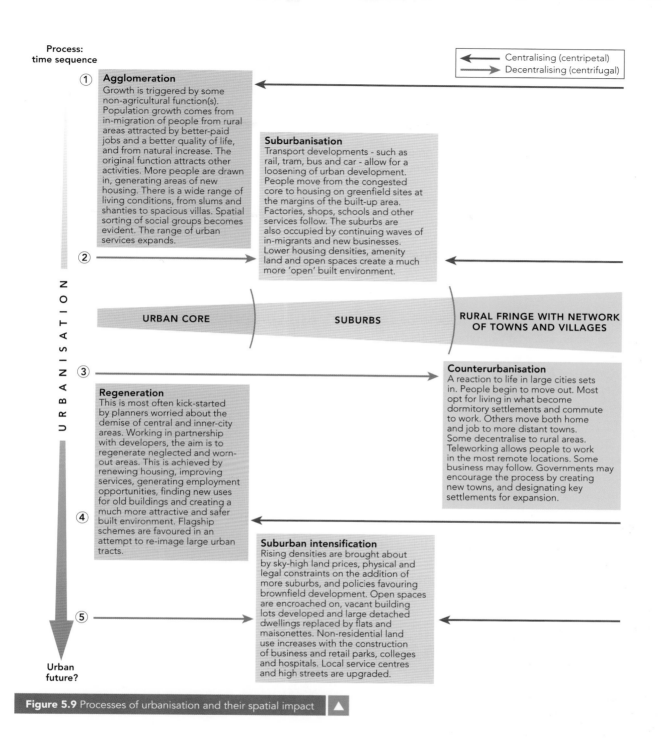

Process: time sequence

Centralising (centripetal)
Decentralising (centrifugal)

① **Agglomeration**
Growth is triggered by some non-agricultural function(s). Population growth comes from in-migration of people from rural areas attracted by better-paid jobs and a better quality of life, and from natural increase. The original function attracts other activities. More people are drawn in, generating areas of new housing. There is a wide range of living conditions, from slums and shanties to spacious villas. Spatial sorting of social groups becomes evident. The range of urban services expands.

Suburbanisation
Transport developments - such as rail, tram, bus and car - allow for a loosening of urban development. People move from the congested core to housing on greenfield sites at the margins of the built-up area. Factories, shops, schools and other services follow. The suburbs are also occupied by continuing waves of in-migrants and new businesses. Lower housing densities, amenity land and open spaces create a much more 'open' built environment.

URBANISATION

②

URBAN CORE   **SUBURBS**   **RURAL FRINGE WITH NETWORK OF TOWNS AND VILLAGES**

③

Counterurbanisation
A reaction to life in large cities sets in. People begin to move out. Most opt for living in what become dormitory settlements and commute to work. Others move both home and job to more distant towns. Some decentralise to rural areas. Teleworking allows people to work in the most remote locations. Some business may follow. Governments may encourage the process by creating new towns, and designating key settlements for expansion.

Regeneration
This is most often kick-started by planners worried about the demise of central and inner-city areas. Working in partnership with developers, the aim is to regenerate neglected and worn-out areas. This is achieved by renewing housing, improving services, generating employment opportunities, finding new uses for old buildings and creating a much more attractive and safer built environment. Flagship schemes are favoured in an attempt to re-image large urban tracts.

④

Suburban intensification
Rising densities are brought about by sky-high land prices, physical and legal constraints on the addition of more suburbs, and policies favouring brownfield development. Open spaces are encroached on, vacant building lots developed and large detached dwellings replaced by flats and maisonettes. Non-residential land use increases with the construction of business and retail parks, colleges and hospitals. Local service centres and high streets are upgraded.

⑤

Urban future?

Figure 5.9 Processes of urbanisation and their spatial impact

U2
5

Managing urban change

Figure 5.9 shows these processes as working in one of two directions or both. Agglomeration, regeneration and suburban intensification are centralising in character. They help to maintain the compactness of the urban area. On the other hand, suburbanisation, like counterurbanisation, is largely decentralising. Its most obvious impact is to cause the urban area to spread outwards.

There is a third group of processes shown in Figure 5.8. These are the internal processes; they too generate change and impact on the urban pattern.

Filtering and gentrification are to do with housing and social class. Filtering involves housing passing down to lower-income inhabitants, whilst gentrification works in the opposite direction with residential areas being improved and upgraded. Exclusion involves particular groups of people being barred from living in certain parts of the urban area, mainly for reasons of ethnicity and level of income. The ghetto and the gated community are two such examples. Re-imaging and regeneration are, as just suggested, part of reurbanisation and are mainly focused on the cores of urban areas.

Key terms

Centralising: the clustering of people and their activities around a particular location.

Decentralising: the outward movement of people and activities from established centres.

Exclusion: the shutting out of particular groups of people (e.g. the poor, ethnic minorities) from particular areas or from full participation in society.

Theory into practice

Look at the suburban ring of an urban area known to you. What is the evidence that suburban intensification is taking place?

Case study | Tales of two cities (1): Newcastle-upon-Tyne and Port Moresby

The purpose of this case study is to describe and analyse the urban areas of two cities located in very different parts of the world – Newcastle-upon-Tyne in north-east England and Port Moresby, the capital city of Papua New Guinea, in South-East Asia. To what extent do these two urban areas differ in terms of their functions, patterns and processes? It is important to remember the significant difference in the development context of the two urban areas. Per capita GNI in the UK in 2005 was $US 37 740, compared with a very modest figure of $US 500 for Papua New Guinea in the same year. It is also necessary to bear in mind the temperate/tropical contrast in the two locations, as well as the difference in the length of their histories. Figures 5.10 and 5.12, plus Figures 5.11 and 5.12, provide support for answers to the questions set out in Figure 5.14.

Newcastle-upon-Tyne

Population: 190 000 (2001)

Location: Latitude 55°N; on the north bank of the Tyne estuary in north-east England.

History: First settlement dates from Roman times and developed at a crossing point of the River Tyne. During the Middle Ages it was England's northern fortress against the Scots. In the sixteenth century it became increasingly involved in the export of coal from the nearby coalfield. To this prosperous activity were added, in the nineteenth century, shipbuilding and heavy engineering. In the second half of twentieth century, all three pillars of the city's economy disappeared and have been replaced by offices and retailing as the major employers. Today, the city functions as a high-order, regional service centre.

Figure 5.10 Aerial view of Newcastle-upon-Tyne

Figure 5.11 Map of Newcastle-upon-Tyne ▶

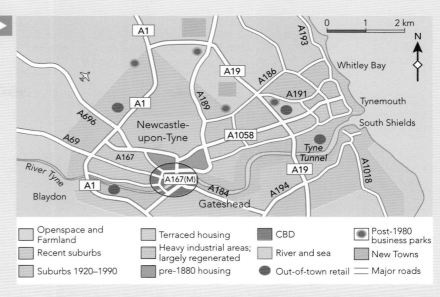

Form: Although physically separated by the River Tyne, it is a vital part of a much larger urban complex – the Tyneside conurbation.

☐ Openspace and Farmland	☐ Terraced housing
☐ Recent suburbs	☐ Heavy industrial areas; largely regenerated
☐ Suburbs 1920–1990	☐ pre-1880 housing
☐ CBD	☐ Post-1980 business parks
☐ River and sea	☐ New Towns
● Out-of-town retail	═ Major roads

Port Moresby

Population: 255 000 (2000)

Location: Latitude 9°N; on the shores of the Gulf of Papua (south-east coast of the island of New Guinea).

History: The settlement was first established by the British in 1880s to act mainly as a port and a defensive base. In 1975, it became the capital of the newly independent state of Papua New Guinea. Since then, its population has more than doubled and its built-up area has spread well beyond its peninsular historic core (Figures 5.12 and 5.13).

Form: It is a free-standing city of two parts – the 'old' and new town'. Growth of the built-up area has engulfed some villages, which have become the centres of recognisable districts.

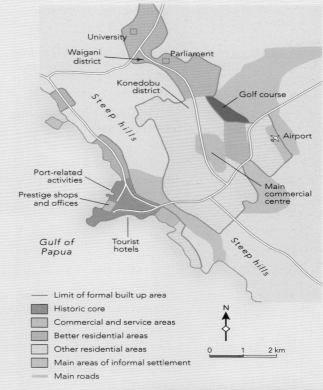

── Limit of formal built up area	
☐ Historic core	
☐ Commercial and service areas	
☐ Better residential areas	
☐ Other residential areas	
☐ Main areas of informal settlement	
═ Main roads	

▲ **Figure 5.12** Map of Port Moresby

Figure 5.13 View of the outskirts of Port Moresby ▼

Question	Newcastle-upon-Tyne	Port Moresby
1. What impact has the physical geography of the site appear to have had on the urban area?	The river separates it from Gateshead. Development downstream has been halted by the coast. Expansion is now northwards and upstream.	Limited growth along the coast. Ridge of steep hills has created a break in the inland spread of the built-up area. In effect, the site has created a two-part urban area.
2. What are the main functions and sources of employment?	Some manufacturing; information technology; commercial and professional services; retail distribution. Recreational space.	National administration; the provision of goods and services; port activities; limited service industry; informal activities.
3. Is it possible to discern a core area, a surrounding suburban ring and an urban-rural fringe?	Yes. The suburban ring is a wide one. The separation of urban and rural land uses within the fringe is fairly clean-cut.	Yes, an historic nucleus and a largely physically-separate suburban ring. There has been some infilling of the gap between the two by informal settlement. The urban-rural fringe is untidy due to the rashes of informal settlement.
4. What are the main features of:		
◆ the core	Much of it now occupied by the central business district.	There are two cores – an historic one and a modern commercial centre. The former contains the port and tourist districts, as well as the best shops and modern offices.
◆ the suburban ring	Much expansion during the twentieth century. The character of residential areas varies with the age of building. Densities generally decrease outwards from the core.	The suburban ring is low-rise and comprises many detached dwellings (many occupied by extended families).
◆ the rural fringe?	Edges of built-up area fairly crisply defined. Retailing complexes and business parks increasingly evident.	Large tracts of informal settlement; little agriculture; some recreational spaces.
5. To what extent is there a spatial segregation of:		
◆ different land uses	Clear segregation of commercial, industrial and residential uses.	Fairly clear segregation of commercial and service, port, tourist and residential uses. The Waigani district contains the Parliament and other public buildings. Most of the government ministries and the national sports stadium are located in the Konedobu district.
◆ different social and ethnic groups?	Strong segregation on the basis of socio-economic class (emphasised by the provision of council housing). Possible to discern some sorting based on ethnicity.	Australian and European residents are conspicuous in the areas of better housing. Active segregation amongst native residents based on tribal origins. The poorest residents, mainly found in the belts of informal settlements, are the recent incomers from rural areas. There are very high rates of unemployment and crime.
6. What processes are currently particularly active and where in the urban area?	**Suburbanisation** – but now tightly controlled by planners. **Suburbanisation intensification** **Counterurbanisation** – some outward movement of people and activities to nearby smaller towns and rural areas. **Regeneration** – particularly evident in areas formerly connected with shipping. **Filtering and gentrification** – most evident in the inner parts of the suburban ring.	**Agglomeration** – the growth of informal settlements prompted by substantial rural–urban migration **Suburbanisation** – inland and along the coast. **Regeneration** – some in the historic core. **Exclusion** – of poor from best residential locations and on the basis of tribal origins.

Figure 5.14 Newcastle-upon-Tyne and Port Moresby compared in terms of their functions, patterns and processes

The main conclusions to be drawn from this comparison of urban characteristics in two very different parts of the world are:

◆ the impact of physical site conditions, particularly of features such as coastlines, hills and rivers, is much the same in both cities

◆ the functions of the two cities may differ in detail, but tertiary activities are dominant in both

◆ it is possible to discern the same structural components – core, suburban ring and rural fringe

– but there are differences in the precise character of each zone

◆ the segregation of people is common to both, mainly on the basis of wealth but also ethnicity

◆ the processes at work differ because of differences in the ages of the two cities and their developmental context – Newcastle, located in an MEDC, has progressed further along the urbanisation pathway.

Discussion points

Why might you expect the influence of site conditions to be less in MEDC urban areas?

Why is the development context of an urban area such an important influence on its processes?

Theory into practice

Using the same classification as in the key to Figure 5.11, make a sketch of a town or city which you know well.

Are you able to recognise the three main zones: core, suburban ring and rural fringe?

Hopefully, now that you have a sound grasp of the process of urbanisation and its global impact, as well as an awareness of the distinctive features of urban areas, you are ready to move on to examine some of the issues created as those areas change over time.

5.2 What are the social and economic issues associated with urban change?

The rise and fall of urban areas

It is commonly believed that all urban areas enjoy uninterrupted and unending growth. In reality, urban growth is often interrupted by periods of stagnation or even decline. Eventually, however, most urban areas manage to bounce back from such setbacks. There are very few examples of urban areas declining to the point of disappearing.

You should already be familiar with the forces behind urban growth from the previous section. The main powerhouse is economic development at a national level, but it is often helped by more local factors such as location, resources, enterprise and progressive government (Figure 5.15). On the other hand, urban decay can be triggered in a range of different ways. Economic globalisation and its associated global shift have recently caused many MEDC cities to lose their manufacturing base (deindustrialisation) as factories have moved to more profitable locations in other parts of the world. As a consequence, many LEDC cities, particularly in China and India, are now booming as manufacturing centres. Cities such as Manchester, Newcastle-upon-Tyne and Sheffield, once renowned as industrial centres, have had to find new functions in order to stave off the onset of decay in the post-industrial period.

Physical processes, such as the silting of rivers, have threatened the livelihood of many ports. This has been especially true of those located at the head rather than the mouth of an estuary. In these instances, it has been necessary to find new functions to compensate for the inescapable loss of the port function. The devastation caused by natural hazards, such as hurricanes and earthquakes, can certainly halt urban growth. In most cases, the setback and its decay are temporary.

Remember too that urban areas and their buildings age. Over time, buildings need more and more money to keep them up to standard. There often comes a point when old buildings can no longer properly accommodate modern uses. In short, physical decay is inherent in the older parts of all urban areas. The main way of stopping it is by urban regeneration.

Growth	Decline
◆ Access to resources	◆ Natural disaster
◆ Opening up of new markets	◆ Change in physical geography
◆ Local enterprise	◆ Exhaustion of local resources
◆ TNC interest	◆ De-industrialisation
◆ Inward investment	◆ Loss of investment confidence
◆ Government support	◆ Political/civil unrest
◆ High rates of natural increase	◆ Poor image and popular perception
◆ Favourable image and popular perception	◆ Net out-migration
◆ Net in-migration	◆ Fall in natural increase rate

Figure 5.15 Reasons for the rise and fall of urban areas

Economic globalisation: the process where the economies of the world are moving closer together and becoming more integrated.

Post-industrial: term used to describe a set of changes and processes at work since the 1970s that have transformed cities, economies and societies. The process involves a shift of emphasis from manufacturing to services.

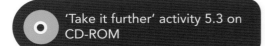

'Take it further' activity 5.3 on CD-ROM

Economic
- Wider range of job opportunities
- Low unemployment
- Inflation of housing prices
- Growing demand for commercial and social services
- Rising cost of living
- Increased consumption of resources

Social
- High rates of immigration and population growth
- Housing shortages
- More social mobility
- Widening gap between rich and poor – multiple deprivation
- Overstretched social services
- Faster pace of living – more stress
- Increased social malaise – crime, terrorism, vandalism, etc.

Environmental
- Hazards – natural and man-made
- Erosion of rural space
- More pressure on rural areas immediately beyond urban fringe
- Dereliction and regeneration
- Rising densities within expanding built-up area
- Increased volumes of traffic
- Traffic congestion
- Increased pollution of air and water
- More noise, dust and visual pollution

 Figure 5.16 Some issues created by urban growth

Issues associated with growth

It might seem odd that urban growth creates issues. You may be thinking – surely, urban growth brings prosperity, not problems. Sadly, this is not the case. Issues or problems arise mainly because those whose job it is to manage urban areas fail to appreciate the scale and speed of growth. The actions they take tend to be reactive rather than proactive. As a result, the provision of such basic things as water and sewage treatment, transport, housing and social services tends to lag, often badly so, behind the rising level of demand. It is in this widening gap between demand and supply that many of the issues associated with urban growth are born.

In trying to identify the issues associated with urban growth, it is useful to refer back to 'What is urbanisation?' (pages 175–76). Figure 5.16 is a classified list of 22 issues or problems associated with the growth of urban areas. The list is far from exhaustive. It should be easy for you to add to this list, perhaps as a result of your first-hand experience of living in an urban area. The case study of Birmingham and Dhaka that follows

will illustrate some of the social and economic issues. The environmental issues will be dealt with in the next section.

Activity

Take four issues (two from each of the top two categories in Figure 5.16) and in your own words explain what makes them issues.

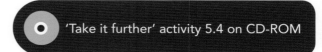

'Take it further' activity 5.4 on CD-ROM

Issues associated with decay

The main challenge associated with urban decay is to identify what is causing it in the first place. Then it is a matter of tackling those causes. The trouble is that once decay sets in, it can easily lead to still more decay. Figure 5.17 identifies probably the more significant contributors to what is, in effect, a vicious downward spiral of decline and decay. With each 'lap', the rate and the degree of decay increase. Once out of control, the downward vortex accelerates.

Each contributor shown in Figure 5.17 might be seen as one 'issue' or problem requiring some sort of remedial action. However, the situation is not as simple as that. Because these issues are interlinked, it becomes difficult to know the best place to break into the downward spiral in order to stop it. Clearly, this makes the task of halting urban decay much more challenging.

An integral part of this vicious circle of decay is the cycle of poverty (Figure 5.18) that traps people in decaying urban areas. Sadly, this cycle tends to be transmitted from one generation to the next. The children of poor parents receive little parental support and are forced to attend inadequate schools. As a result, they leave school at the earliest possible opportunity and with few qualifications. This, in turn, means that they have difficulties in finding work and that they can only expect to earn low wages for doing

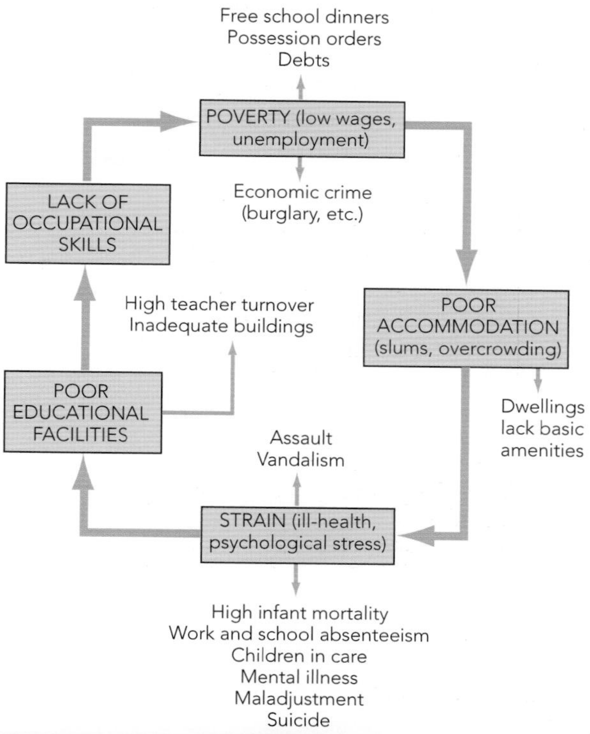

Figure 5.18 The cycle of poverty ▲

menial tasks. Thus they remain trapped in an unending cycle of poverty and are largely powerless to improve their lot. Their impeded access to good housing, a secure and well-paid job and adequate services means that they are the victims of multiple deprivation (see below).

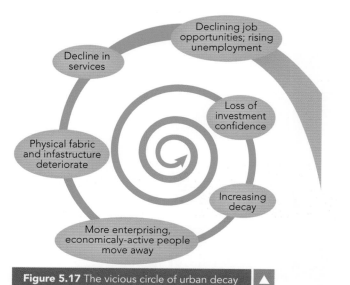

Figure 5.17 The vicious circle of urban decay ▲

Key terms

Cycle of poverty: the idea that poverty and deprivation are transmitted from one generation to the next, thus creating a self-perpetuating vicious circle.

Multiple deprivation: a term signalling the fact that deprivation is usually a matter of lagging behind in a number of related aspects of life, such as employment, housing, services and health.

Theory into practice

Of the contributors shown in Figure 5.18, which do you think is the most effective one to tackle first? What sort of remedial action would you recommend?

Case study | Tales of two cities (2): Birmingham and Dhaka

These two cities illustrate the issues created by urban change. It is more about urban growth than decay, but it might be useful to bear in mind the saying 'in the midst of growth, there is decay'.

Birmingham

Birmingham is the UK's third largest city. Its popular image is of a modern and progressive city (Figure 5.19). But it is a city that has stared decay in the face. For well over a century, Birmingham was known as 'the workshop of the world' or 'a city of a thousand trades'. It was the first industrial city in the world to be created in the Industrial Revolution that started in England in the late eighteenth century. During the second half of the twentieth century, however, it became one of the first victims of the global shift in manufacturing, as its industrial base was rapidly eroded and removed elsewhere. Today, Birmingham has transformed its economy; nearly 80 per cent of its economic output is generated by the tertiary sector. It is a major centre of banking, finance and insurance. Tourism is an increasingly important part of the city economy. With major facilities such as the International Convention centre and the National Exhibition Centre, Birmingham now accounts for over 40 per cent of the UK conference and exhibition trade. The city's sporting and cultural venues attract large numbers of visitors.

> **Key term**
>
> **Image:** the perceived reputation and/or appearance of a town or city.

the media to promote and spread the image of the 'new' post-industrial Birmingham. It has also required much regeneration in the city's core area and inner ring to provide the modern offices and shops that are an essential part of its new economic role and the city's 'new face'.

Although the city has successfully changed its image and has recently enjoyed strong economic growth, there are problems. Growth has placed a great strain on the city's transport system, with many roads and the central railway station overloaded during peak periods. Adding to the transport issue is the fact that many of the higher skilled jobs have gone to commuters from the surrounding West Midlands. At the same time, high rates of unemployment exist in inner-ring districts. In short, the benefits of Birmingham's successful transformation have not been distributed evenly throughout the city. This will be illustrated in the next section, which looks at urban deprivation. The sad fact is that the city's now prosperous core is almost surrounded by a ring that, despite much regeneration, still contains old and inadequate housing, poor services and poverty. Many of the residents of such lagging areas feel increasingly disaffected. Birmingham's population still shows strong polarisation.

> **Key terms**
>
> **Disaffected:** feeling unjustly treated by, and isolated from, society.
>
> **Polarisation:** a widening gap between extremes, as for example, between rich and poor.

How has this change in identity come about, this re-imaging from decaying, dirty industrial city to dynamic slick service centre? Much credit must go to the determination and vision of the city's decision-makers and managers. There has been a heavy reliance on the financial support of companies, great and small. It has required the help of

Figure 5.19 Birmingham – a progressive city

Dhaka

Dhaka, the capital of Bangladesh, may be on the other side of the world to Birmingham and be much larger (nearly 7 million compared with just over 1 million), but there are links between them. Birmingham, as we shall see in the next section, is home to one of the largest Bangladeshi communities in the UK. Both cities are involved in the global shift in manufacturing. Birmingham has been one of its 'losers', while Dhaka has been one of its 'winners'. Transnational corporations (TNCs) have been drawn to Dhaka by its large supply of very cheap labour. The city has become particularly involved in the clothing or garment industry. Its 'sweat shops' now produce clothes that are sold in the shops of most MEDC cities. There is no doubt that the city has benefited from the growth of this branch of manufacturing, but there is still much unemployment (the rate is around 25 per cent). This results from two factors. First, there is a high rate of population growth (around 4 per cent per annum) created by huge volumes of rural–urban migration and high levels of fertility. Secondly, this rate of population growth is far outstripping the rate at which new jobs are being created. This is a common situation in LEDC cities. Although willing to work long hours for very low wages, most people are forced to find other ways of making a living outside the formal job market. This may involve selling in the street, shoe-shining, rubbish collecting or scavenging bottles and other types of waste for recycling. Begging, petty crime and prostitution are other, less legal ways of scratching a living. These activities make up what is known as the informal economy.

> ## Key terms
>
> **Transnational corporations (TNCs):** A large company operating in more than one country and typically involved in a wide range of economic activities.
>
> **Informal economy:** made up of activities that are not officially recognised but are undertaken by poor people in order to survive. They include selling on the street, low-cost transport and jobs in small workshops.

In Dhaka, the driving of rickshaws (by pulling, pedal power or motor) is the most common informal activity, with nearly half a million people involved. With little by way of public transport, rickshaws play a vital part in keeping the city moving. But as such they greatly add to the general congestion on busy and inadequate roads. Half a million children are estimated to be involved in other activities in the informal economy. Most of them work from dawn to dusk, earning on average the equivalent of about 12p per day to help support their families. The jobs vary from begging and scavenging to domestic service and collecting money on motorised passenger carriers such as minibuses. These children work in vulnerable conditions. They are exposed to hazards, such as traffic accidents, street crime, violence, drugs, sexual abuse, toxic fumes and waste products. These extremely poor working conditions lead to serious health and developmental problems.

By LEDC standards, Dhaka is a thriving city (Figure 5.20). It is the commercial core of Bangladesh. The city has a growing middle class that is increasing the demand for consumer and luxury goods. But the majority of its residents are living below the poverty line. Many are surviving on less than £1.50 a day. They,

like some of the residents of Birmingham's inner ring, feel they are missing out on the economic success and growth that are coming the city's way. Decay is perhaps not an appropriate word to use here because things can only get better – or rather it is difficult to imagine that they could get any worse for so many people.

Activity

1   This case study has identified a number of issues associated with urban change. Make a list of them and then classify them as either social or economic.

2   Identify and comment on any discrepancies between your list and that in Figure 5.16.

Urban deprivation

The final part of this section focuses on an issue or characteristic that is common to both scenarios – growth and decay. The case study on Birmingham and Dhaka has already hinted that, just as there can be decay in growth, so there can be poverty in prosperity. For the latter situation, the term deprivation is widely used. Deprivation is said to occur when a person's well-being falls below a level generally regarded as an acceptable minimum. This minimum applies not just to one, but to a range of different aspects of daily life. In the UK, a multiple deprivation index (MDI) has been developed to assess this multi-faceted situation across the whole country. It is based on seven different quality of life indicators that are given different weightings in the calculation of the index (Figure 5.21). The MDI

Quality of life indicator	Weighting (%)
Income	22.5
Employment	22.5
Health and disability	13.5
Education, skills and training	13.5
Access to housing and services	9.3
Crime	9.3
Living environment	9.3

Figure 5.21 Components of the multiple deprivation index ▲

values for local authority areas (districts) are calculated by averaging the MDI scores of the 'super output areas' (SOAs) into which each district is divided. Local authorities are then ranked on the basis of these average scores, with the highest scores indicating the most deprived areas. The thousands of SOAs are also ranked and then classified on a percentage basis as being among the 10 per cent most deprived areas, the 20 per cent most deprived, and so on.

Key term

Deprivation: when a person's well-being falls below a level generally regarded as an acceptable minimum.

Case study | Tales of two cities (2): Birmingham and Dhaka (continued)

Birmingham

In 2004, Birmingham was ranked sixteenth out of 354 districts as the most deprived in England. However, Figure 5.22 shows that the city contained many Super Output Areas (SOAs) ranked as being among the 10 per cent most deprived in England. They form a very conspicuous broad ring in the areas of old housing immediately surrounding the city centre. But there are also pockets of high deprivation towards the urban fringe in areas of more recent residential development. This is particularly the case in the south and south-west of the city. Many of these pockets coincide with areas

of social housing built during the second half of the twentieth century. Clearly, there are many residents in Birmingham who are not sharing in the city's current prosperity.

Key term

Social housing: housing for poorer households, usually provided by local authorities or housing associations and rented by tenants.

Figure 5.23 tells us something more about the people who are missing out. Around 30 per cent of the city's population is made up of people from Bangladesh,

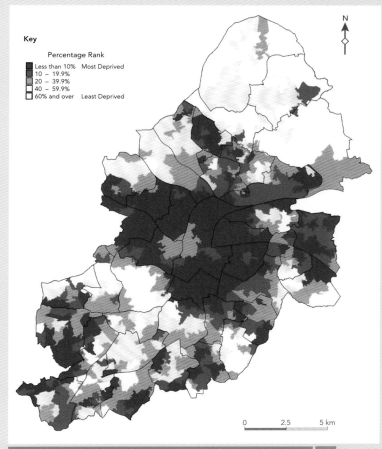

Figure 5.22 Distribution of multiple deprivation in Birmingham (2004)

Key

Percentage Rank

- Less than 10% Most Deprived
- 10 – 19.9%
- 20 – 39.9%
- 40 – 59.9%
- 60% and over Least Deprived

0   2.5   5 km

India, Pakistan, Africa and the Caribbean. The great majority of these non-white people live in the inner-city ring of deprivation. In some areas, they account for over 60 per cent of the population. It seems clear that deprivation in the UK, at least, is partly to do with ethnicity and discrimination.

Dhaka

Like most LEDC cities, there are few if any statistics to indicate how widespread deprivation is in Dhaka. But there is plenty of visible evidence that it is a serious problem. The malnourished and poorly clothed people seen crowding the streets are one indicator (Figure 5.24). As in Port Moresby (page 183), there are large tracts of informal, substandard housing ('bostis' – illegal shanty settlements) scattered around on land regarded by developers as being unsuitable for development mainly because of the flood risk. The city residents also suffer a very high crime rate and frequent outbreaks of political and religious violence.

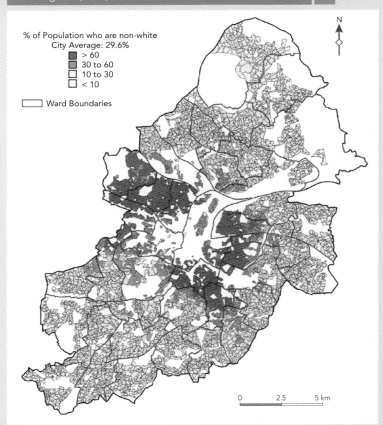

% of Population who are non-white
City Average: 29.6%

- > 60
- 30 to 60
- 10 to 30
- < 10

Ward Boundaries

0   2.5   5 km

Figure 5.23 Distribution of non-whites in Birmingham (2001)

Figure 5.24 Profusion of poverty in Dhaka

'Take it further' activity 5.5 on CD-ROM

Whilst the symptoms of deprivation are widely recognised, ranging from poor housing to ill-health, from unemployment to low income, from illiteracy to crime, its causes are poorly understood. We know where it occurs, but why is it that in all populations there are winners and losers? Why do the losers lose? Is it through lack of education, as suggested in Figure 5.18? Is it because of a lack of opportunities? Or is it something genetic that is passed down from one generation to the next? If we are unclear about its causes, what hope is there of us ever removing it?

Discussion point

How do you explain the existence and persistence of deprivation?

Managing service provision

Figure 5.21 has suggested that in the UK, at least, an important part (over one-third) of multiple deprivation is related to services, particularly social services such as nurseries, schools, medical centres, hospitals, dentists, care homes and libraries. Either these services are lacking or they are inaccessible to people by virtue of distance or cost.

In many countries, it is the responsibility of government (local and national) and planners to forecast the type and scale of services that will be necessary as a resulting of ongoing urban growth. Making accurate forecasts is no easy matter. These managers also have to match the provision of services to the spatial pattern of demand within the urban area. In short, the challenge is to provide what is wanted, where it is wanted. Funding apart (i.e. finding the necessary money), this apparently simple task is in practice a highly complex one. There are four other major complications (Figure 5.25):

◆ *Thresholds* Every service needs the support of a minimum number of people (threshold). This threshold varies from service to service. For example, the number and location of primary schools will depend on the number of children aged 5 to 11 years and where they live within the urban area, as well as the size of the school. But demand distributions change (see below).

Key term

Threshold: the minimum number of people needed to sustain a service.

◆ *Life-cycle changes* Our service needs change as we age.

◆ *Areas and populations age* Think of a new suburban housing estate attracting first-time buyers and young families. The needs of those families are mainly to do with schools. Eventually, the children leave home and the area, but their parents remain. The demand for schools falls, but the demand for day centres, medical and home support services increases. What happens to those schools as their enrolment numbers fall? Where are those services for the elderly to be located? And still later, what happens to those services that have been provided for the elderly when that generation gradually dies out and the area is gradually occupied once again by younger people?

◆ *Physical access* It is inevitable that people have to travel to take advantage of a particular service. Not every service can be provided within easy walking distance. But some people are more mobile than others – they have cars. So should the planners locate services in areas of low car ownership and expect the car owners to do the travelling? Or do they provide subsidised public transport to the service delivery locations? This leads us to the wider issue of traffic and transport that will be discussed in the next section.

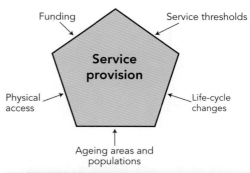

▲ **Figure 5.25** The challenges of service provision

Theory into practice

Give examples of how your own service needs have changed, and how you expect them to change over the next 50 years.

Birmingham

The inadequacy of service provision is certainly a feature of Birmingham's inner-ring of deprivation (Figure 5.22). That is partly explained historically. The housing there was built during the nineteenth century when the city was booming as an industrial centre. Much of it was built for factory workers, mainly in the form of small, back-to-back terraced units with only the most basic of water and sewage services. The lack of compulsory education meant that few schools were ever built. There were, however, select areas of the ring where the more wealthy people lived, mainly in detached villas. By the end of the Second World War (1945), the inner-city belt had become very rundown – little had been done by way of house maintenance and better-off households were moving out to the suburbs. The ring was ripe for regeneration. However, the sudden influx of immigrants from the Caribbean and the Indian subcontinent put a temporary brake on this – even the worst housing became occupied by people desperate for cheap accommodation. Villas were subdivided into small flats. Eventually, wholesale redevelopment did take place. Local residents (both native and immigrant) were re-housed either in replacement high-rise blocks or on suburban council estates (Figure 5.26). However, the top priority of the city at the time was to meet the acute housing need. As a consequence, the provision of services, particularly in the inner ring, took second place.

Sadly, it does look as if the whole re-housing exercise has failed to eliminate deprivation – reduced it, maybe. Much deprivation remains in its original locations, but some has been relocated in the suburban ring in so-called sink estates (Figure 5.22).

Dhaka

The speed of Dhaka's growth (it has quintupled in 30 years to 12 million) and the poverty of Bangladesh as a whole have meant one thing. The provision of services has lagged very far behind the rise in demand. Only two-thirds of households in Dhaka are served by the city water supply system. It is the only city in Bangladesh with a water-borne sewage system, but this serves only 25 per cent of the population, while another 30 per cent are served with septic tanks. Most solid wastes are often dumped untreated in nearby low-lying areas and water bodies. Dhaka has one of the highest rates of death from infectious disease of any city in Asia. The worst provided areas are the 'bostis' (Figure 5.27). There are only 45 000 hospital beds in the whole country and there over 4000 people per doctor. Despite this, life expectancy has risen from 37 to 60 over the last 50 years. While there may be as many as 16 000 primary schools in Dhaka, 20 per cent of children still do not receive even compulsory education. The literacy rate is a meagre 45 per cent.

The key to better service provision lies in the economic development of the country as whole. Being the capital means that of all the cities in Bangladesh, Dhaka is likely to be the first to benefit. However, its mega-city size and the speed of its recent growth mean there is a huge backlog in service provision to be made good before many people begin to enjoy the same quality of life as even those in the most deprived areas of Birmingham.

▲ **Figure 5.26** Redevelopment of Birmingham's inner ring

Figure 5.27 A Dhaka bosti ▲

Of the two broad categories of issue considered in this section, it seems that there are more issues of a serious nature falling under the social than the economic heading. That may seem rather strange bearing in mind that the states of urban growth and urban decay are primarily the outcome of economic circumstances. However, the incidence of deprivation suggests that it is commonplace in both scenarios. It also appears that when it comes to service provision, economic circumstances largely call the tune. Finally, remember that there are economic and social issues shown in Figure 5.16 that have not been discussed at all. This does not mean that the 'missing' issues are any less important or serious.

Discussion point

How far do you agree that the provision of social housing is the key to solving the problem of multiple deprivation?

5.3 What are the environmental issues associated with urban change?

The Brown Agenda was launched by the UN in 1992 at the same time as the Earth Summit was being held in Rio de Janeiro. It has done much to focus global attention on the built environment. It has drawn attention to the consequences of development in general and of urban growth in particular. The major issues raised on the Agenda include (Figure 5.28):

◆ the lack of safe water supply, sanitation and drainage (sewage disposal)

◆ the inadequate management of solid and hazardous waste

◆ the incidence of accidents – traffic, industrial and environmental

◆ uncontrolled emissions into the atmosphere

◆ the occupation of unsafe land by shanty towns (informal settlements).

Figure 5.28 highlights some of the key concerns expressed at the summit.

'Our roads are a major killer. We need to do much more to ensure the general safety and well-being of our citizens.'
India

'Recycling should be a top priority. We can't go on burying our mountains of rubbish.'
Finland

'Whether we like it or not, shanty development happens. Better to 'site and service' than demolish and create even more homeless poor.'
Brazil

'Nothing spoils our image more than derelict land and buildings. Finding new uses and funding are the real challenges.'
Wales

'Too much air pollution - too many smogs. No wonder our citizens have breathing problems.'
Chile

'We need to be sure that there is safe piped water for everyone.'
Zambia

'We take a pride in our green spaces. Residents enjoy them and they do a bit to combat global warming.'
Australia

'Our ecological footprint is too deep. We need to be much more energy efficient.'
Malaysia

'The private car is our worst enemy. Congestion is a waste of time, money and fossil fuel. Getting people to leave their cars at home is not easy.'
England

Figure 5.28 Environmental issues of urban change ▲

This section looks a little more closely at the first four issues. The previous section has already touched on the fifth issue. Instead of squatter settlements, the issues of brownfield sites and flood-plain development will be considered. Before moving on, you may have noticed that the first four issues are more related to urban growth than urban decay. Equally, you will sense from Figure 5.28 that the environmental issues of urban change vary from country to country.

Key terms

Brown Agenda: Part of the international drive to make cities and urbanisation more sustainable. Mainly concerned with the living environment and the quality of life of slum dwellers. It was one of the outcomes of the Earth Summit held in Rio de Janeiro in 1992.

Earth Summit: the first United Nations Conference on Environment and Development, held in Rio de Janeiro in 1992.

Activity

Identify the specific issues raised in the speech bubbles in Figure 5.28.

Water and waste

Along with food, water is vital to human survival. Furthermore, the quality of water is crucial to human health. It goes without saying, therefore, that water is essential to urban growth. Water is needed for:

- domestic consumption
- manufacturing use
- cooling purposes in electric power stations
- sewage treatment and waste disposal.

Worldwide, the demand for water continues to rise. Demand in England and Wales has risen to 40 billion litres per day. The key thing about water is not just ensuring that supply meets demands – ensuring the safety or purity of that supply is also critical. The problem is that urban water supplies are readily contaminated, as for example, by the discharge of urban sewage and industrial wastes into rivers and lakes and by the pollution carried in urban runoff (Figure 5.29). Agricultural practices, such as applying manure and artificial fertilisers, can lead to lethal chemicals seeping into reservoirs and groundwater stores. Then there is the problem of collecting and storing water to ensure a continuous supply, even during long dry seasons.

Key term

Waste disposal: the act of getting rid of, or making safe, the products of an urban area's metabolism.

Discussion point

Why is per capita domestic water consumption increasing in MEDC urban areas?

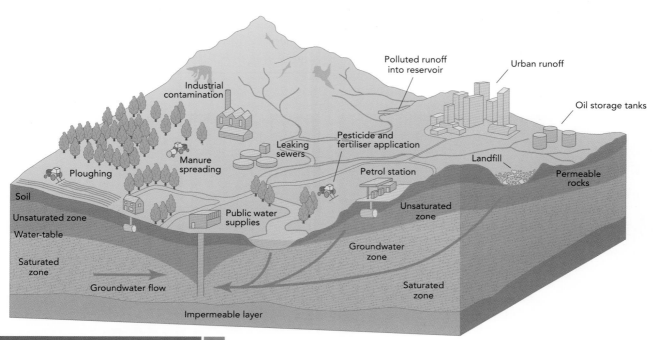

Figure 5.29 Sources of water pollution

Urban areas create huge quantities of solid waste, particularly those in MEDCs with their high levels of consumption and their throwaway societies. Recycling waste materials is the obvious way forward, but at the moment, for cost and technological reasons, recycling is mainly limited to paper, glass and plastics. Much could be done by composting biodegradable waste. At present, not all local authorities in the UK are processing this waste. The challenge remains as what to do with un-recyclable waste, particularly toxic waste. Should it be burned, buried or dumped at sea? All three options are distinctly environmentally unfriendly.

> **Key term**
>
> **Consumption:** the increasing consumption of goods and services that undermines the environment and exacerbates inequalities.

 'Take it further' activity 5.6 on CD-ROM

Traffic and transport

The movement of people and goods is a vital aspect of the urban area. The more efficient this movement, the better it is for the environment, the inhabitants and many of their activities. Three specific issues may be identified here: traffic congestion, oil consumption and public versus private transport.

Traffic congestion

Traffic congestion is the bane of most urban areas today (Figures 5.30a and b). It is the outcome of a number of factors, such as outdated and inadequate road systems, the great rise in private-car ownership and the concentration of services and jobs in the core areas of towns and cities that encourages an acute convergence of traffic. Traffic congestion is not just a waste of time and fuel, there are costs in terms of pollution, noise and stress. It is being shown, however, that levels of congestion can be reduced by various forms of traffic management.

◆ Segregation – rerouting heavy traffic, establishing ring routes for through traffic, pedestrianisation.

◆ Parking – pitching prices at the right level, rationing parking permits, limiting the amount of parking space.

◆ Vehicle control – traffic light systems, congestion charging, restricting vehicle use to alternate days.

◆ Public transport – making it an attractive alternative to the car.

◆ Traffic information systems – advising travellers about routes, hold-ups, black-spots and alternative modes of transport.

These sorts of interventions may prove effective in MEDC cities but they are unlikely to work in many LEDC cities. In the latter, much of the congestion is caused by the great volumes of paratransit traffic. Minibuses, hand-drawn and motorised rickshaws, scooters and pedicabs (pedal tricycles used as taxis) carry both passengers and goods along narrow, crowded streets (Figure 5.30). Paratransit is one of the main activities of the informal economy that prospers in all LEDC urban areas (see page 189).

> **Key term**
>
> **Paratransit:** an alternative mode of flexible passenger transport that does not follow fixed routes or timetable schedules. Typically vans, minibuses or rickshaws are used to provide paratransit services.

Figure 5.30a Traffic congestion in MEDC urban area

Figure 5.30b Traffic congestion in LEDC urban area

Figure 5.31 Transport choice?

Oil consumption

Much of the energy used by various modes of transport comes from the burning of oil-based products. There are two major worries here, namely that oil stocks are rapidly running out and that burning oil products is a major source of air pollution (see below). The use of bio-fuels is just beginning. This perhaps offers some hope in helping to solve the problems associated with the use of oil.

Public transport

One obvious way of reducing traffic congestion and the consumption of oil is to provide cheap and efficient public transport. But breaking the downward spiral of public transport (Figures 5.31 and 5.32) will require getting really tough on the use of the motor car. People must be persuaded to leave their cars at home as much as possible.

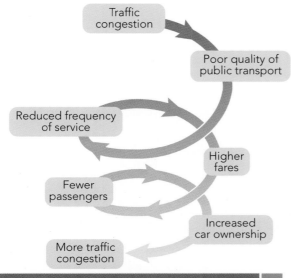

Figure 5.32 The downward spiral of public transport

Activity

Outline possible ways of discouraging car use.

'Take it further' activity 5.7 on CD-ROM

Air pollution

Air pollution has been greatly reduced in MEDCs as a result of deindustrialisation and the global shift in manufacturing, together with clean air legislation. In fact, MEDCs have been accused of 'exporting' their pollution to LEDCs.

Ever-increasing use of the motor vehicle represents a major challenge in all parts of the world. Although countries have toughened up on exhaust emissions, the motor vehicle remains, as just pointed out, a major polluter of the atmosphere along with coal-, oil- and gas-fired electric power stations. Two aspects of air pollution are of particular concern – its effects on human health and wildlife, and its contribution to global warming.

There are three main air pollutants that are serious hazards to human health.

◆ Suspended particles – made up of soot, smoke, dust and liquid droplets. Cause chronic respiratory illnesses and worsen heart conditions.

◆ Sulphur dioxide – produced by the combustion of fossil fuels by factories, motor vehicles and domestic boilers and fires. This causes acidic rain and can be extremely detrimental to the health of the young and elderly.

◆ Nitrogen dioxide – caused by fuel combustion, aerobic decomposition and nitrogenous fertilisers. Also causes acid rain.

Figure 5.33 on the next page shows that even in cities with generally low levels of air pollution, the level of one particular pollutant can be quite high and a threat to human health. Amsterdam and Tokyo are two such examples.

Activity

Compare Beijing, Milan and Mumbai in terms of the data given in Figure 5.33. Suggest reasons for any differences that you identify.

U2

5

Managing urban change

	Population (millions)	Suspended particles (mg m³)	Sulphur dioxide (mg m³)	Nitrogen dioxide (mg m³)
WHO recommended maximum levels		90	50	50
Amsterdam	1.5	40	10	58
Athens	3.1	178	34	64
Beijing	12.0	377	90	122
Berlin	3.4	50	18	26
Brussels	1.9	78	20	48
Copenhagen	1.6	61	7	54
London	7.6	–	25	77
Los Angeles	13.1	45	9	74
Mexico City	18.1	279	74	130
Milan	4.3	77	31	248
Montreal	3.4	34	10	42
Moscow	9.3	100	109	–
Mumbai	18.0	240	33	39
New York	16.6	–	26	79
Singapore	3.6	–	20	30
Sydney	3.7	54	28	–
Tokyo	28.0	49	18	68

Figure 5.33 Air pollution in selected cities (2000) ▲

'Take it further' activity 5.8 on CD-ROM

Urban dereliction and brownfield sites

In 'The rise and fall of urban areas' (pages 185–86), the comment was made that decay can occur in areas of urban growth. You need to remember that urban areas are highly dynamic and frequently changing. One 'negative' product of this dynamism is dereliction. It has three main contributory causes.

◆ The inevitable ageing and decay of buildings with the passage of time which cause maintenance costs to spiral.

◆ The movement of urban activities to better and more profitable locations.

◆ Basic changes in an urban economy brought about by the collapse of notable activities.

The result of these causes is to leave behind a residue of abandoned buildings and derelict land. The term brownfield site or land is now more commonly used. There are two related issues here. What new uses can be found for this abandoned land in order to bring about its regeneration? And when it comes to urban growth, should this brownfield land not be used in preference to building on greenfield sites at the urban fringe (Figure 5.34)?

Key term

Derelict land: an old term for a brownfield site.

Environmental Agency
Brownfield Land Redevelopment: Position Statement

Key issues

THE ENVIRONMENT AGENCY is committed to bringing more land into sustainable use. Concentrating development on brownfield sites can help to make the best use of existing services such as transport and waste management. It can encourage more sustainable lifestyles by providing an opportunity to recycle land, clean up contaminated sites, and assist environmental, social and economic regeneration. It also reduces pressure to build on greenfield land and helps protect the countryside.

Brownfield land is often more expensive to develop than greenfield. Moreover, with the shifting patterns in population and demographics, the location of large amounts of brownfield land in the former industrial heartlands of the country matches poorly with the areas of highest current land demand in the south and south-east.

Some brownfield and derelict land can represent important wildlife habitat, public green space or a core part of urban green networks. These are important in providing good quality of life, and brownfield reuse must strike an appropriate balance in the interests of sustainable development.

Many brownfield sites are former industrial sites and much of this land may be contaminated. ... At the present time there is confusion concerning the treatment of flood risk when brownfield sites are redeveloped. ... It is expected that the area of land at risk of flooding will increase substantially by 2050 due to climate change. Areas of industrial and brownfield land currently not considered to be at risk of flooding may be at risk in the future. This means that some of the engineering solutions used to remediate contamination, such as encapsulation of contaminated soil left in situ, permeable reactive barriers, or bentonite walls, may not be appropriate if the site is to be subject to inundation by flood waters.

Figure 5.34 Extract from recent Environmental Agency position statement about brownfield land ▲

In England, it is now government policy that 60 per cent of all new homes should be built on brownfield land. This sounds all well and good, particularly if it helps to cut down on the use of greenfield land, but there are some real difficulties.

◆ Not all local authorities have large amounts of brownfield land.

◆ There are disagreements about the precise definition of 'brownfield' and therefore how much of it there really is. For example, the government view is that old quarries are not brownfield land, but residential gardens are. Garden grabbing is becoming commonplace. Developers buy up houses with gardens, knock down the houses and build two or three new houses crammed on the same site. One new home in five in south-east England is built on garden land. This practice is a major contributor to suburban intensification.

◆ Across the UK there are thousands of sites which have been contaminated by previous industrial uses and may present significant risks to human health and to the wider environment. Cleaning them up so that they are fit for housing can be very expensive and a challenge to current technology.

◆ Not all brownfield sites have the physical access necessary for residential development.

◆ The problem of 'bad' neighbours – it is no good building new homes on a brownfield site next door to a sewage treatment works or a heavy industrial plant.

> **Key term**
>
> **Garden grabbing:** a process undertaken by developers where the gardens of existing houses are sold off as building plots or an old house is demolished to make way for several new dwelling units. The net outcome is a raising of housing densities and suburban intensification.

Discussion point

Is the case for using brownfield land really stronger than that for not using greenfield sites?

Building on the floodplain

Over the last few years, much of England has experienced some degree of flooding (Figure 5.35). In the worst hit areas, thousand of homes have been ruined and people's lives thrown into despair. Most of these ruined properties are in urban areas located on floodplains – Doncaster, Hull, Worcester

Figure 5.35 Flood headlines

WATER EVERYWHERE BUT NOT A DROP TO DRINK

THINK BEFORE YOU BUILD

BROWN PLEDGES NEW RULES FOR BUILDING

Ministers Warned Three Years Ago Over Flood Defense Failings

Floods Claim Another Victim

After Deluge, Britain Will Be Swamped by a Tidal Wave of Costs

Housing Can be Built on Flood Plains, Minister Insists

Warning Over Flood Plain "Ghettos"

Come Hell and High Water

Flood Plain Homes Uninsurable

and Tewkesbury. Clearly, these periodic extreme and catastrophic floods are a warning about the risks of allowing urban development on such potentially hazardous areas (Figure 5.36). Yet, at the same time, the government, faced with a need to build more and more affordable houses, is saying that inevitably much of the new housing will have to be built on such land. Planning applications for developments on floodplains in Britain have been going up every year for the last five years. Over five million people are now living or working in flood-risk areas in England and Wales. On the other hand, it is easy to see why it is so tempting to build on floodplains. Floodplains offer level and fairly cheap sites; they have relatively little agricultural value, and because they have often been ignored by early urban growth, they can provide sites close to urban cores. But remember that their original value is the part they play in natural systems, namely as 'safety valves' that help dissipate flood waters.

> **Key term**
>
> **Floodplain:** the part of a valley floor over which a river spreads during seasonal or short-term floods.

Living with the flood **risk** is not just the UK's problem. The risk levels are probably even higher in densely populated LEDCs, and nowhere more so than in Bangladesh. There is very little of that country that is not floodplain. The heavy monsoon rains and the convergence of major rivers greatly heighten the flood risk. Every year thousands are drowned and homes are destroyed.

"It isn't on a flood plain, is it?"

Figure 5.36 Floodplain concerns!

> **Key term**
>
> **Risk:** an estimate of the likely outcome of some decision or future event. In most cases, the estimate is subject to error.

Discussion point

Is it really worth allowing urban areas to spread onto floodplains?

Case study | Tales of two cities (3): London and Mumbai

The two cities chosen here are not only different in terms of their locations and the economic development status of the countries in which they occur. They are also different with respect to two different aspects of size. The population of Greater Mumbai is much larger than that of Greater London (over twice as large), but London occupies a much larger area (over three times as large). The net outcome is a staggering difference in population densities. Places do not come much more densely populated than some districts in Mumbai.

London

Site: The Thames floodplain and gently rolling hills to the north and south of the river.

Area: 1579km^2

Population: 8.5m (2005)

Density: 4761 persons per km^2

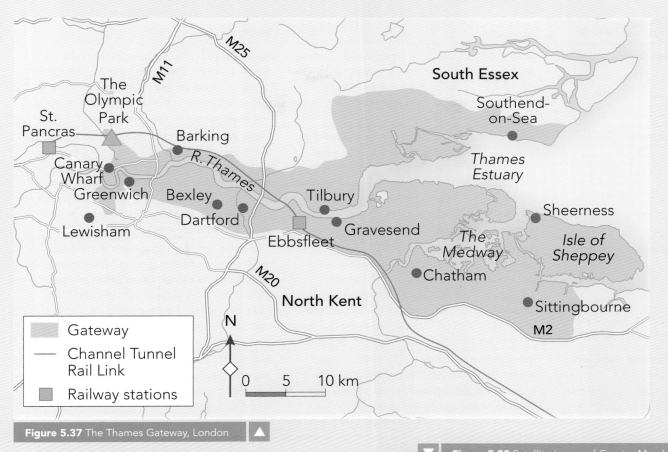

Figure 5.37 The Thames Gateway, London

Legend:
- Gateway
- Channel Tunnel Rail Link
- Railway stations

Map labels: M11, M25, St. Pancras, The Olympic Park, Barking, R. Thames, Canary Wharf, Greenwich, Bexley, Lewisham, Dartford, Ebbsfleet, M20, North Kent, Tilbury, Gravesend, The Medway, Chatham, M2, South Essex, Southend-on-Sea, Thames Estuary, Sheerness, Isle of Sheppey, Sittingbourne

N 0 5 10 km

▼ **Figure 5.38** Satellite image of Greater Mumbai

Mumbai

Site: Originally an archipelago of seven islands, but reclamation has welded them into one (Salselle Island). More recent growth has spread onto the mainland at New Mumbai.

Area: 438km²

Population: 18m (2005)

Density: 29 000 persons per km² (Figure 5.38)

Environmental issue	London	Mumbai
Water supply	All dwellings have safe, piped water. The major management issue is one of meeting the level of demand. London is located in the driest part of the country. Per capita consumption of water continues to rise.	Nearby lakes collect rainwater that falls on Western Ghats. Present supply is outstripped by rapid urban growth. 60 per cent of the 18 million inhabitants live in slums; and many do not have access to safe, piped water.
Waste management	Most properties are linked to piped sewage systems; the remainder have sceptic tanks. The main management problem is how to dispose of the increasing amounts of domestic refuse mainly related to the rising consumption of packaged food and drink. Campaigns to recycle paper, plastic and glass.	More than 5000 metric tons of solid waste are generated every day. The municipal corporation is in charge of the disposal of this waste, and evidently cannot cope with the task. Much waste simply dumped by the roadside and left to rot. Besides the smell, there are risks to health.
Atmospheric pollution	Levels have been significantly reduced by a series of Clean Air Acts, de-industrialisation and tougher controls on motor vehicle emissions. The incidence of respiratory diseases much reduced during second half of twentieth century. The management challenge is to lower pollution to even lower levels.	There is a very high incidence of chronic respiratory problems, arising from extreme air pollution. The causes of pollution are mainly industries in the eastern suburbs and New Mumbai, garbage burning and insufficient control over emission levels from vehicles.
Traffic and transport	Congestion charging has helped to ease traffic congestion in Central London. The main management problem is persuading people to use public transport, particularly for commuting. Can public transport be made a sufficiently attractive option to tempt people to leave their cars at home?	Nearly 90 per cent of commuters use public transport (suburban trains, buses and ferries) because it is cheap and reliable. Car ownership is low and distances to work can be vast. However, the roads are heavily congested (591 vehicles per km as opposed to international standard of 300) and there are over 60 000 road accident deaths a year.
Brownfield sites	Extensive use of brownfield sites in the inner-city ring. Flagship schemes, such as Canary Wharf, have given much publicity to the potential of regeneration. Management problems include dealing with contamination risk and reconsidering whether gardens really are brownfield sites.	There is plenty of derelict land – redundant docklands and sites of former cotton mills, but there is little investment for regeneration. Some of the land is occupied by squatter settlements and as such may be regarded as an acceptable temporary use until proper housing can be provided.
Floodplains	Much of London is built on the Thames floodplain. The Thames Barrage offers a good degree of protection, but a huge expansion of the city (Thames Gateway proposal, Figure 5.37) will involve building on the floodplain downstream of the barrage. The 2012 Olympics site in east London is mainly floodplain.	Very little of the urban area is built on flood plains. The concern is the coast and the clearance of the mangroves that have acted as natural barriers to the sea. Urban development in such areas is at particular risk when there is a combination of high tides and monsoonal cloudbursts.

Figure 5.39 London and Mumbai compared in terms of environmental issues and their management

Activity

Summarise the similarities and differences between London and Mumbai shown in Figure 5.39 in terms of the six environmental issues. You might refer to Figure 5.14 as a possible model.

This section has looked at five environmental issues associated with urban change. It is inescapable that urban change in general and urban growth in particular will impact on the natural environment in a variety of ways. In most of these impacts, there is an element of risk – to human health and safety as well as to the natural world. The whole question of how to minimise those impacts and risks leads us into the final section of this chapter.

5.4 How can urban areas be managed to ensure sustainability?

What is a sustainable urban area?

Sustainability is a bit of a buzzword these days. Many people use it, but does it mean the same thing to everybody? Most would agree that it is about improving situations today in a way that does not have to be paid for tomorrow. The Brundtland Report (1987) was the first to coin the phrase sustainable development. It was defined as:

'To meet the needs of today without compromising the ability of future generations to meet their own needs'.

> **Key terms**
>
> **Sustainability:** Improving the quality of life while living within the earth's carrying capacities.
>
> **Sustainable development:** Meeting the needs of today while not compromising the capacity of future generations to meet their needs.
>
> **Need:** Something one cannot do without, such as food and water.

When applying this definition to urban areas, three questions need to be asked:

1. What are the 'needs of today'? The most important are probably (Figure 5.40):
   - health and welfare needs – decent, affordable housing; medical provision for all who need it; protection from environmental hazards
   - social needs – access to education; personal security; equal opportunities
   - economic needs –an adequate livelihood; secure job; range of employments
   - political needs – freedom of speech; civil rights; participation in the decision-making process
   - environmental needs – food and water; raw materials; freedom from pollution.

2. Is it possible to meet these needs today? To be realistic the answer must be 'no'. There is a huge backlog of unmet needs to be cleared. Urban areas throughout the world are, and have been for centuries, failing on all or most of the five counts.

3. Will it still be possible to meet these needs tomorrow? The most likely answer is 'no'. The reason is quite simple, as suggested at the end of the last section. Urban areas will always consume non-renewable resources, pollute the environment and embody risk. They will do all these things, despite people's best efforts to avoid them.

Improving sustainability

Once it is recognised that urban areas cannot, and will never become wholly sustainable, it is more realistic to focus attention instead on ways that can make urban areas *more* sustainable than they are present. Perhaps the following definition of sustainability is helpful here:

'Improving the quality of life while living within the earth's carrying capacities.'

To move in this direction depends on taking five courses of action that relate to the needs set out in Figure 5.40.

- Minimising the ecological footprint.
- Improving the quality of the living environment.
- Waging war on deprivation and discrimination.
- Raising public participation in government and decision-making.
- Ensuring a sound economic base.

Goal
Meet the needs of today without compromising the ability of future generations to meet their needs

Health and welfare needs
- decent, affordable housing
- medical provision for all who need it
- a caring society
- quality of life

Social and cultural needs
- access to education
- equal opportunities
- freedom from discrimination
- personal security

The needs of today

Political needs
- freedom of speech
- civil liberties
- participation in decision-making process
- democratic government

Economic needs
- secure jobs
- range of employments
- a fair wage
- advancement opportunities

Environmental needs
- food and water
- raw materials and energy
- freedom from pollution
- protection from hazards

Figure 5.40 Needs of today

Input actions
- Conserve natural resources
- Ensure efficient use of resources
- Protect biodiversity
- Respect environmental capacity

Internal actions
- Recycle waste
- Provide 'green' infrastructure
- Make living space healthy and secure
- Reuse brownfield sites
- Make urban areas more compact
- Reduce use of private car
- Create a fairer society
- Encourage wide participation in decision making

Output actions
- Minimise emissions and pollution
- Restrict use of greenfield sites
- Meet leisure and recreation needs

Figure 5.41 Making urban areas more sustainable ▲

Many consider that the first of these is perhaps the single most important action. But what is an ecological footprint? This is best understood if we think of the urban area as an open system, with inputs, outputs and internal actions (Figure 5.41). The inputs of an urban area are made up mainly of those things that are brought in to support urban life. They range from food and water to building and industrial raw materials. The problem is that some of those resources are non-renewable and therefore exhaustible. Our present use of resources is coming close to exceeding the earth's carrying capacity. The outputs of urban areas include waste, pollution and the spread of the built-up area into the countryside. The ecological footprint of an urban area is the sum total of its inputs and outputs.

The ecological footprint is measured in terms of the amount of productive land area necessary: 1) to produce the resources that a population consumes and 2) to assimilate the wastes it produces. So it measures a range of things needed by that population. They include the amounts of farmland, forest, water, energy and other material inputs. Ultimately, the ecological footprint is expressed in terms of the number of hectares of land needed to meet all the needs of one person (Figure 5.42).

Key terms

Ecological footprint: how much land it takes to provide the resources used by, and to dispose of the waste produced by individuals or groups of people.

Carrying capacity: the number of people who can be supported in a given area within natural resource limits, and without degrading the natural, social, cultural and economic environment for present and future generations.

▼ **Figure 5.42** Ecological footprints of selected countries

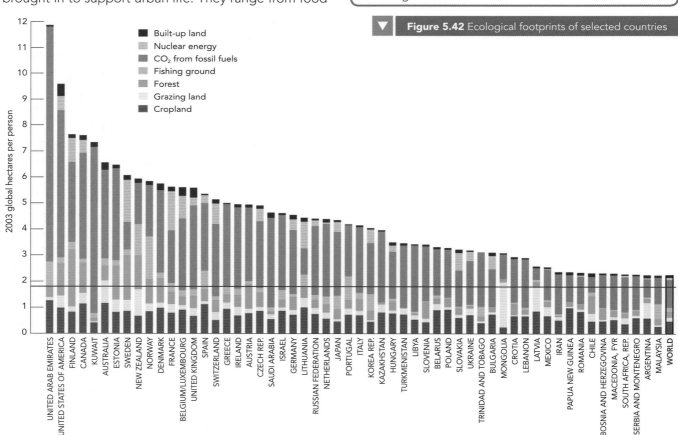

Unfortunately, there is as yet little footprint data for LEDCs and their cities. We do know that most European cities have footprints approaching 3 global hectares, whereas in the USA the figure is around 4 hectares. The total ecological footprint of London (the average footprint multiplied by number of people) is around 125 times the city's built-up area (Figure 5.43). In ecological terms, we need to understand that nearly all the land 'used' by urban areas in fact lies outside their boundaries. Perhaps we should think of this space as urbanised land.

> **Key term**
>
> **Urbanised land:** the land required to meet the needs the population of a town or city, that is, land used to supply food, water and other resources, as well as recreational space.

The ecological footprint is largely, but not exclusively about inputs. Figure 5.41 shows that there are two other 'arenas' where actions should be taken to improve urban sustainability. These are reducing the pollution of water and air (by treating raw sewage to much higher standards and by controlling more tightly emissions from motor vehicles and factories) and by restricting the use of greenfield sites to accommodate urban growth. In addition, there is a whole range of actions that can be taken within the urban area to improve sustainability. These include:

- ◆ recycling waste
- ◆ regenerating brownfield sites
- ◆ creating a fairer and more open society
- ◆ improving the quality and security of the built environment
- ◆ keeping urban areas compact and reducing use of the motor vehicle.

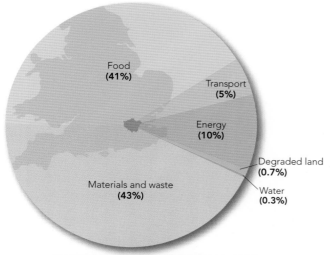

Figure 5.43 Ecological footprint of Londoners ▲

Food (41%)
Transport (5%)
Energy (10%)
Degraded land (0.7%)
Water (0.3%)
Materials and waste (43%)

Sustainability in practice

When it comes to making places and activities more sustainable, there is a variety of things that we can do and in doing so perhaps ultimately influence decision-makers. Some of these are shown in Figure 5.44. But while individuals can contribute much, there is still a need for pioneering projects to show others what is possible.

Buy appropriate products – e.g. fair trade goods, FSC certified wood, high efficiency electrical goods, recycled paper, etc.

Become involved – join national or global pressure groups.

Recycle – e.g. paper, tins, plastic, glass, clothes, books, organic waste, etc.

Sustainability

Become informed – watch documentaries and read reliable reports on sustainability issues.

Support active NGOs – e.g. RSPB, WWF, Oxfam, Christain Aid, Water Aid, etc.

Regularly check your personal ecological footprint.

Figure 5.44 Making it happen ▲

There are a few cities in different parts of the world that have pioneered moves to become more sustainable (see Take it Further 5.4). These include Curitiba (Brazil), Chattanooga (USA) and Putrajaya (Malaysia). They and others serve as beacons of hope for a more sustainable future. Agenda 21 has been another significant step forward. At the Rio de Janeiro Earth Summit in 1992, the United Nations agreed that the best starting point to achieve sustainable development is at the local level. In fact, two-thirds of the 2500 action items of Agenda 21 relate to local councils. Each local authority has had to draw up its own Local Agenda 21 (LA21) strategy following discussions with its inhabitants about what they think is important for their area and their quality of life. The principle of sustainable development must form a central part of every strategy.

> **Key term**
>
> **Quality of life:** the general state or condition of a person or a population living in a given area. It is a broad concept including the quality of health, housing, educational attainment, employment and public services, etc., which may be measured to help inform social policy.

As part of LA 21 in Liverpool, for example, the public and the city council have drawn up their own indicators to discover the success of sustainable development. These indicators include the number of parks and the number of people living close to them, education standards and crime figures. In Cheshire, the local authority has set up a transport task group as part of its LA21. This aims to set up commuter plans to discourage people from travelling by car. Kirklees, in West Yorkshire, has encouraged partnerships between themselves, businesses, charity groups and members of the public to help them to achieve a sustainable level of development.

Case study | Tales of two countries: the coming of eco-friendly urban areas?

In this case study, the spotlight falls on recent proposals to pioneer the building of eco-friendly urban areas in the UK and China.

UK

Although planned in the 1960s and at a time when sustainability was not a word in the planners' vocabulary, Milton Keynes was perhaps a pioneering venture. There were elements in its design and construction that hinted at trying to be environmentally friendly. They were the inclusion of much parkland; the interspersing of jobs and residential areas to reduce commuting, and the construction of energy-efficient housing (Figure 5.45).

▼ **Figure 5.45** Milton Keynes during its early development

In July 2007, town halls and developers in England and Wales were invited to bid for cash to build five new eco-towns in which the houses and infrastructure would be carbon neutral. What the government had in mind were small stand-alone settlements, each containing between 5000 and 20 000 homes. These new eco-towns would be expected to have good transport links with existing towns and cities. In contrast to the high-density apartments and small dwelling units that now characterise many of our urban areas, the eco-towns would have plenty of green spaces. The dwellings would be family houses rather than apartments, and have gardens. It was expected that up to 50 per cent of the accommodation would be affordable housing with a mix of owner-occupied, social and rented homes. The houses would be built using timber, solar thermal panels, double-glazing, insulation and biomass boilers that do not use fossil fuels. In a further effort to reduce carbon emissions, shops, primary and secondary schools would all be within walking distance. One eco-town, Northstowe, is already being built north of Cambridge, and others near Bristol and Peterborough are expected to start construction soon.

Key terms

Carbon neutral: counteracting and balancing emissions of carbon dioxide by growing plants to act as fuels or planting trees in urban areas to offset vehicle emissions.

Affordable housing: low-cost and subsidised housing that is available to people who cannot afford to rent or buy houses generally available on the open market.

While any attempt to make urban areas more sustainable is to be welcomed, serious difficulties lie ahead.

◆ Most of the working population of these eco-towns will commute to nearby towns and cities. Such travel is unlikely to be carbon neutral.

◆ The experience of the Bedzed Project in south-east London (Figure 5.46) has clearly shown that while homes can be made carbon neutral, it is much more difficult to do likewise with the urban infrastructure.

◆ The building of new towns, presumably on greenfield sites, is in conflict with present government policy of giving priority to the use of brownfield land.

Figure 5.46 Bedzed housing, south-east London ▼

U2

5

Managing urban change

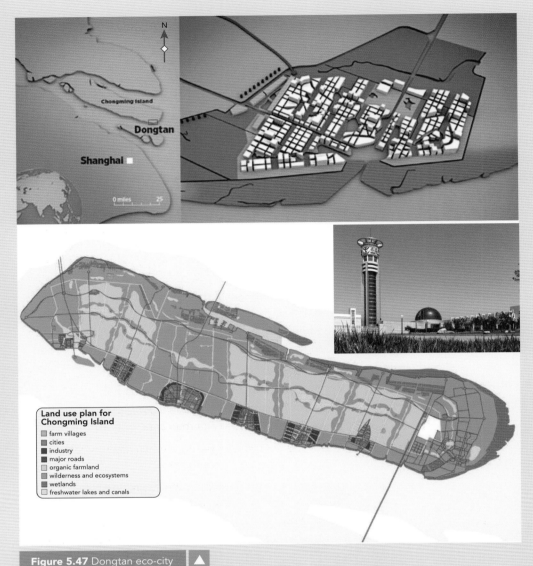

Land use plan for Chongming Island

- farm villages
- cities
- industry
- major roads
- organic farmland
- wilderness and ecosystems
- wetlands
- freshwater lakes and canals

Figure 5.47 Dongtan eco-city ▲

◆ It is necessary to radically change the lifestyles of residents, from a total commitment to recycling to living without the car.

On the positive side, however, we know that we can build carbon-neutral homes. That fact alone is very encouraging bearing in mind that housing is currently responsible for over a quarter of the UK's carbon dioxide emissions.

China

In order to promote the urbanisation of China, there are plans to build 400 new cities and 10 000 new towns over the next two decades. These will accommodate some of the 300 million or more people that China wants to move from rural to urban areas. Shanghai, situated on the southern shores of the Yangtse delta in Eastern China, is the largest city in a rapidly urbanising China and the seventh largest in the world.

Widely regarded as the flagship of China's modern economy the city is now one of China's most important cultural, commercial, financial, industrial and communications centre. It is also one of the world's busiest ports, and in 2005 became the largest cargo port in the world. Since 1992, when China started its move to become a more market-oriented economy, Shanghai's population has grown from 12 to 16 million. Shanghai's awesome skyscrapers and modern lifestyle have become symbols of the 'new' China and its recent economic development. Shanghai is now planning for still more growth. This will involve the construction of a new city the size of Manhattan (New York) on its doorstep. Dongtan (Figure 5.47), as it is to be called, will be built on the eastern end of Chongming, a large island in the Yangtse River delta, a short distance to the north-east of Shanghai's CBD and close to Shanghai's new airport. The site is one of the last big undeveloped spaces in the Shanghai area. It will be one of a number of new cities to be built on the island.

The new city will eventually extend to cover 8800 hectares and is expected to house several million people. It is claimed that Dongtan will be the world's first genuinely eco-friendly city, powered by renewable energy resources (mainly HEP). It will be as close to carbon-neutral as possible. The first phase of the development will cover 630 hectares, roughly three times the size of the City of London. It will include a transport hub and a port to accommodate fast ferries from the mainland and the new Shanghai airport, a leisure facility, an education complex, space for high-tech industry and housing. It has been designed by a British company and will be completed by 2010, in time for Shanghai's hosting of the World Expo.

The fact that the city will be built on agricultural land and that it will inevitably disturb the huge wetland and wildlife refuge adjacent to it may question the eco-friendly nature of the project. However, this attempt to put sustainability high up the agenda should be applauded and its is hoped Dongtan sets the example for China's other proposed new cities.

 'Take it further' activity 5.9 on CD-ROM

Discussion point

Debate the following proposal: 'Sustainable urban areas are little more than wishful thinking'.

It may be impossible to achieve a truly sustainable urban area. However, pioneering projects in cities as far apart as Chattanooga and Curitiba, together with the above examples, clearly show that there is a growing wish to improve the general sustainability of urban areas. Doing this certainly promises a better urban future as well as creating urban areas that are pleasant places to live and work in.

Knowledge check

1 In what ways do urban areas differ from rural areas?

2 Explain what is meant when it is said that 'urbanisation is a multi-strand process'.

3 Describe the three main features of urban change at a global scale.

4 How and why do the processes of urbanisation change along the urbanisation pathway?

5 Why do land values affect the spatial patterns of urban areas?

6 Identify possible ways in which urban areas in MEDCs differ from those in LEDCs.

7 With the aid of examples, examine the reasons why the fortunes of urban areas fluctuate.

8 'Urban growth gives rise to more issues than urban decay.' To what extent do you agree with this statement?

9 Examine ways of breaking the vicious downward spiral of urban decay.

10 'Deprivation is a relative term.' Explain what this means.

11 Summarise the main challenges associated with providing services in an urban area.

12 Using examples, suggest how urban areas might 'lead the way to a greener world'.

13 Examine what you think are the main threats to human health in urban areas.

14 Are the use of brownfield sites and floodplains just issues for MEDC urban areas? Justify your viewpoint.

15 Explain why transport is so important to the prosperity of urban areas.

16 Explain why it is unlikely that urban areas will ever become completely sustainable.

17 What is meant by an ecological footprint? Why is it used, and what are its limitations?

18 Which of the internal actions shown in Figure 5.41 is likely to contribute most to making urban areas more sustainable.

19 'Eco-city and eco-town projects are simply showpieces that will do little to make existing urban areas more sustainable.' Do you agree? Give your reasons.

A list of useful websites accompanying this chapter can be found in the Exam Café section on the **CD-ROM**

Exam Café
Relax, refresh, result!

Relax and prepare

What I wish I had known at the start of the year…

Roger

"I find mind-maps and spider diagrams useful as they produce very visual images which are easier to remember. These help me to recall information but also help in planning answers. They link my thoughts and provide the connections that are key in geography."

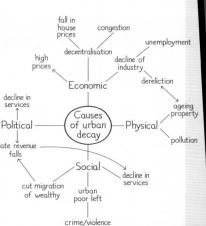

Katy

"I sometimes use colour coding to prioritise or classify information or different shapes or different-sized words or even pictures to help increase my visual memory. Once I've drawn a diagram, I keep reading it from the centre outwards, so each time I remember more of it. Once I think I've memorised it, I try re-drawing it from memory."

Hot tips

Said

"When revising, I often work with a friend to answer key questions such as, 'Why is the demand for housing in the UK rising when the birth rate has been falling?'. Together we were able to come up with the following possible explanations:

◆ The number of households are rising because divorce, life expectancy and people choosing to live on their own is on the increase and this leads to a demand for more but smaller housing units.

◆ People are becoming more affluent and this can raise their aspirations and they may seek larger houses in nicer areas.

◆ More people are able to afford a second home.

◆ There has been an increase in immigration, especially recently from eastern Europe, further increasing demand."

Common mistakes – Alex

"Sometimes I miss the link between sections in a question e.g. in the following example part i) and part ii) are linked. If I get part i) wrong, then ii) will be wrong as well.

Q1

i) Describe the changes in the pattern of urban land use in a specified city in a MEDC.

ii) Suggest reasons for the changes described in (i).

I now know that the use of ii) means it is the same data as i). My reasons must relate to the changes I've describe in part i) and not be general ones about changes in urban land use. When a) and b) are used then b) doesn't depend or link to the answer in a)."

Refresh your memory

5.1 What are the characteristics of urban areas?

Physical	Climate, relief, drainage, rock type
Economic	Transport routes (sectors – Hoyt) Ability to out bid for sites (Burgess) Size of site available Mutual attraction/repulsion (Harris and Ullman)
Social	Historical – inertia, conservation, reputation Mutual attraction/repulsion Land ownership – estate development Religious factors
Political	Planning controls – land-use zoning Need for centrality or safety
Characteristics	Shape, form, layout, density, building types Population features e.g. number, age, ethnicity, socio-economic Land uses – industrial, housing, recreational, commercial, retail

5.2 What are the social and economic issues associated with urban change?

Population	Migration, birth rate, ethnic mix, age structure
Mobility	Private and public transport
Income	Increased or decreased inequalities in wealth or income
Political	Planning initiatives and their impacts, local tax, land-use zoning
Employment	Structure – industrial versus service employment, changing technology, unemployment
Public services	Cost and viability – type, location, 'post code lottery'

5.3 What are the environmental issues associated with urban change?

Pollution	Air, water, noise, visual, solids
Water	Scarcity (transfer schemes, use of aquifers, reservoir building)
Traffic	Congestion, pollution, land use (roads, car parks, garages, etc)
Land	Pressure on marginal land and ecosystems (parks, conservation, green belts)
Micro-climate	Heat island, wind channelling, higher rainfall, smog
Dereliction	Visual pollution, safety, health

5.4 How can urban areas be managed to ensure sustainability?	
Physical	Climate – creation of urban micro-climate Relief – building on unsuitable sites e.g. steep slope Drainage – lack of, water shortage Vegetation – loss of habitats Pollution – air, water, land, noise, visual
Economic	Settlement – housing quality and quantity, cost Power – shortage, reliability, cost Industry – lack of jobs or low pay, exploitation Services – lack of sufficient schools, clinics etc Transport – congestion, poor public transport
Social	Wealth inequality and deprivation Ethnic tensions High cost of social support Inner-city decay Slums and shanties Crime and security

Top tips . . .

Geography is ideal for the use of diagrams or maps to help revision. A well labelled (annotated) diagram or map helps you remember but can also act as your case study. So often a quick sketch diagram or map can give a sense of place far quicker than writing. Remember to pick examples that could serve to answer a number of questions. Again, remember to use colour coding to prioritise or classify information.

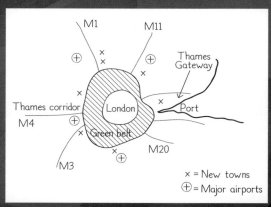

x = New towns
⊕ = Major airports

Get the result !

Sample question

In what ways can urban change cause increased atmospheric pollution? [6 marks]

Examiner says

This answer scores maximum marks because – (1) it links cause and effect and gives an example, (2) it elaborates on forms of pollution, (3) provides another clear change linked to an impact, (4) and another clear change linked to an impact, (5) further example of a clear change linked to an impact, (6) other clear changes linked to an impact.

Student answer

The increase in traffic and congestion leads to more photo-chemical smog e.g. Los Angeles (1) and other air pollutants such as particulates and ozone (2). The vertical growth of many cities has created high-rise buildings that trap fumes (3) and the buildings themselves may pollute the air via air conditioning or heating exhausts (4). Demolishing buildings and replacing them creates dust that pollutes the air (5). Industrial growth and the need for more power can also increase pollution from burning of fossil fuels (6).

Examiner says

There is more than enough here for full marks.

Examiner's tips

How to deal with data response – charts and figures. This is usually the first part of a question and is designed to see if you can interpret (read the implications of) a diagram. Sometimes you are expected to describe the trend or state the values of something. This you do by reading off the values on the diagram using the scale on the axes. You must quote real values. Saying 'it is high' isn't good enough. Trends can be positive or negative, strong or weak. Always look for odd values (anomalies) that don't fit in. In other cases you are asked to compare or describe the relationship between two or more variables. In this case you *must not* describe one then the other but write in sentences that combine both variables. Again relationships can be positive or negative, linear or non-linear and strong or weak.

The use of maps and diagrams in answers can convey a lot of information and so save a lot of writing and time. If maps and diagrams are used:

◆ keep them simple – don't try to draw the UK exactly
◆ draw them quite large – it is easier to read larger diagrams and they look neater
◆ only use colours if it helps and remember certain colours mean something in geography, e.g. blue = water, brown = upland
◆ number them as Figures and refer to them in your writing
◆ put them with the appropriate text
◆ label clearly and briefly
◆ remember conventions, e.g. North on a map, labelled axis on a graph.

Chapter 6

Managing rural change

Rural areas are constantly changing. At a global level, people are leaving rural areas in LEDCs and heading for the cities. In contrast, in many MEDCs people are leaving urban areas for rural areas. Yet within rural areas there is great variety – some areas are very successful whereas other settlements are declining. Why should this be so? And what are the implications for people who live in these settlements and the planners who try to find solutions for these problems settlements. Can rural areas survive – can they be forms of sustainable development?

Questions for investigation

- What are the characteristics of rural areas?

- What are the social and economic issues associated with rural change?

- What are the environmental issues associated with rural change?

- How can rural areas be managed to ensure sustainability?

Consider this

Rural areas are changing quickly.

Many areas no longer have farming and food production as their main economic activity. Is this a good thing? What are the implications for people around the world if rural areas change and no longer provide the food that we have been used to receiving?

6.1 What are the characteristics of rural areas?

A rural settlement is a place where people live and carry out a variety of activities, such as trade, agriculture and manufacturing. Most but not all rural settlements are hamlets and villages.

> **Key term**
>
> **Rural settlement**: in the UK, this is defined as a settlement characterised by a small population, low population densities and primary industries.

Settlement patterns/morphology

There are two main aspects of pattern that can be examined. First, there is the pattern of a group of settlements in a given area. Are there any overall patterns – for example, are they located evenly or unevenly, are they random or clustered? Are the settlements located in lowland or upland areas, in lines such as along river valleys? Then, there is the morphology (shape or pattern) of the individual settlement. Is it linear or cruciform? Is it dispersed or nucleated? Dispersed settlement is more common in areas of low population density whereas nucleated settlements have often developed around a resource such as a spring or a defensive site.

Settlement hierarchy

A settlement hierarchy is the organisation and structure of settlement based on size and the number of functions of a settlement. At the top of the hierarchy are cities and conurbations. At the base are dispersed individual farmsteads (Figure 6.1).

Hamlets

At the next level are hamlets. This is a small collection of farms and houses, but generally lacks all but the most basic services and facilities. The trade generated by the population, which is often less than 100 people, will only support low order services such as a general store, a sub post-office or a pub. Low order services generally have a small threshold population and serve a small catchment area.

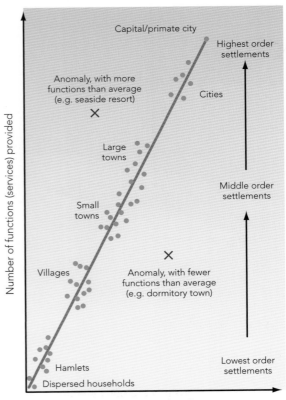

Figure 6.1 Settlement hierarchy ▲

> **Key terms**
>
> **Threshold population**: the minimum number of people necessary before a particular good or service will be provided in an area.
>
> **Catchment area**: the area from which people travel to obtain a particular service or product, such as the area from which a school draws its pupils. Catchment areas are also known as spheres of influence, hinterlands or urban fields.

Villages

By contrast, a village is much larger in population. Hence it can support a wider range of services, including school, church or chapel, community centre and a small range of shops (Figure 6.2).

Small market town

Small market towns are characterised by larger populations than villages and by a greater variety of high and low order goods. Their catchment area is bigger than that of hamlets and villages and people will range much further to obtain the goods and services that they offer.

Small market town	General store, post office, pub, butcher, garage, grocer, hardwear, primary school, baker, bike shop, chemist, confectionery, dry cleaner, electrical, TV/radio, furniture, hairdresser, laundrette, local government offices, off-licence, photo shop, restaurant, shoe shop, solicitor, supermarket, undertaker
Village	General store, post office, pub, butcher, garage, grocer, hardwear, primary school
Hamlet	General store, post office, pub

Figure 6.2 A simple rural hierarchy

Key terms

High order goods: comparison goods such as electrical goods and furniture that the shopper will buy only after making a comparison between various models and different shops. A high threshold population is needed to sustain a shop selling comparison goods, and people are prepared to travel some distance (range) to obtain the goods.

Low order goods: goods that are purchased frequently, such as milk or bread. People are not prepared to travel far to buy a convenience good and there is no real saving in shopping around. The extra cost of 'shopping around' outweighs any savings that may be made.

Range: the distance that people are prepared to travel to obtain a good or service.

Rural settlement functions

Rural settlements offer certain functions and services. Only basic or low order functions are found in smaller hamlets whereas the same functions and services are found in larger settlements (villages and market towns) together with more specialised ones – high order functions. High order services require a greater threshold population to support them, and also draw their custom from a much larger sphere of influence than low order services. The market towns will draw custom from the surrounding village and hamlets as well as serving their own population. The definition of hamlet, village and town is not always very clear-cut and these terms represent features which are part of a sliding-scale (continuum) rather than separate categories.

Processes that operate in rural areas

There are many processes that operate in rural areas. These include population growth and decline, the growth and decline of settlements, and changes in the rural economy. In addition, there are impacts on rural society and the physical environment.

In some rural areas, especially those close to large urban areas, there is in-migration of population and growth of settlements. This growth is called counter-urbanisation, that is, the growth of smaller settlements. Many residents, however, may live in the countryside but continue to work in the cities and towns. Thus, commuter or dormitory settlements have developed. In the areas of growth there is also much social change – the composition of the population changes in terms of age, occupation and class. Moreover, the growth of settlement and transport infrastructure can have a negative impact on the physical environment.

In contrast, some remote rural areas are experiencing decline of economic opportunity and of population. This affects the age-structure of the area, and the attractiveness of the region for further investment. As people move out, a cycle of deprivation and decline may develop.

Activities

1   What is meant by the terms a) hierarchy; b) low order goods; and c) high order goods?

2   What is the relationship between high order goods and a) range and b) threshold?

Factors affecting rural functions and opportunities

The function of rural areas varies spatially and temporally. Rural areas that are close to large urban areas have greater opportunities for economic activity. There is a larger market for the goods produced and there is a better transport infrastructure. In contrast, remote isolated rural areas offer fewer opportunities, partly because of poor accessibility and partly because of a smaller market or population. Opportunities include those in farming (for example, lowland areas close to urban markets have the opportunity to offer Pick Your Owns or diversify into recreational activities). In contrast, upland farmers have fewer alternatives. There are also opportunities for mining, forestry, economic activity (lowland rural areas generally have better access and a larger potential workforce), recreation and housing. Accessible areas have more potential.

Functions may change over time; farming is now much less important in terms of employment compared with 50 years ago, and most rural residents now work in services in a nearby urban area. As functions change, so too does the nature of rural settlements. Many rural settlements are now far more important for residential purposes than for farming, and in many rural settlements, modern industrial estates or office parks mean that industry may be as important, or more so, than farming as a form of employment.

Case study | Settlement functions and opportunities in West Oxfordshire

Rural settlements in West and North Oxfordshire show great variety within a small area. This reflects differences in location, physical, socio-economic and cultural factors. The settlements offer a range of opportunities. For example:

- Yarnton is close to Oxford and is a dormitory settlement
- Begbroke has a residential function but also contains a large high-tech firm, drawing workers from the surrounding region
- Bladon has tourist function (being the burial place of Sir Winston Churchill) and used to contain many workers' cottages on the Blenheim Estate: many of these are now rented
- Woodstock is also an important tourist location: Blenheim Palace and Estate is located at Woodstock; the village is an important regional centre containing a secondary school, doctors' surgery, chemist and other higher order functions
- settlements such as Long Hanborough and Combe contain railway stations and therefore have an important commuter function.

Key term

Dormitory settlement: a settlement in which people live (sleep) but do not work. It is also called a commuter settlement.

The importance of the Duke of Marlborough as a landowner and an employer in the region cannot be underestimated, for example, a number of settlements such as Combe and Bladon are found just outside the Estate and provide homes for workers on the estate. It is a good example of how a cultural factor (hereditary ownership of land) has affected settlement development. In most of the villages described, farming is no longer an important factor. The farmers generally live on their farms, away from settlements.

Hierarchy and functions

The West Oxfordshire Settlement Sustainability Report categorised the 120 settlements within West Oxfordshire according to their size, character and role, in particular the number of services and facilities they offered (Figures 6.3 and 6.4). This created four distinct groups.

- Group C – service centres (nine towns and villages)
- Group B – 12 medium-sized villages
- Group A – 21 smaller villages
- Group D – small villages, hamlets and open countryside – about half of the total number of settlements are not included within Groups C, B

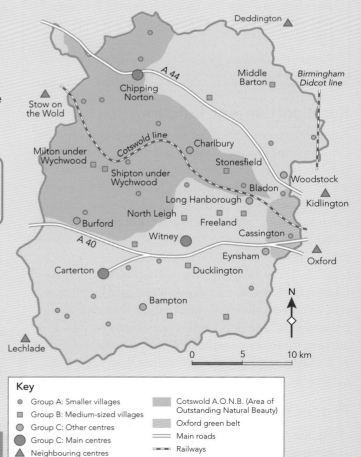

Figure 6.3 Settlement hierarchy and location in West Oxfordshire

Key
- Group A: Smaller villages
- Group B: Medium-sized villages
- Group C: Other centres
- Group C: Main centres
- Neighbouring centres
- Cotswold A.O.N.B. (Area of Outstanding Natural Beauty)
- Oxford green belt
- Main roads
- Railways

SETTLEMENT CATEGORIES	RESTRAINT INDICATORS							POSITIVE INDICATORS															
	Green Belt	AONB	Conservation Area	<500 people (2001 Census)	> 8km from a main service centre	No direct access to principal road	No shops open on daily basis	>1000 people (2001 census)	<4km from a main service centre	Post Office	Shops*	Other non-food shops	Primary school	Secondary school	Community building	Public house/hotel	Place of worship	Library, doctor's, fire station, police station	Playing fields	Built sports facilities, available for public use	Petrol filling station	Range of employment opportunities	Access to principal road	Good bus service
GROUP A: Villages	N1	N2	N4	N5	N6	N7	N8	P1	P3	P4	P5	P6	P7	P8	P9	P10	P11	P12	P16	P17	P18	P19	P20	P22
Ascott-under-Wychwood																								
Cassington																								
Stanton Harcourt & Sutton																								
GROUP B: Medium-sized Villages																								
Ducklington																								
Minster Lovell (S. of B4047)											3													
Shipton-under-Wychwood																								
GROUP C: Service centres																								
Witney											3													
Burford											3													
Woodstock											3													

* One/two food shops (including Post Office where applicable) or three or more food shops
(3) Three or more shops

Figure 6.4 Example of settlement functions in West Oxfordshire ▲

and A because of their size, lack of facilities and individual characteristics.

The settlements were classified using a set of 30 indicators. These included:

◆ 'positive factors' which show the availability of services and facilities such as schools, doctors, shops and public transport

◆ 'restraint factors' which show a lack of local facilities, significant distance from local service centres and locations within an area of constraint such as the Cotswolds Areas of Outstanding Natural Beauty (AONB) or the Oxford Green Belt.

Parish Survey

The 30 indicators were reviewed through the use of Parish Surveys, which have been undertaken by the District Council every three years since the mid-1970s and through monitoring of planning applications. The most recent survey was undertaken at the beginning of 2007.

Profile of the Service Centres – Group C

There are nine towns and villages in this category. It contains the District's key service centres of Witney (pop. 25 000), Carterton

(pop. 14 000) and Chipping Norton (pop. 6000) and other settlements which act as local service centres, such as Woodstock (Figure 6.5).

Services and facilities

Each place has a wide range of services, facilities and employment opportunities. Many are linked to Oxford by relatively good bus services. Most have experienced significant amounts of house building, with associated population growth.

Factors influencing the development of service centres

Most of the service centres have developed as a result of good accessibility, a good economy (the

▼ **Figure 6.5** Woodstock – a local service centre

wool and blanket industry in Witney goes back to the Middle Ages and has only recently collapsed), as well as political and environmental factors. Economic factors, such as accessibility and job opportunities are important in the recent growth of settlements in the area. In contrast, political factors, such as land ownership have influenced the distribution and location of settlement. The major land holding of Blenheim Palace and Estate has helped the growth of Woodstock as a tourist location. Environmental factors, such as potential for flooding, affect location of settlements – see page 31. The floodplains of the main rivers in the area – the Thames, Evenlode and Windrush – have restricted where settlements have grown and how their expansion has been influenced. Social factors, such as the availability of a skilled workforce, are important in attracting industry to the area, which in turn encourages more skilled workers to the area.

Profile of medium-sized villages – Group B

There are 12 villages within Group B. With populations over 1000, Group B villages have larger populations than Group A. Many of the villages are located within environmentally sensitive areas.

Services and facilities

In terms of facilities and services, all villages in Group B have a school, community hall and at least one pub. Many have experienced significant house building over the previous decades, but all experienced the closure of shops and the withdrawal of other services and facilities since the last Parish Survey was carried out in 2003.

Since 2003, Freeland has lost its post office and village shop. Ducklington and Freeland are the only villages in Group B not to have a shop open on a daily basis. Other closures include petrol filling stations in Freeland, Middle Barton, North Leigh and Stonesfield (Figure 6.6). Only six villages in Group B now have their own petrol stations. Stonesfield and Milton under Wychwood have both lost full-time GP surgeries.

Since 2003, five villages in Group B have seen an improvement in bus services. All villages now have a daily bus service. Ducklington, Middle Barton and Milton under Wychwood are the only villages in this group that do not have a bus service which is more frequent than one an hour. Shipton under Wychwood has seen the closure of the local residential home for older people. However, North Leigh now has a new village hall offering facilities for community use, including a library and youth centre.

Factors influencing the development of medium-sized villages – Group B

Group B villages generally offer few job opportunities and are located in areas not as accessible as Group C centres. They are also located away from the C centres. Socially they are more likely to have an older population and have fewer new housing estates. Consequently, the environmental impact on these settlements has been less than in larger settlements. In a similar way to the larger settlements, the growth and location of these settlements is affected by important landowners such as the Duke of Marlborough at Blenheim, and the floodplains of the main rivers.

Profile of the smaller villages – Group A

There are 21 villages within Group A of varying characteristics. All villages have populations less than 1000, with some less than 500. Two villages, Bladon and Cassington, are located within the Oxford Green Belt (Figure 6.7), nine villages are located within the Cotswolds Area of Outstanding Natural Beauty and all but six villages include a Conservation Area. Due to their location and limited range of local facilities provided within Group A villages, they are not viewed as suitable locations for significant new development.

Services and facilities

The number of local facilities varies within each village in Group A. Each village has a place of worship. Just over 60 per cent of the Group have a post office. Most have a school and all but six have

Figure 6.6 Shops in Stonesfield – a category B settlement

Figure 6.7 Empty shop premises in Bladon – a category A settlement

at least one shop. However, none have a GP surgery or library.

Since the Parish Survey was carried out in 2003, there have been a number of changes within the Group A villages. While some have seen an improvement in facilities and services, others have lost them. A replacement village hall has opened in Cassington. Cassington has also gained a post office, bringing the number of Group A villages with a post office to 13.

Only four villages in Group A do not have playing fields available for public use. Since 2003, Bladon and Over Norton no longer have such facilities available (the playing fields are privately owned).

There has been a significant improvement to the local bus service with Group A villages. Since 2003, over three-quarters of Group A villages have seen a more regular bus service. This is primarily due to the introduction of community bus services for the elderly and people with limited mobility.

Factors influencing the development of smaller villages – Group A

The smallest settlements offer few job opportunities. They are, in effect, dormitory settlements. Many of the settlements, such as Cassington, come under the sphere of influence of Oxford and Witney for employment. Expansion of the settlement is limited by the River Evenlode and its floodplain. Cassington has a relatively small young population, due to the lack of job and social opportunities, but an above average number of adults, due to it being a good environment in which to raise a family.

Opportunities within rural communities in West Oxfordshire

West Oxfordshire offers many opportunities in its rural communities. Although farming does not employ many people, it is still important for food production and an important factor in environmental issues. Many rural settlements, such as Woodstock, Long Hanborough and Begbroke have industrial estates and science parks. In addition, there are opportunities with respect to recreation and tourism (e.g. Blenheim Palace and Estate), as well as in quarrying and services.

Activities

1 Study Figure 6.3. This shows the number and location of Group C (other centres), medium-sized villages, and small-sized settlements. Comment on the number and relative distribution of each of the three types of settlement.

2 Suggest reasons for the location of the 'main centres'.

3 In what ways have the main transport routes affected the location of settlements in the region?

4 Study Figure 6.4. Choose any two of the 'restraint factors' and suggest why they limit the growth of a settlement.

5 Study the 'positive indicators'. Select and justify one factor which is a good indicator of settlement hierarchy, and one which is not a good indicator of settlement importance.

 'Take it further' activity 6.1 on CD-ROM

Discussion point

'Very small rural areas are doomed!' To what extent do you agree?

Case study | Rural settlement in the Eastern Cape of South Africa

Functions

Welcomewood is a rural resettlement camp of c. 2000 people. Housing quality is mixed; plot sizes permit some subsistence agriculture; the road network and quality are poor; water is provided by means of a tap along the roadside and there is no provision of either electricity or refuse collection. Local sources of employment are very limited. Welcomewood was resettled with Africans from Upington (a few hundred miles away) during the 1970s. Yet over 30 years later there is still social differentiation between the original population, who speak mostly Xhosa, and the relocated population, whose primary language is Afrikaans.

Peelton is located c. 12km north-east of King William's Town (Figure 6.8) it is a largely pastoral area, owing to its rather limited water supplies. Again, like Welcomewood, it lacks a viable agricultural basis. Some small-holdings are worked but the majority of the population look to secondary and tertiary activities for a livelihood. The population size is estimated at c. 5000 and the quality of the housing varies from mud and sticks to more elaborate stone buildings. Water has to be collected from nearby wells and electricity is not provided. Local sources of employment are limited to the schools and clinics. Unemployment and poverty are widespread.

> **Key terms**
>
> **Resettlement**: a process carried out by the former South African government in which certain population groups (notably black and coloured people) were forcibly removed from their homes and relocated elsewhere – often hundreds of kilometres away.
>
> **Xhosa:** the name given to one of South Africa's most populous black tribes and to their language.

> **Key terms**
>
> **Secondary activities**: manufacturing industries which convert raw materials into finished and semi-finished goods.
>
> **Tertiary activities:** services which include educational, health, financial, retail and all activities that are not manufacturing, primary (fishing, forestry, farming and mining) and quaternary (research and development).

Figure 6.8 Location of Peelton and Welcomewood

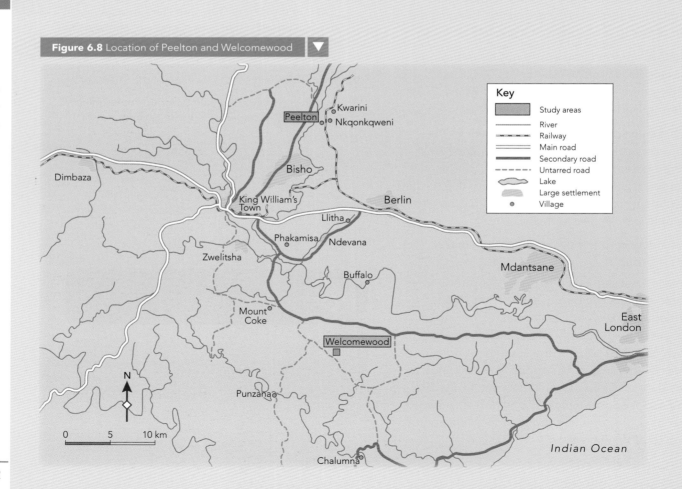

Key
- Study areas
- River
- Railway
- Main road
- Secondary road
- Untarred road
- Lake
- Large settlement
- Village

Opportunities

The Eastern Cape is the poorest region in South Africa. It does not have mineral resources such as gold or diamonds, but it has limited amounts of fertile soil and water. Farming is one of the main opportunities in the region. In addition, there is a large manufacturing sector in East London, and many of the rural areas provide workers for this sector. Increasingly, tourism is being developed as an economic sector.

Use of vegetation

Natural vegetation is utilised in a number of ways. The most frequent uses are building material for fences, fuel and forage. Results from a survey in Welcomewood suggest that the collection of woods from woodlots near Mount Coke, c. 6km away, was quite frequent.

Fuel

Most householders in Welcomewood use wood as a fuel compared with only 13 per cent in Peelton. In Peelton the collection of wood was important, in the households with lower income. Some were even able to make money through the sale of wood. Mainly mothers and children travelled up to 3km to collect wood. The importance of wild vegetation as forage can be gauged by the potentially hazardous presence of large numbers of cattle, sheep and goats grazing on the roadsides.

Factors affecting development

There are a number of factors affecting development. Accessibility is one – the region is remote and isolated from the main centres of economic activity in South Africa. Economic structure is another. The region has an important manufacturing base in areas such as East London and Port Elizabeth. Farming is the most important activity in the surrounding rural areas. The workforce is generally unskilled – out migration of the younger, skilled, innovative workforce creates a skills shortage in the region.

Farming in the region

Cultivation is primarily a subsistence activity. A number of factors combine to make agricultural productivity low and variable. Both Welcomewood and Peelton have one borehole each, providing a limited supply of water, some of which is used for irrigation. Fields close to the river in Peelton are used for arable agriculture, although this land is privately held. In Welcomewood, situated on the top of an exposed ridge (Figures 6.9 and 6.10), the water situation is more serious given the higher evapotranspiration rates associated with exposed windy places.

Factors affecting agricultural development

Rainfall

In the Eastern Cape region, the low and variable rainfall adds to the problem of diminishing land per capita, caused by resettlement and population growth.

> **Key term**
>
> **Per capita:** per head of population.

Soils

Soils vary a great deal within rural areas. In Peelton, owing to the presence of alluvial soils, a large proportion of the land near the river is suitable for cropping, whereas in Welcomewood, the quality is much poorer and only a limited amount can be used for cropping. Instead, most is used for livestock.

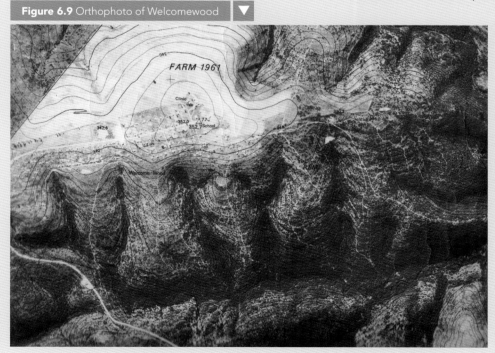

Figure 6.9 Orthophoto of Welcomewood

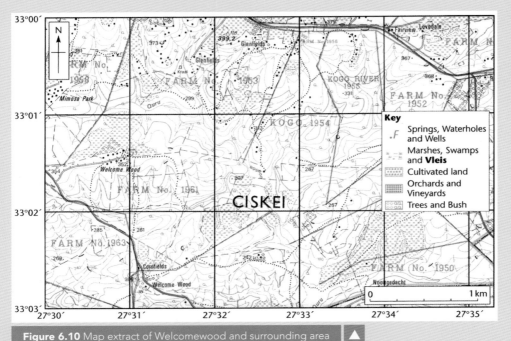

Figure 6.10 Map extract of Welcomewood and surrounding area ▲

Access to the land

Output is also affected by access to land. This depends, in part, on the tenure system, sharing and renting. Many of the Welcomewood residents have land holdings (34 per cent), although these are mostly quite small, 1.7 hectares, the amount given to them on resettlement. Land-use analysis shows that 77 per cent produce vegetables, 23 per cent maize, and 27 per cent graze animals on communal ground. By contrast, land holdings in Peelton vary a great deal. The houses are mostly on 1 hectare but go up to as much as 6 hectares. Use of holdings varies also, with 87 per cent using them for growing vegetables, 10 per cent for maize and 3 per cent not making any use of them at all. A further 15 per cent make use of communal land.

> **Key term**
>
> **Tenure:** the rights that people have on the land; for example, ownership or rental.

Other constraints

The ability to produce is affected by many other constraints such as the high cost of water, high cost of seeds and the lack of fences to deter domestic animals from eating crops. The cost of developing a vegetable garden is relatively high, Rand 600–700, depending upon the area to be fenced. Irrigation could be useful, although given the lack of water and the long distances between many of the plots and the boreholes, this was not a common practice. Likewise, the use of manure does not appear to be either regular or widespread.

Shortage of labour

Shortage of labour also affects agricultural production. The large number of female-headed households, migrants and commuters in the two areas is characteristic of low productivity. In fact, some women claimed that the absence of males had caused the household to abandon agriculture altogether.

Thus the picture of farming that emerges is one in which the land is unable to support the rural population. Nevertheless much of the land remains unused, a result of many factors. These include the practice of fallow, the absence of labour, low yields, high risk of operations, theft by other villagers, lack of land rights, access to land and the inability to afford the necessary inputs and equipment. Those who do engage in farming do so mostly as a means of supplementing food supplies, as the size of holdings are not large enough to support viable economic units. Farming is largely a secondary activity to supplement other jobs. However, farms are too small to support a decent herd of animals.

Unemployment

While seasonal unemployment has long been considered a feature of the rural economy, in Peelton permanent unemployment is more frequent. The Welcomewood population would appear to be more vulnerable than the Peelton population given its remoteness and lack of alternative opportunities. Its distance from sources of employment and history as a resettled community mean that there are probably fewer contacts with people with whom it may be possible to secure employment. Peelton, on the other hand, has a regular bus service to Bisho, and from there to King William's Town and Zwelitsha, making the search for employment easier.

The rural areas do not provide a living for most of their inhabitants, at least not from agriculture. There are other sources of employment, such as in the service sector, notably clinics, schools, and police departments, although these provide only about 20

Figure 6.11 Welcomewood ▲

jobs in both Peelton and Welcomewood. There are also shops, one in each village, and an informal sector.

> **Key term**
>
> **Informal sector:** the untaxed, unregulated sector of the economy – often called the 'black market' in the UK.

Services

Water

Local services are lacking. In Welcomewood, one water tap provides the whole of the village with its water

supply. None of the houses have piped water. Carrying a 20kg container of water is physically demanding, taking up a large number of hours daily and will effect the women and children who were mostly responsible for its collection.

Similarly, in Peelton, residents on the Nkqonqweni and Kwarini sites have to travel distances of 1km and 3km respectively to reach the borehole. The local topography, valley slopes of, on average, about 5°, does not make the collection any easier. Indeed, the return journey with the container full is uphill. In some of the settlements on the edge of the Welcomewood area households collect water from the dam. The risk of contamination from animals is high and schistosomiasis (bilharzia) has been reported at the local clinic.

> **Key term**
>
> **Bilharzia:** also known as schistosomiasis – an infectious disease caused by penetration of the schistosome worm into humans, via the feet in stagnant water, and their subsequent migration to growth in the human bladder.

Banks and post office facilities

The provision of services in the rural areas is limited. Neither village has a bank or a post office. The general procedure for people posting or collecting mail (or indeed receiving their pension) is via the local shopkeeper who sorted post on visits to Bisho, King William's Town or Zwelitsha.

Recreational facilities

Recreation spaces are few. Welcomewood has a field with a goalpost in it, whereas Peelton has nothing. School fields were not used at all for any sports.

Figure 6.12 Housing quality in Welcomewood

Figure 6.13 Peelton environment

Figure 6.14 Housing quality in Peelton

'Take it further' activity 6.2 on CD-ROM

Activities

1 Study Figure 6.9.

  a What does the orthophoto suggest about the relief of the area?

  b How might the relief of the area affect the potential for farming?

2 Suggest ways in which human factors have affected the development of the rural environment in Welcomewood.

Theory into practice

For a rural area you have visited and studied, describe its site and situation, form (shape) and function.

6.2 What are the social and economic issues associated with rural change?

Recent changes to rural settlements since the 1970s in the UK, have resulted in depopulation of remote rural areas, fewer farms, changing agricultural land use, an increase in population in more accessible rural areas, counterurbanisation, and therefore an increase in long-distance commuting. In addition, new roads, airports, housing schemes and theme parks have been constructed.

Key term

Counterurbanisation: an increase in the population of smaller settlements such as villages and towns. It involves migration from larger urban settlements to smaller ones.

Key factors leading to growth or decline in rural areas

Improvements in transport

One of the greatest changes has been brought about by improvements in transport, both public and private. There is a definite relationship between the type and rate of change that is occurring in rural settlements and distance from large urban areas (Figure 6.15). The most accessible villages have grown the most. Many villages have grown at alarming rates and have lost their original character, form and function (Figure 6.16). These are often described as dormitory, commuter or suburbanised villages (Figure 6.17). Figure 6.17 shows the development of the suburbanised village. Growth occurs initially around the original village core. Much of the growth is along main roads (the ribbon development) but there is also some infilling. Some farms and buildings that were outside the original settlement (isolates) become incorporated into the main village. At a later stage, large-scale infilling causes the

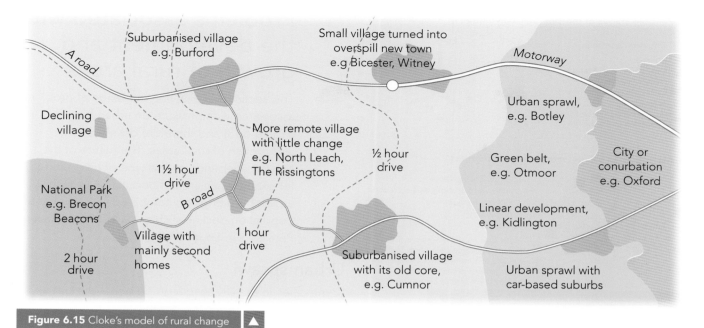

Figure 6.15 Cloke's model of rural change ▲

village to appear far more compact and nucleated than at first.

Characteristics	Original village	Suburbanised village
Housing	Detached, stone built houses with slate/thatch roofs; some farms, most over 100 years old; barns	New, mainly detached or semis; renovated barns or cottages; expensive planned estates, garages
Inhabitants	Farming and primary jobs; labouring or manual jobs	Professionals/executives; commuters; wealthy with families or retired
Transport	Bus service; some cars; narrow winding roads	Decline in bus services as most families have one or two cars; better roads
Services	Village shop, small junior school, public house, village hall	More shops, enlarged school, modern public houses, and/or restaurant
Social	Small, close-knit community	Local community swamped; village may be deserted by day
Environment	Quiet, relatively pollution-free and open space	More noise and risk of more pollution; loss of farmland

Figure 6.16 Changes in the suburbanised village ▲

Key terms

Dormitory settlement: a rural settlement that has a high proportion of commuters.

Commuter settlement: a settlement in which people live but don't work.

Suburbanised village: a settlement where there has been recent growth changing its character from primarily farming orientated to residential with the majority working in nearby towns.

Increased standards of living

As people become better off and have access to one or more cars, they are able to live further away from their place of work. If they believe that the quality of life is better in small settlements, and that they can afford a bigger house, they may be tempted to move.

Decreased size of households

The traditional nuclear family – with grandparents, parents and children – all living in the same household, has all but disappeared. Instead, there are more people living alone – whether due to lower death rates, increased rate of divorce, or children leaving the parental home to live in their own home – which means that there is a much greater demand for housing, and there are more homes with fewer people in them. Consequently, there has been an increased demand for housing in rural settlements as well as in urban areas.

Growing dissatisfaction with urban lifestyles

Many large urban areas are considered to be unsafe, polluted, expensive, unfriendly and a poor place to

(a) Possible stages of morphological evolution of a suburbanised village

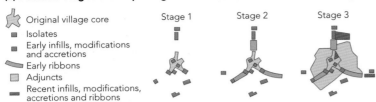

Stage 1   Stage 2   Stage 3

- ✦ Original village core
- ■ Isolates
- ■ Early infills, modifications and accretions
- ⌐ Early ribbons
- ☐ Adjuncts
- ▬ Recent infills, modifications, accretions and ribbons

(b) Metropolitan village; morphological features

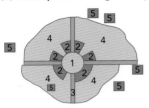

1. Original village core
2. Infills, modifications and accretions
3. Ribbon development
4. Adjuncts
5. Isolates

Note: This model diagram indiactes all the morphological elements likely to be present in a metropolitan village. The arrangement of these elements is likely to vary considerably between villages.

Figure 6.17 A model of suburbanised villages ▲

bring up a family. In contrast, rural areas are often perceived to be the opposite – clean, safe and a good environment to raise a family. Whether either is true is debateable. Nevertheless, the belief (perception) exists and it causes some people to move from larger urban areas to smaller rural settlements.

Increase in car ownership

Increased levels of car ownership allow people and families to live further away from their place of work. Commuting is increasingly easy as transport developments improve.

Improving technology

With increased technology, such as the Internet, it is possible for some workers to work from home. This is called telecommuting. Work is sent electronically to the company office.

Problems associated with the development and growth of rural areas

This section will examine the range of problems associated with pressure on urban fringes (Figure 6.18) and therefore development and growth of rural areas. The nature of the pressure depends on the type of urban fringe, for example, an area of population growth such as southern England compared with an area of population decline such as parts of the Gower Peninsula in South Wales.

Urban sprawl

Some cities in the UK have a green belt – but not all. One of the functions of the green belt is to prevent urban sprawl. In areas that lack a green belt, such as Swindon, the potential for urban sprawl is much greater since there is not the planning legislation to prevent it.

The need for more housing

There are great pressures on certain rural areas for housing developments. There are also pressures for economic and transport developments. The increase in the demand for housing occurs throughout the country. Between 1991 and 2001, a further 460 000 new homes were needed.

Increased demand for housing is generated by: longer life expectancy; young people leaving home earlier; more families splitting up and moving into separate homes; and more people preferring modern houses with good facilities.

Large urban areas are no longer desirable because they are seen as expensive, polluted, and unsafe. By contrast, in a rural area or a small town, property costs

Figure 6.18 Issues in the rural–urban fringe ▼

Greenfield sites		Brownfield sites	
Advantages	*Disadvantages*	*Advantages*	*Disadvantages*
Land may be accessible	Habitat destruction	Redevelopment of derelict land	Land may be contaminated
Cheaper land	Reduction in biodiversity	Does not harm the environment	Widespread air and water pollution
People prefer more space and pleasant environments	Increased pollution	Creates jobs locally	Congestion
Allows planners more freedom	Increased impermeability leads to flooding	Provides a boost to local economies	Overcrowding
Easier to plan for infrastructural developments	Increased traffic on the road, and cross city commuting	May use existing infrastructure	Land is expensive

less and larger plots of land are available. Much of the new housing is being built in new towns, overspill towns and small towns.

Key terms

Greenfield site: an undeveloped site usually on the edge of an urban area or in a rural area.

Brownfield site: a derelict inner-city site, usually the site of former industrial activity.

Where should new homes be built?

The location of West Oxfordshire and its high environmental quality has helped sustain a high demand for housing. Since 1981 West Oxfordshire's population has grown by 25 per cent.

The present housing stock in West Oxfordshire is estimated to be just under 40 000, of which 14 000 are in Witney and Carterton. The West Oxfordshire local government's objective in relation to housing is to:

◆ identify sufficient sources of new housing

◆ locate new housing (Figure 6.19) where it will have the least adverse impact on the character and resources of West Oxfordshire

◆ seek a range of new residential accommodation which provides a variety of sizes, types and affordability.

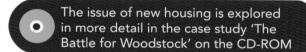

The issue of new housing is explored in more detail in the case study 'The Battle for Woodstock' on the CD-ROM

Figure 6.19 New housing development in West Oxfordshire ▼

Industrial growth and urban services

Some services such as reservoirs or cemeteries located in the urban fringe, may be attractive as visitor destinations whereas mineral workings, sewage works and landfill sites can be unattractive and polluting.

Recreational pressures for golf courses and sports stadia

Some recreational activities in the urban fringe like stock-car racing and scrambling can erode ecosystems and create localised litter and pollution. However, country parks, sports fields and golf courses can lead to conservation. The golf course at South Hinksey in the urban fringe of Oxford is a good example. It keeps a large area of land 'semi-natural' with trees, grassland and shrubs, providing a range of habitats and therefore helping maintain a varied biodiversity. On the other hand, there are many that claim that golf courses waste a large amount of water in the watering of greens and destroy natural habitats. Country parks provide a more direct way of conserving the environment – for example, Shotover Country Park in east Oxford.

Transport and infrastructure development

Motorways destroy countryside that they pass through where a lot of land is used for slip roads and roundabouts, especially near junctions. For example, the development of the M40 motorway to the east of Oxford. Often the building of major new roads through

environmentally sensitive areas meets with much local and national opposition, for example, Twyford Down near Winchester. Some transport infrastructure can have a positive impact, for example, cycle ways can improve access and promote new development.

Examples of new developments linked to motorway development include the service station near Junction 8 of the M40 (near Thame) and Bicester Village Designer Retail Outlet, which has good motorway access.

Agricultural developments

Whilst there are many well-managed farms and smallholdings in the urban fringe, they can suffer from litter, trespass and vandalism; some land lies derelict in hope of planning permissions for development. Pear Tree farm to the north of Oxford sold some of its land to make way for a park and ride scheme (Figure 6.20).

Service development

There are some well-sited, well-landscaped rural developments but there are incidences of other development, for example, some out of town shopping areas and some unregulated businesses like scrap metal and caravan storage, which can be unsightly and detrimental to the environment.

Case study | Rural change around Oxford

Factors leading to the growth of rural areas around Oxford

There is little doubt that Oxford's green belt is in crisis. Superstores and science parks used to be seen as the biggest threat but this has now changed to housing, roads, golf courses and recreation facilities. Recent developments on Oxford's green belt include an all-weather athletics tracks at Horspath, and a sports fields at Kidlington (Stratfield Brake). Oxford United's new sports ground is located on green belt land close to Blackbird Leys, and large areas of the Cherwell Valley

▼ **Figure 6.20** Issues in the rural–urban fringe

are seen as particularly vulnerable to the expansion of college sports grounds. Park-and-ride schemes were intended to reduce traffic problems in the city centre. However, they add to the pressure on rural areas around the city.

Developments

Examples of rural land that has been developed.

◆ The Pear Tree park-and-ride and service station covers 30 acres of green belt (Figure 6.20).

◆ There is a substantial housing development at Blackbird Leys.

◆ A golf course at Hinksey Hill opened in the mid-1990s.

◆ The University has built a world-class business centre The Oxford Business School.

◆ The University has developed a nano-technology centre at Begbroke.

◆ There is a business park at Long Hanborough.

Any or all of these developments in the urban-fringe lead to: increased road congestion causing noise and air pollution; building development on land causing environmental damage to natural habitats; and change and conflict in rural communities as villages become subsumed by the encroaching urban area.

Oxford Science Park

The Oxford Science Park is an on-going development within the green belt. It is a 30-hectare site located on the edge of Oxford near Kassam Stadium. It has been developed as a partnership between Magdalene College, Oxford and the Prudential Assurance Company Limited. During the construction phase, the use of heavy machinery and the removal of vegetation has meant that soil erosion is a particular problem in the area.

Industrial development

In addition to the Science Park, there are a number of new developments. Oxford University have created a new science park at Begbroke. Similarly Unipart built a new factory at Horspath. The crucial factor for the local residents is that the factory is just 230m from the nearest house and is 14m high. Villagers claim that a process of 'step-by-step industrialisation' has destroyed the character of Horspath. This has angered local residents and others who see it as another example of green belt land being used for development purposes.

Key term

Industrialisation: the growth of manufacturing industry in an area.

Development of recreational facilities

Oxford United's stadium has been developed on a rural area on the edge of Oxford. The new stadium has nearly 2000 parking spaces and spaces for 15 coaches. Other leisure facilities include a hotel, cinema, bowls complex, and leisure pool. These will be built on neighbouring land. In addition, as part of the planning agreement, Oxford United have to contribute to:

◆ the upgrading of the A4074 Sandford interchange

◆ parking controls in Littlemore and Blackbird Leys

◆ a new all-weather pitch near the new stadium.

Development of transport services/ infrastructure

A fifth park-and-ride service into Oxford was built near Kidlington at Water Eaton in 2005. It has spaces for 400 cars and was built on land adjacent to a derelict grain silo.

Economic and social problems arising from development

Development can also lead to a range of problems. For example, the increased demand for housing sends house prices up. This results in a lack of affordable housing for young people. In addition, new housing is generally targeted at the wealthy. There is also a change in employment structure. There is a decline in primary activities (farming, forestry, mining and fishing) as well as manufacturing, and an increase in services. Many of the service jobs are skilled jobs, and this results in fewer, less well-paid jobs for some local people. This can lead to social differences and problems between the original residents and the newcomers.

Activities

1 Explain why there is pressure to develop the rural–urban fringe.

2 Suggest how, and why, some of the pressures on the rural–urban fringe have had negative impacts.

Decline in rural services

The Rural Development Commission's 1991 survey of rural services throughout the UK revealed that:

◆ 39 per cent of parishes had no shop

◆ 40 per cent had no post office

◆ 51 per cent had no school

◆ 29 per cent had no village hall

◆ 73 per cent had no daily bus service

◆ fewer than 10 per cent had a bank or building society, a nursery or day-care centre, a dentist or a daily train service.

Where they existed, services tended to be concentrated in larger villages and market towns. Overall, the services which most people take for granted are few and far between in the countryside. With the lack of public transport, rural people are heavily reliant on cars to reach essential services, recreational or other facilities.

Population changes have led to most rural areas having above average numbers of elderly people and below average numbers of 16–20 year olds.

Loss of local shops

In 1991, only 61 per cent of settlements under 300 population had one or more weekly mobile shops. A significant minority of people – about a quarter – still depend on a local shop for their everyday needs, but the proportion varies widely in different areas.

Primary schools

Many rural primary schools closed in the 1960s and 1970s, especially in smaller villages of under 200 people. By 1991 only 40 per cent of parishes had a primary school and these usually had a population of over 500 people.

Healthcare services

By 1991, only 16 per cent of parishes had a permanent GP surgery (they are most common in parishes of more than 1000). Consultation rates for GPs are lower than those in towns, even for those villages with a surgery, and are lowest in remoter villages with no surgery.

Transport services

In 1991, only 27 per cent of parishes had a daily bus services and only 7 per cent had a rail service.

Although by necessity car ownership is higher in the countryside than in the towns, significant numbers of people, especially amongst the elderly and women, have no or limited use of a car. Norfolk is a very good example of where rural transport has declined and had a negative impact on settlements.

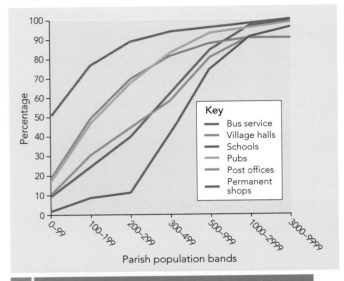

▲ **Figure 6.21** Population size and availability of services

Activities

1 Study Figure 6.21. What proportion of settlements with less than 100 people have access to: a) a bus service; b) a village school; and c) a post office. How does this compare with settlements which have 500–999 people and those with 1000–2500 people.

2 What appears to be the critical size of settlement to sustain a range of services? Support your answer with evidence.

Discussion point

Choose any two of the following groups: a) low income groups; b) the elderly; c) the infirm; d) people without access to a car; e) youth. What are the implications of declining rural services for the two groups that you have chosen?

 'Take it further activity' 6.3 on CD-ROM

Case study | Rural change in the Eastern Cape

Factors leading to rural change and decline

Forced population relocation

Apartheid

Forced population movements were a feature of the former government of South Africa, especially during the apartheid era (Figure 6.22). Forced resettlements generally referred to black (African) people who were removed from white-owned farms and villages to the so-called homelands. Black spot removals referred to African villages where the black residents were forced to leave and resettle in a homeland. Homeland consolidation referred to the redrawing of the political boundaries of the homelands so as to incorporate black communities and remove them from South African responsibility. Group area removals, in contrast, referred to population removals at an urban area, in which each population group was to live in racially segregated zones. Betterment planning, or 'rehabilitation', has given rise to the most widespread form of resettlement in South Africa. It differs from other forms of forced resettlement, in that it took place entirely within the former homelands.

> **Key terms**
>
> **Apartheid**: a form of separate development for different races, which allocated different areas at a national urban and local scale for different population groups. It was a form racial capitalism.
>
> **Homelands**: territory that was set aside for black inhabitants of South Africa as part of the policy of Apartheid.

Betterment

In the rural areas, betterment was a key process. Betterment has been officially portrayed as an attempt to combat erosion, conserve the environment and improve agricultural production in the homelands. However, betterment planning was unable to achieve these goals because it was unable to deal with the basic problem that these areas were already overcrowded. Betterment resulted in a rearrangement of the land-use system within many rural areas in the homelands, with the land being divided into newly proclaimed arable, residential and grazing areas. To enable this, people were moved into new residential areas. Betterment planning occurred in about 80 per cent of the former Ciskei homeland. As many as 375 villages in (former) Ciskei were dispossessed of land rights through the implementation of betterment. Betterment occurred very widely and involved significant loss of land. People generally found themselves further from resources such as wood, water and fields, and were significantly impoverished.

Although the Apartheid era is over, there are still areas of very high population densities in the former homeland areas. Much of the population is impoverished, socially as well as economically, and unable to migrate to the large urban areas of South Africa, where there are more jobs and the standards of living are better.

Poverty and unemployment

Poverty is not uniformly spread across South Africa. The standard of living for many black people in rural areas is very low. There is also a regional dimension to this problem. Regions such as the Western Cape and Gauteng, that are home to cities such as Cape Town and Johannesburg, are the economic core of the country and therefore relatively rich, while those that are rural in character exist on the margins and are poor.

It is well known that the Eastern Cape region is one of the poorest regions in South Africa. Nearly 70 per cent of people in the region lived in poverty in 2002. The situation has deteriorated rapidly since 1996, when 54.3 per cent of people were classed as poor. The Eastern Cape is currently on a sharply downward trend (Figure 6.23). Unemployment levels in the Eastern Cape are also higher than in any other province in the country.

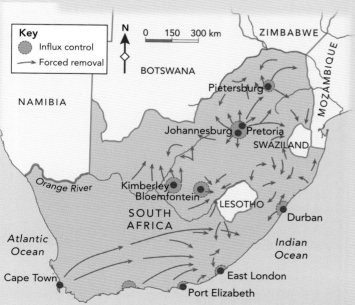

 Figure 6.22 Forced resettlement and the homelands of South Africa

	1999	2000	2001	2002
National – male	28.8	28.9	30.9	32.6
Eastern Cape – male	46.2	46.3	49.2	51.5
National – female	43.6	44.5	47.8	50.6
Eastern Cape – female	53.3	54.1	58.3	62.0

Source: Border Rural Committee, Annual Plan 2005

Figure 6.23 Unemployment change (per cent) in the Eastern Cape

Most of the wealth in the Eastern Cape occurs in the west, in and around the industrial city of Port Elizabeth. Most of the poverty occurs in the rural areas of the former homelands, Ciskei and Transkei. For example, between 1996 and 2001, the percentage of households in the nine Keiskammahoek villages that had no source of income more than doubled, from 17 per cent to 41 per cent. Keiskammahoek was part of the former Ciskei homeland. This compares with the total Amahlati average in 2001 of 18 per cent. Amahlati covers some areas that fell into former 'white' South Africa. In contemporary South Africa poverty is concentrated in the former homelands. In every village without exception, fewer people were employed in 2001 than in 1996. This points to serious economic decline in the area over the past decade.

The former homeland areas represent the greatest single development challenge facing South Africa. They are areas that have been excluded and marginalised from mainstream South Africa – economically, socially, environmentally and politically. If anything, the homelands are now more marginalised from South African society than they were in 1994. Economic activity has declined, unemployment has risen and poverty is even more endemic.

Social and economic problems associated with the decline of the homelands

Social problems

Standard of living

The Human Development Index (HDI) is a measure of development which includes life expectancy, literacy and income. It thus provides a composite index of development (Figure 6.24). Overall the HDI for the Amathole District suggests that there have been some improvements in the quality of life of residents within the district between the period 1996–2005 with increases recorded in all local municipalities. However

Human Development Index		
	1996	2005
Mbhashe	0.37	0.42
Mnquma	0.46	0.50
Great Kei	0.42	0.44
Amahlathi	0.46	0.50
Buffalo City	0.56	0.59
Ngqushwa	0.41	0.46
Nkonkobe	0.45	0.49
Nxuba	0.46	0.50

 Figure 6.24 Human Development Index for Amathole, 1996–2005

the number of people in poverty (Figure 6.25), that is, households where the total income of all members of the household falls below a particular level (Figure 6.26) has increased and the area is clearly in decline.

> **Key term**
>
> **Human Development Index:** a composite measure of development comprising life expectancy, purchasing power parity (gross national income at local purchasing price and levels of literacy).

The average Human Development Index for the UK is 0.94 (18th out of 177 countries) and the average for South Africa is 0.653 (121st out of 177 countries).

Figure 6.25 Number of people in poverty

	1996 (%)	2005 (%)
Mbhashe	68.6	77.9
Mnquma	62.3	75.0
Great Kei	56.1	79.5
Amahlathi	59.0	75.3
Buffalo City	41.5	50.1
Ngqushwa	64.1	80.2
Nkonkobe	55.9	69.2
Nxuba	56.6	80.3

Access to services

There are low levels of household access to basic water, sanitation and energy in the former homeland areas of Mbhashe and Mnquma and levels of access to emergency health services (ambulance) are appallingly low throughout the district. However, the levels of access to other services appears to be better. Only Nqushwa (44 per cent) and Mnquma (53 per cent) reported significantly less than two-thirds of households with access to clinics (Figure 6.27).

Figure 6.26 Poverty in Peelton

HIV and AIDS

Amathole District had an antenatal prevalence rate of 27.4 per cent in 2004 of HIV, up from 21.7 per cent in 2002. Current data on HIV is unreliable. It is estimated that 10 per cent of the Eastern Cape population is HIV positive, and 19.2 per cent of adults (20–64 years). In the Eastern Cape those at greatest risk are women, who have higher HIV prevalence rates than men. The

African population is at greatest risk, and the epidemic is growing fastest among youth (15–25 years).

> **Key term**
>
> **Antenatal prevalence rate:** the prevalence (predominance in an area) rate in unborn children.

Economic problems

Transport

Transportation in Amathole is constrained both by infrastructural issues and by limited access to public transport. Despite significant investments in new roads

Figure 6.27 Household access to basic services ▼

	% of population access to clinics	% of population access to primary schools	% of population access to roads in good condition	% of population access to ambulance service
Mbhashe	79.1	78.9	77.1	20.9
Mnquma	53.2	85.6	51.4	39.9
Great Kei	65.0	90.1	0.8	12.4
Amahlathi	84.2	93.5	2.5	31.8
Buffalo City	80.3	82.3	46.3	45.1
Ngqushwa	43.6	82.9	5.7	5.4
Nkonkobe	65.0	80.5	6.4	21.3
Nxuba	99.1	84.2	29.9	82.0

in the district, many local communities remain trapped in isolated and disconnected local communities often with very poor road infrastructure. Although the district has an extensive network of roads, the bulk of these roads are in poor condition. Currently, the process of maintenance and upgrade is severely hampered by the lack of clarity with regards to roles and responsibilities between various roads role players as well as the lack of dedicated funding to deal with backlogs.

In the Eastern Cape, it is estimated that while 1800km of roads are in a good condition, 435km are in a poor condition with weak ride-ability, and 50 per cent of gravel roads are in a reasonable condition. In 2002/2003, Rand 1 282 473 million was allocated to the maintenance of rural roads: six bridges were constructed; 3100 road signs were replaced; 110km of road reserve fencing was replaced; and 1200km of road markings were repainted. Many rural areas remain relatively remote and inaccessible, thereby reducing the potential for development.

Skills

Most working people (29 per cent) in the Amathole District have either elementary skills or are unskilled workers. Only 4 per cent of Amathole's working people fall into the skilled category. Craft and trade (14 per cent) are significant fields of occupation, followed by a proportion of technical skills and professionals, and a significant number of people in service-oriented areas of work.

Development and the ability of the economy to provide job opportunities is strongly correlated to the availability of a suitably educated and skilled workforce, hence it is critical that the local educational institutions should deliver appropriate and high-quality education.

Unemployment and household income

Unemployment has increased between 1996 and 2005 in all areas (Figure 6.28). This is a clear reflection of the currently limited local economic activities and livelihood options within these areas.

Activities

1 Study Figure 6.23.
   a Explain how unemployment in the Eastern Cape changes over time?
   b How does it compare with the national total?

2 Study Figure 6.24. Describe the changes in the HDI over time and comment on the differences in the HDI with different parts of the Eastern Cape.

3 a How has the percentage of people in poverty (Figure 6.25) changed over time?
   b How does it vary between different parts of the Eastern Cape?

4 Comment on the access to services as shown in Figure 6.27.

Problems in the farming sector

In the period before black people were forced into reserves and, later, homelands, tribal groups were not confined to small areas. The loss of their traditional lands led to the decline of the black rural economy. Increasing poverty prevented black farmers from affording the inputs necessary to improve yields. As the reserves were unable to feed the needs of the black population, many black people resorted to the only means possible – they became migrant labourers and entered the cash economy (subsistence farming was replaced by commercial farming in which farmers were forced to pay taxes and thereby enter the cash economy). Thus, migrant labour was a result and a cause of low productivity in black agriculture. The failure of many migrants to send much of their wages back to homeland areas further weakened the agricultural base (Figure 6.29).

Hence, the decline in productivity and profitability of black agriculture is a direct outcome of the nature of capitalist development in South Africa. Total food production in the homeland areas is sufficient for about one-third of the homeland needs. The failure to modernise and to increase output is related to a

Figure 6.28 Unemployment and household income for the Eastern Cape and Amathole District

	Unemployment rate (%)		Household income less than Rand 1500/month (%)
	1996	2005	
Eastern Cape	48.4	53.5	65.2
Amathole	47.8	52.7	67.0
Mbhashe	72.9	75.8	71.6
Mnquma	61.2	65.4	76.0
Great Kei	28.7	38.2	76.0
Amahlathi	50.8	59.4	73.5
Buffalo City	39.9	44.8	55.0
Ngqushwa	70.0	76.5	66.8
Nkonkobe	62.1	65.9	77.8
Nxuba	44.1	57.4	61.8

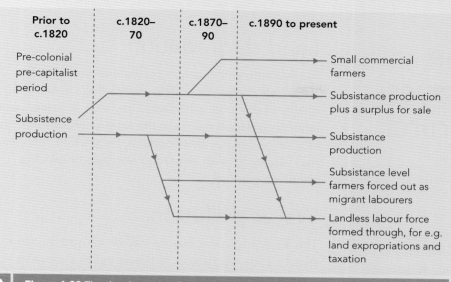

	Prior to c.1820	c.1820–70	c.1870–90	c.1890 to present

Pre-colonial pre-capitalist period

Subsistence production

Small commercial farmers

Subsistance production plus a surplus for sale

Subsistance production

Subsistance level farmers forced out as migrant labourers

Landless labour force formed through, for e.g. land expropriations and taxation

▲ Figure 6.29 The development of underdevelopment in black agriculture in South Africa

number of factors: the limited size of plots; yields up to five times less than on white farms; and up to 20 or 30 per cent of the land being left unused.

Factors explaining the low levels of food production in homelands can be classified under two broad categories: those operating at a farm level and those which are largely state ones. Factors operating at a farm level include: availability and use of land; labour supplies; and availability of capital.

The homeland areas accounted for just 13 per cent of the land area of South Africa but they contained 44 per cent of the population. Population increased rapidly in the twentieth century. The forced removal of black people from 'white' South Africa and their relocation in the homeland areas increased the population pressure. Tightening of influx control, that is, preventing black people from moving to white South Africa further increased population pressure and declining amounts of land per person (Figure 6.30). Despite the large numbers of people present in the homelands the quality of labour is very poor. The homelands have often been described as a dumping ground for surplus people and so there are a disproportionate number of women, children, elderly and infirm. The most able labour is usually migrant labour. Fifty-nine per cent of rural households were headed by women. African women in rural areas face oppression from four sides, because they are: African, women, poor and live in rural areas. Most domestic chores are labour intensive,

which reduces the female labour availability for agriculture. The position of women in South African society is extremely low and so female-headed households are doubly hit when it comes to agricultural production.

Capital availability is very low in black farms. Wages sent by migrants tend to be small and irregular, and therefore unable to provide for farming inputs. One of the paradoxes of farming in the homeland areas is that there appears to be a labour shortage. There is an excess of population but up to 30 per cent of the land remains uncultivated in any one year. The main reasons for under cultivation include shortage of finance to purchase inputs such as seeds and fertiliser, and lack of access to land in terms of land rights. Low productivity is also due to poor transport, lack of agricultural and managerial skills, and better returns from waged labour.

Activities

1 Define the following terms: homeland; migrant labour; population pressure.

2 Explain how the migrant labour system could be both a cause and an effect of low productivity in black agriculture.

3 Using any three of the factors mentioned as reasons for low productivity in black agriculture, explain in detail how they led to the underdevelopment of black agriculture.

Theory into practice

For a rural area you have studied, describe the social, economic and environmental changes that have occurred there.

Discussion point

How are the changes and pressures in rural areas in South Africa different from those in the UK?

 'Take it further activity' 6.4 on CD-ROM

	1970	1980	1990	2000	2020
Cultivated	0.6	0.5	0.4	0.3	0.2
Other	5.5	4.2	3.2	2.4	1.5

Figure 6.30 Land per person 1970–2020 (hectare) ▲

6.3 What are the environmental issues associated with rural change?

Land-use change

Changes in land use can have many unforeseen consequences. With any building developments there is an increase in impermeable surfaces, in drainage density, in overland flow, in flooding and a decline in water quality. Similarly, in the change from pastoral to arable farming there may be increases in soil erosion during winter months when the soil is free of vegetation. The removal of vegetation may decrease biodiversity.

Traffic congestion and pollution

With more traffic on the road there is greater congestion and more air pollution. This is particularly important around major junctions and at certain times, such as in the morning and evening rush hour. The pollutants include gases such as carbon monoxide, oxides of nitrogen, metals (such as lead) and volatile organic compounds.

Land degradation

In areas where population pressure is great, land degradation may occur. This means there is declining quality of the land, and declining yields. In some cases, land degradation may take the form of accelerated soil erosion and gulley development. Vegetation changes may also occur with degraded scrubland replacing improved pasture.

Water pollution

The decline in water quality may be caused by many factors: increased use of fertilisers in fields close to water courses; illegal dumping of material; increased rates of soil erosion due to land use changes; accelerated runoff near new buildings and building sites. The result is a reduction in water quality and the biodiversity associated with natural streams and rivers.

Rural dereliction

In areas of population decline, rural areas may show signs of dereliction. Farmhouses become derelict and gradually decay; farm buildings are abandoned and their remains litter the countryside; farm equipment is abandoned and pollutes the area. This creates areas of countryside that can look run down and neglected.

Case study | Environmental issues associated with rural change in Oxfordshire

In much of southern England, rural areas are under considerable strain. This is especially true in rural areas close to expanding urban areas. As we have already seen on p 231, the pressures are clearly shown in the rural areas around Oxfordshire.

Extension of Oxford's green belt

Oxford' green belt is set to be substantially redrawn north and south of the city, opening the way for major housing schemes (Figure 6.31).

Land-use change

One of the main changes is that of land-use. In 2007, planning inspectors in Oxfordshire gave their backing to a 4000 home urban extension to the south of Oxford on green belt land (Figure 6.32). The 370-acre site was chosen because planners believe the new homes could be integrated into the city's public transport system. In addition, the area is close to sources of employment such as BMW, the Oxford Business Park and the Oxford Science Park. The proposed settlement would

Planning inspectors are expected to pinpoint specific pockets of land that could be removed from Oxford's 50-year old green belt, without causing substantial environmental damage to the city's setting. However, in many cases environmental damage is widespread. This includes an increase in the impermeable surfaces, an increase in flooding, a decrease in biodiversity, an increase in traffic and consequently, air pollution, declining water quality in Oxfordshire's rivers and streams, and the pollution of groundwater resources.

Land off Grenoble Road (Figure 6.32), where Magdalene College wants to build 4000 homes, is to be the most significant section of the green belt to be reviewed. Magdalene College and Thames Water, owners of the site near the Kassam Stadium, made a strong submission to the public inquiry arguing a settlement on the edge of the city could help address the city's chronic shortage of affordable housing.

Shipton Quarry, where developers want to build 5000 homes, is another key site that the inspectors want to be examined. The site, north of Kidlington, was formerly a cement works. The recommendations are also expected to boost Oxford University's hopes of building thousands of new homes on a 368-acre greenfield site between Kidlington and Yarnton, next to the main Woodstock to Oxford Road. Independent planning inspectors spent four months examining housing numbers proposed by the South East England Regional Assembly (Seera).

Inspectors appeared to have rejected the idea of any wholesale review of Oxford's green belt on the grounds that it would result in years of debate, involving six councils.

Figure 6.31 Changes to Oxford's green belt – adapted from the *Oxford Times*, 24 August 2007 ▲

stretch from Sandford-on-Thames to Greater Leys and would mean building a multi-million pound sewage system and a new secondary school. Residents in nearby villages, such as Garsington, and the South Oxfordshire District Council are opposed to the plan and intend to contest it.

Another proposed development is a 368-greenfield site between Yarnton and Begbroke (Figure 6.33). Oxford University owns Begbroke Science Park at the centre of the site (Figure 6.34), as well as much of the surrounding farmland. The University have made an outline proposal for more link roads and housing development in the area but in addition more schools and facilities would be needed. Local residents complain the access to Oxford from Begbroke and Yarnton is already poor, due to the large volumes of vehicles entering Oxford on a daily basis. Any extra vehicles would only make the situation worse. Arguably the planned housing developments will offset the rural dereliction evident in these areas.

Elsewhere, changes in farming are having an impact on the environment. The switch from pastoral to arable farming in parts of the Cherwell Valley have led to increased rates of soil erosion, especially in wet winters. Hedgerows and trees have been cut down to make larger fields which are easier for machines to work in. The result has been termed the 'prairiesation of Oxfordshire'

▼ **Figure 6.32** Location of the proposed extension to Oxford

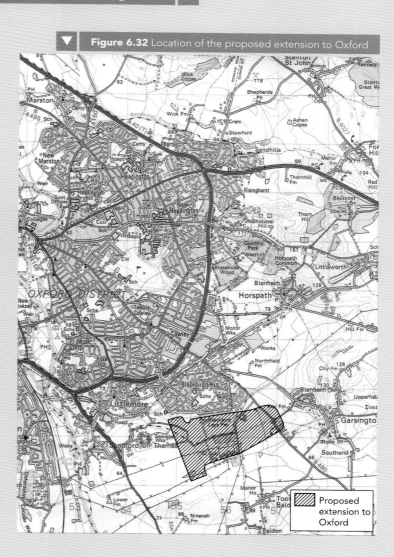

Proposed extension to Oxford

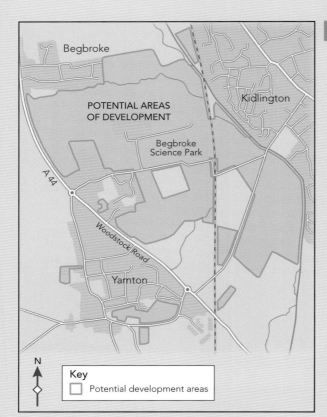

Begbroke

Kidlington

POTENTIAL AREAS
OF DEVELOPMENT

Begbroke
Science Park

A 44

Woodstock Road

Yarnton

N

Key
☐ Potential development areas

─ **Key term** ─

Eutrophication: growth of algae as a result of fertiliser leaching from farmland.

Recreation and leisure

Recreation and leisure also has a major impact on the rural environment. For example, Shotover Country Park attracts 600 000 visitors a year. Some of the land in the Country Park is still owned by the University, but it is now all managed by Oxford City Council's Countryside Service. As further protection against development, the City of Oxford has undertaken to preserve its own farmland on the south side of the hill 'in its natural state' as a condition of receiving a large gift of land from the Oxford Preservation Trust into the Country Park.

A large proportion of visitors to the Country Park arrive by car, and this in itself is a problem for the future. Large numbers of visitors are eroding the paths nearest the car park, and this has become a management problem, along with the other problems of access for horse riders, cyclists, and disabled people. There is also great pressure to develop the surrounding land for more housing. In contrast, Wytham Woods are owned by the University of Oxford and visitor access is controlled by the University. Here there is less impact from visitors and less pressure for development.

and led to land degradation. Increased application of fertilisers has led to eutrophication in streams and watercourses. Nevertheless, water quality in many of the streams and rivers in West Oxfordshire is of a good quality. Figure 6.35 shows how many kilometres of Oxfordshire's rivers are good/poor quality.

Figure 6.34 Begbroke Science Park ▲

Year	Category	A (km of water quality)	B km	C km	D km	E km	F km	Total km	Good %	Fair %	Poor %	Bad %
2005	Biology	150	79	36	5	0	0	260	84	16	0	0
2005	Chemistry	154	81	20	7	8	0	270	87	10	3	0
2005	Nitrates	0	0	1	3	172	94	270	–	–	–	–
2005	Phosphates	5	55	24	34	139	14	270	–	–	–	–

Key: A and B – good quality; C and D – fair quality; E – poor quality; F – bad quality

Figure 6.35 Water quality in West Oxfordshire in 2005 ▲

▲ **Figure 6.36** Yarnton Nurseries

Even small changes such as the development of garden centres leads to an increase in the amount of traffic on the road and the construction of car parks at the expense of farmland (Figure 6.36).

Key term

Guided Transit Express (GTE) system: a tram system operating on a predetermined course.

Traffic congestion

A transport plan to solve traffic congestion problems between Witney and Oxford is to be carried out over the next 15 years. The three-phase plan was unveiled yesterday (15 October) at a meeting of Oxfordshire County Council's executive.

Schemes to introduce bus lanes or a dual carriageway on the A40 between Oxford and Witney or reviving a railway were finally ruled out by councillors.

Phase one will involve improving the Wolvercote and Eynsham roundabouts and Cassington junction signals to improve traffic flow. A link road will be built between the A40 and A34 north of the Pear Tree inter-change. This will take five years. The government was never going to be convinced by the A40 bus lane. The real beauty of the new scheme is that the first stage of improvements can begin relatively quickly.

Phase two, in six to seven years' time, will include a Witney park-and-ride scheme and a Guided Transit Express system to connect the Pear Tree park-and-ride to the city centre. No estimate has been given for the total cost of the 15-year scheme. Buses would run on concrete guideways, at first in Oxford city, but they would eventually go all the way to Carterton and use part of the disused Witney to Oxford railway route.

Figure 6.37 Plan to tackle A40 congestion – adapted from *Oxford Mail*, 16 Oct 2002 ▲

The GTE is also included in phase three of the new strategy, which could be as far away as 10 to 15 years, during which time the population of both Witney and Carterton will have grown by 25 per cent. County councillors agreed to put the proposals out to consultation with local councils and agencies such as bus operators.

The whole of the Oxford road network is operating very close to capacity and the only way that substantial growth in travel demand between West Oxfordshire and Oxford can be met is by encouraging public transport.

The heavy volume of traffic causes major problems with congestion and air quality. In particular the roads converging on Oxford are very slow moving and reduce the air quality, especially during morning and evening rush hours. Even parking the many coaches that take passengers into and out of Oxford is causing areas of the countryside to be given over to coach parks (Figure 6.38). This increase in impermeable surface can lead to localised runoff and flooding.

Funding is a key issue. The dualling of the A40 all the way from Witney to Headington will cost at least £100m. The revival of the Witney to Oxford railway has been estimated at over £100m and was rejected by consultants but the majority of funding would come from the government's annual transport grant to the county council.

Water pollution: Oxfordshire's new reservoir

Another large-scale project is the new reservoir in south-west Oxfordshire. The plan is for the reservoir to store water from the Thames during high flows and supply the Upper Thames area and the Thames during low flow. It could also help supply water to the proposed nuclear power stations in Oxfordshire – a possible plan to convert Didcot Power Station – if either of these materialise. However, there are a number of potential impacts to consider. These include on-site and operational impacts to the environment as listed below.

On-site impacts include:

◆ pollution risks during construction
◆ the effect of the reservoir on flood risk and drainage
◆ diversion of water courses
◆ the effect on groundwater levels (leakage from reservoir).

Figure 6.38 Coach park at Yarnton, Oxfordshire ▼

Operational impacts include:

◆ the physical, chemical and ecological implications of abstraction from the Thames

◆ maintaining reservoir water quality in terms of oxygen, algae and temperature characteristics.

Nevertheless, there are a number of potential benefits of the scheme. The main one is the security of water resources in the region. The proposed reservoir is impressive.

Critics claim that 'Lake Oxford' will swamp a vast area of precious English countryside near Abingdon, destroying wild animals and plants.

Construction is likely to start in 2010 and finish in 2020 and involve a massive fleet of lorries and even a new train line to bring materials and people to the site. At the same time, vast new pipelines will be built to link it to the Thames and there will be a new water treatment and pumping centre. There will also be a boat park, water sports club house, beach, jetty, pier and slipway. The impact on the environment is likely to be considerable.

Activities

1 Briefly explain why there are pressures to develop rural areas in Oxfordshire.

2 How can changes in either farming or recreational activities impact rural environments?

3 Visit the website of the county you live in, or if you live in a large urban area such as London or Birmingham, one of the counties around you. What are the main issues for rural settlements in the area that you have chosen?

 'Take it further' activity 6.5 on CD-ROM

Discussion point

Are environmental problems an inevitable result of growth in rural areas?

 For a contrasting case-study refer to 'Environmental issues associated with rural change in South Africa' on the CD-ROM.

6.4 How can rural areas be managed to ensure sustainability?

The sustainable development of rural areas requires a careful balance of socio-economic and environmental needs. This requires detailed planning and management.

What is sustainability?

Sustainable development is defined as a form of development in which basic standards of living are raised without compromising the needs of future generations. That means there must be some form of management of the resource base so it can continue to be used in years to come.

How can it be applied to rural areas?

In many rural areas, especially in LEDCs, poverty is widespread. Sustainable development is needed in order to take these people out of poverty and to raise their standards of living. At the same time, if the development is sustainable, it will ensure that the standard of living is maintained for the next generation. In rural areas, there are options for sustainable farming, sustainable use of water, of energy resources and of transport.

How can sustainability be improved?

Sustainability can be improved only if there is commitment from the local people all the way up to the national and international community. Many rural residents can see clearly the benefits of managing resources, such as soil and water that will guarantee their livelihood. However, if there is not the support at a national level, many small-scale, local projects may be doomed to fail. Much of the government input is in the form of funding, but there are other practical measures, such as support, educational programmes, field workers, and the development of rural credit schemes.

Implementing sustainability

Sustainable schemes can only be implemented after there has been discussions and planning between local residents and officials from government and other institutions, such as universities, local businesses and schools. Detailed plans need to be drawn up, and the results of any scheme need to be monitored, so that the plans can be modified if need be.

Case study | Planning for a sustainable future in the Amathole District Municipality, South Africa

In order to plan for a sustainable future, the key strategic focus areas for the Amathole District Municipality include:

◆ reducing unemployment by half

◆ poverty eradication

◆ investing in sustainable infrastructure development including improvement of roads and an integrated public transport system

◆ enhancing the economy of the district

◆ reducing the impact of HIV and Aids.

To facilitate this, the Eastern Cape Provincial Government will be investing some Rand 2.5 billion in infrastructural development within the district by 2008/09.

Priority economic sectors in Amathole

In the context of the very high unemployment and poverty rates there is clearly an urgent need for new investments to create jobs and improve livelihoods in the Amathole District. Based on existing economic activity, market opportunities and present resource/assets/skill bases the particular industries offering potential include: livestock farming; irrigated horticulture; forestry; manufacturing; construction; trade and business services/ICT; and tourism.

The skills base of the population is low, and only 16 per cent of the district's working age population have attained the Matric or above. The Matric is required for the study of university degrees.

Key term

Matric: a formal South African school qualification taken at age 18.

Livestock farming

Animals and animal products are the dominant farming activity in the District, accounting for 72 per cent of agriculture value added. There is potential to improve livelihoods from livestock through livestock improvement programmes (marketing, branding, genetic improvement etc). There is scope for improved cooperation between communal livestock farmers and commercial farmers.

Agriculture's competitive advantage in the district is enhanced by:

◆ proximity to markets, notably East London (Buffalo City)

◆ product diversity (across a variety of agro-ecological zones)

◆ access to business services/ICT/logistics network.

The Amathole District has quite significant untapped agriculture, forestry and water resources, and these resources are at the heart of any sustainable growth potential and economic activity in the district. Figure 6.39 highlights the areas where there are potential opportunities and a sustainable agricultural scheme will target a range of these areas. Given this existing natural resource base, the Provincial Department of Agriculture plays an important catalyst role in promoting subsistence farming (food security) and small- and large-scale commercial agriculture.

Another key element of required infrastructure development in the district is the road network. The Amathole District suffers from quite significant road and rail infrastructure backlogs given the rural character of large parts of the district. Amathole District also suffers from acute backlogs in social and economic infrastructure.

Enterprise	Product	Potential exists in local municipal	Management and technical enterprise	Jobs	Processing infrastructure required	Markets	Market potential
Sheep (extensive)	Mutton Wool	Nxuba, Nkonkobe, Amahlathi, Mbhashe Mnquma	Medium	1/100ha	Existing agents and abattoirs	Local (mutton) Export (wool)	Good
Beef (extensive)	Beef	Mbhashe, Mnquma, Great Kei, Buffalo City, Nkonkobe, Nxuba, Amahlathi, Ngqushwa	Medium	0.8/100ha	Existing agents and abattoirs	Local	Good
Dairy (irrigated)	Fresh milk	Buffalo City, Great Kei, Ngqushwa	High	1/35 cows	No	Local	Good (seasonal surpluses)
Game	Hunting, Live game, Tourism	Nxuba, Great Kei, Nkonkobe, Ngqushwa, Amahlathi	Medium	0.3/100ha	No	National Export	Good (Foreign Exchange related)
Poultry	Broilers	Buffalo City, Amahlathi, Great Kei, Mnquma, Mbhashe	Very high	1/2000 birds	Abattoir	Local	Good
Poultry	Eggs	Buffalo City, Amahlathi, Great Kei, Mnquma, Mbhashe	High	1/5000 birds	Packaging	Local	Good
Field crop Maize	Grain	Mnquma, Mbhashe Amahlathi	Low	8/100ha	Existing milling	Local	Good
Field crop dry bean	Dry sugar beans	Mnquma, Mbhashe Amahlathi	Low	8/100ha	Packaging	Local	Good
Vegetables (open irrigated)	Vegetable	All (isolated areas)	Medium	2/ha	No	Local	Good (limited and seasonal)
Hydroponic production (tunnels)	Tomatoes	Buffalo City, Great Kei, Mbhashe, Mnquma Ngqushwa	Very high	12/ha	Packaging existing (EL)	Local National	Good (limited local)
Citrus	Fresh fruit	Nkonkobe, Ngqushwa, Great Kei, Amahlathi	High	0.8/ha	Pack shed	Local Export	Medium (quality and foreign exchange related)
Pineapples	Fruit for canning	Buffalo City, Ngqushwa, Great Kei, Mnquma	Medium		Existing (East London)	Export	Medium (foreign exchange related)

Figure 6.39 Opportunities for agriculture in Amathole

Key term

Broilers: broiler chickens (often referred to as 'broilers') have been selectively bred and reared for their meat rather than eggs.

Options for sustainable agriculture

There are many possibilities for improving agricultural productivity in the region. Irrigation schemes can be expensive but there are cheaper schemes for example:

◆ *establishing small-scale gardens and subsistence farms* These could provide a very useful extra supply of food in just a few weeks. If such gardens are integrated with improvements in harvesting and storage there ought to be a much larger supply of quality food at very little cost.

◆ *developing farming cooperatives* There is a pool of labour and experience, the latter perhaps somewhat limited, which could share the cost of tools and seed, and use the produce for their subsistence needs.

245

◆ *erection of fences or barbed wire*   This would help prevent theft and trampling of crops by livestock. It could become a cottage industry.

◆ *using drought-resistant fodder crops*   For example, the American aloe (Figure 6.40). Pastureland is especially fragile owing to a combination of drought, overgrazing, population pressure, and the absence of land ownership policies. Trying to decrease herd size has proved unpopular and unsuccessful. Drought-resistant fodder crops are a good alternative.

Essential oils

An essential oil is any concentrated liquid containing aromatic herbs or plants. They have been used for medicinal purposes and interest in them has been revised in recent decades especially with the growth of aromatherapy. The production of essential oils holds considerable potential as a form of sustainable agricultural development. Not only are the raw materials present but it is also a labour intensive industry and would utilise a large supply of unemployed and underemployed people.

The essential oils industry has a number of advantages.

◆ It is a new or additional source of income for many people.

◆ It is labour intensive and local in nature.

◆ Many plants are already known and used by the people as medicines, and are therefore culturally acceptable.

◆ In their natural state the plants are not very palatable or of great value and will therefore not be stolen.

◆ Many species are looked on as weeds – removing these regularly improves grazing potential as well as supplying raw materials for the essential oils industry.

Forestry

Forestry activity in the district presently exists in three local municipal areas: Amahlathi, Mnquma and Mbhashe. Although the forestry areas are limited, there is nevertheless scope to increase the areas under forestry, particularly in communal areas. The nature of the forests in the Amathole mountain areas

Figure 6.40 The American aloe ▼

include both commercial pine and indigenous forests. Constraints to the development of this sector lie in the slow pace of land and forestry sector reform processes, the limited size of the available resource, and the lack of skills and investment for beneficiation. It is also much harder to run a sustainable forestry industry because of the longer yield time. Where forests are not suitable for commercial exploitation, they have huge potential benefit in resources for traditional and alternative medicines, and for tourism.

Construction

In the district formal employment in construction is low, accounting for only 3.4 per cent of all formal/regular jobs (Figure 6.41). This sector is likely to become a net jobs creator on the basis of both increasing investment in public infrastructure investment (water, sanitation, housing, roads, schools etc) and increased private property development (retail, residential, tourism and industrial developments etc).

Tourism

Tourism in the district has grown considerably over the last decade and now accounts for a significant portion of district value-added and employment, perhaps as much as 10 per cent district value-added and 20 000 existing jobs/incomes. New tourist facilities, such as B&Bs, backpacker hostels and game farms (Figure 6.42) have sprung up in many parts of the district. The district is fortunate in having a diversity of attractions including heritage sites and the Amathole Mountains Escape Route. Tourism will benefit from the investment in a sustainable infrastructure.

The challenge of job creation

The diversity of the Amathole economy (Figure 6.43) and the range and depth of the socio-economic challenges in the district mean that there is no single sector of focus that can transform the economy and create the requisite jobs needed to halve unemployment by 2014.

Transport

To make transport in the region more sustainable the quality of roads will need to be improved. All settlement will be connected to the road network, bridges will need to be built and public transport expanded.

▼ **Figure 6.41** Brick making in the Eastern Cape

Figure 6.42 Rhinoceros on private game park in the Eastern Cape ▲

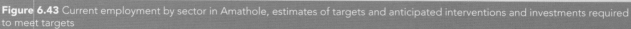

Figure 6.43 Current employment by sector in Amathole, estimates of targets and anticipated interventions and investments required to meet targets ▼

Sector	Required interventions and investments
Agriculture and agro-processing	Improved road and rail linkages (R63, R72, N2)
	Commercial production on communal land
	Settling of land claims such that production is maintained
	Investment capital
	Skills development and retention
	Security improvement
	Improvement in competitiveness and linkages to markets
	Investment in agro-processing facilities
Tourism	Investment in locality development and tourist facilities
	Packaging of tourist products
	Reduction in travel and accommodation costs
	Increase in competitiveness with KwaZulu Natal KZN province, overseas localities
	Increase in foreign and leisure tourists
	Quality and service improvements
	Sector collaboration and coordination
	Investment in facilities and systems for sports tourism

Moving towards a sustainable growth and development strategy

The Amathole District Municipality is clearly grappling with very entrenched problems of poverty and unemployment, and progress seems to have been quite slow so far in addressing these problems. While public and private sector investments are clearly increasing, these levels are currently insufficient to address the growth and development challenges faced by the district. The wide diversity of agricultural products in Amathole, its close proximity to key markets and investment in infrastructure will be key factors in the areas drive towards a sustainable future.

Case Study: Planning for a sustainable future

U2

6

Managing rural change

Activities

1 Briefly outline some of the options for sustainable development in the Eastern Cape region of South Africa.

2 Why is it difficult to achieve sustainable development in the area?

Discussion point

How might management issues in West Oxfordshire differ from those in the Eastern Cape?

Theory into practice

Find out what your nearest county council is doing to promote sustainable development in the region.

Knowledge check

1 What are the characteristics of rural areas?

2 How do rural areas vary in terms of functions, processes and opportunities?

3 What are the social and economic issues associated with rural change?

4 How does structural change lead to economic and social differences within and between rural areas?

5 How does lack of economic opportunities lead to depopulation and decline?

6 What are the environmental issues associated with rural change?

7 How can the changing use of rural areas put increasing pressures on the environment including:

a land-use change

b traffic congestion and pollution

c land degradation

d water pollution

e rural dereliction?

8 How can rural areas be managed to ensure sustainability?

A list of useful websites accompanying this chapter can be found in the Exam Café section on the **CD-ROM**

ExamCafé
Relax, refresh, result!

Relax and prepare

What I wish I had known at the start of the year…

Keisha

"In class yesterday we were asked '*At what stage does a rural area cease to be rural and then what does it become?*', and this led to a lengthy debate. Supposedly different countries have different indicators. Some argue it is when population density passes a certain point or when economic activity is no longer dependent on agriculture. The percentage of rural versus non-rural land use can also be used. The term 'rurban' has been used to describe the mid or transition stage whilst others prefer 'suburbanised village' which indicates the process of rural change. I suppose my point is that you shouldn't always assume that a question has one clear answer, sometimes you need to show a range of opinions!"

Tracy

"I go through the following routine the night before the exam. First, I make sure I know the time of the exam, as you can always get confused between am and pm exams. I then pack my equipment in its see-through container – pencil, two pens, eraser, ruler, some colouring pencils (red, green, blue, brown) and a calculator. I include some correction fluid, but not all centres allow it so check first. I then check the final version of my revision notes. I try not to cram but if I need to do some last minute revision, I check my revision checklists. When I am unable to sleep I go for some gentle exercise rather than doing any more revision."

Common mistakes – Graham

▷ "In an examination I sometimes make the mistake of leaving a section unanswered either by getting a question wrong and crossing it out, or leaving it blank to come back to and forgetting about it. My teacher has advised me not to cross out any answers I am not sure about as I may just be right and to at least write something valid from my memory aids as I can't get any marks for a blank section."

6.1 What are the characteristics of rural areas?

Physical	Climate, relief, drainage, soil type, rock type
Economic	Transport routes Accessibility to large urban area Ability to out bid for sites Size of site available Mutual attraction/repulsion
Social	Historical – inertia, conservation, reputation Mutual attraction/repulsion Land ownership – estate development e.g. National Trust Religious factors
Political	Planning controls – National Parks, SSSIs, Nature reserves Controls on farming e.g. CAP, set-a-side Development schemes e.g. release of green belt Government uses e.g. military ranges
Characteristics	Land use – crops, pasture, woods, reservoirs, moorland, recreation Population – number, age/sex, socio-economic, ethnicity, migration Activities – farming, mining, industry, parks, conservation, water

6.2 What are the social and economic issues associated with rural change?

Population	In and out migration, birth rate, ethnic mix, age structure ageing
Mobility	Increase private and cost and availability of public transport
Income	Increased or decreased income/wealth inequalities
Political	Impact of planning initiatives/controls conservation areas etc
Employment	Decline of agricultural services, changing technology
Public services	Growth or decline, type, location – loss of rural culture

6.3 What are the environmental issues associated with rural change?

Pollution	Air, water (eutrophication), noise, visual, solids
Water	Scarcity (transfer schemes, use of aquifers, reservoir building)
Traffic	Congestion, pollution, land use (roads, car parks, garages etc)
Habitat destruction	Direct and indirect, species diversity
Disease introduction	Accident e.g. Foot and Mouth or planned
New species introduction	Accidentally or planned
Dereliction	Visual pollution, safety, health

6.4 How can rural areas be managed to ensure sustainability?	
Physical	Relief – building on unsuitable sites e.g. steep slope
	Drainage – water shortage, pollution
	Vegetation – loss of habitats
	Pollution – air, water, land, noise, visual
Economic	Settlement – housing quality and quantity, cost
	Power – shortage, reliability, cost
	Industry – lack of jobs or low pay, migrant labour
	Services – lack of sufficient schools, shops, clinics etc
	Transport – cost, poor public transport
Social	Wealth inequality and deprivation
	Cultural change
	Age profile – dependency ratio, birth rates, social services
	Migration – depopulation
Sustainability	In terms of culture, employment, population, environment, viability
	Idea of green technology, integrated management, key settlements

Top tips . . .

▷ Photo-based questions are attractive as the answer is usually there if you can read the photo correctly. Remember to scan the photo carefully working from the centre outward. Don't be afraid to label the photo with items relevant to the question as you do this. If it is a photo for a physical geography question, then look for Structure, Processes and Human impacts; if a human geography question (SPAMIST), then don't ignore the physical geography. If you are asked to sketch the photo, remember the accuracy and detail of your labels (usually at the end of an arrow pointing to the relevant bit on the photo – annotation) are more important than your artistic ability. You may be able to trace the outline or sometimes it is provided for you. Single word labels are of little worth so give some detail as asked for in the question. Do not do a key and write extensively away from the photo.

Get the result !

What are the main characteristics of rural areas on the fringe of large urban areas? [6 marks]

Examiner says

This answer would gain maximum marks because (1) good understanding of basic concept, (2) good attempt to structure answer, (3) explanation of characteristic with resulting secondary impact, (4) explanation of a clear characteristic with cause-effect, (5) another clear characteristic with cause-effect, (6) clear characteristic with cause-effect.

Student answer

The main characteristic is one of rapid change (1). Economically, house prices are rising (2) and there are more affluent populations often commuting to the urban area (3). Socially the community is becoming more mixed and diverse as the wealthy move in (4). Infra-structure is improving as the increased population leads to more schools etc (5) but environmentally there is a loss of habitats as woodland and farming are replaced by housing and roads. (6)

Examiner says

An equally effective way would have been to use an example to illustrate the possible characteristics.

Examiner's tips

There are marks for the quality of your communication and the correct use of geographical terminology, but what does this mean?

Quality of communication refers to the presentation of a logical and coherent discussion. Does the introduction and conclusion have a clear progressive structure? One of the commonest problems is the use of the paragraph – too many candidates either do not use them or produce short, often single-sentence paragraphs. Each paragraph should have points that are linked – the most obvious example would be to have one each for Physical, Economic, Social and Political.

Spelling is an issue but even good candidates can make slips in the pressure of the exam room. What isn't acceptable is the persistent misspelling of geographical terms and places. Learn them or only use them if you are fairly confident of their spelling. Too many can't spell Mississippi, Spurn Head etc. More exotic spellings may be difficult but if examiners can't understand it they will not credit it.

Try to use the correct geographical terminology where appropriate – we use many technical words and it is this vocabulary that you should know and use. Use words such as 'infrastructure' instead of listing aspects such as roads, power stations, schools etc.

Handwriting can be a problem as writing fast in examinations impacts on legibility. Most examiners can read most handwriting. If it is too bad, ask in advance for a scribe or permission to word process.

Chapter 7

The energy issue

'America faces a major energy supply crisis over the next two decades. The failure to meet this challenge will threaten our nation's economic prosperity, compromise our national security and literally alter the way we lead our lives.'

US Secretary of Energy, 2001

Following the 1973 Arab–Israeli war, the Arab nations reduced the supply of oil to the USA and Western Europe in an effort to lessen their support for Israel. This led to a serious energy shortage which became known as **'the Energy Crisis'**. Other less serious shortages of supply have occurred since then which have pushed energy prices up and reminded us that we cannot take energy for granted. The key energy issues for individual countries now are the three 'S's: Sustainability, Security and Strategy.

Questions for investigation

- What are the sources of energy and how do these vary in their global pattern?

- What is the relationship between energy use and economic development?

- What are the social, economic and environmental issues associated with the increasing demand for energy?

- How can energy supply be managed to ensure sustainability?

Consider this

Energy crisis America: a preview

As the energy crisis deepens, people will make fewer trips, plan trips better, drive less and carpool. High mileage cars will be in demand and the SUV market will die. With less and slower driving due to high petrol prices, the number of accidents will go down. Park and Ride schemes will become more prevalent and the government will encourage energy savings with tax breaks. Businesses will turn off advertising lights and more people will wash their clothes in cold water. In poor areas there will be more deaths from heat in the summer and more deaths from cold in the winter due to energy costs. Fuel riots may occur.

The energy crisis will increase illegal immigration from Mexico where there is increasing unemployment. This will add to social tensions. Chronic heating oil shortages may cause a greater migration to the Sun Belt. The middle class may shrink with more people sinking into poverty. There will be more homelessness. Eventually energy efficiency measures unthinkable today will become the law of the land. Extreme right-wing politicians will call for military control of all Middle East oil for the good of the world.

Source: Gasprices-usa.com/EnergyCrisisUSA

7.1 What are the sources of energy and how do these vary in their global pattern?

Renewable and non-renewable (finite) sources of energy

Non-renewable (finite) sources of energy are the fossil fuels and nuclear fuel. These are finite, which means that as they are used the supply that remains is reduced. Eventually, these non-renewable resources could become completely exhausted.

Renewable energy can be used over and over again. These resources are mainly forces of nature that are sustainable and that usually cause little or no environmental pollution. Renewable energy includes hydroelectric, biomass, wind, solar, geothermal, tidal and wave power.

At present, non-renewable resources dominate global energy. The challenge is to transform the global energy mix to achieve a better balance between renewables and non-renewables.

Key terms

Fossil fuels: fuels consisting of hydrocarbons (coal, oil and natural gas), formed by the decomposition of prehistoric organisms in past geological periods.

Renewable/sustainable energy: sources of energy such as solar and wind power that are not depleted as they are used.

Energy mix: the relative contribution of different energy sources to a country's energy production/consumption.

Figure 7.1 shows the big gap in energy consumption between rich and poor countries. Wealth is the main factor explaining the energy gap. The use of energy can improve the quality of life in so many ways. That is why most people who can afford to buy cars, televisions and washing machines do so. However, there are other influencing factors, with climate at the top of the list (cold climates have high energy usages through heating).

The demand for energy has grown steadily over time. Figure 7.2 shows a global increase of over 60 per cent between 1981 and 2006. The fossil fuels dominate the global energy situation. Their relative contribution in 2006 was: oil – 36 per cent; coal – 28 per cent; natural gas – 24 per cent. In contrast, hydroelectricity and nuclear energy accounted for about 6 per cent each. Figure 7.2 includes commercially traded fuels only. It excludes fuels such as wood, peat and animal waste which, though important in many countries, are unreliably documented in terms of consumption statistics.

The availability of renewable and non-renewable sources of energy

The availability of fuel types is one of the key factors affecting the wide variation in fuel consumption seen worldwide (Figure 7.3):

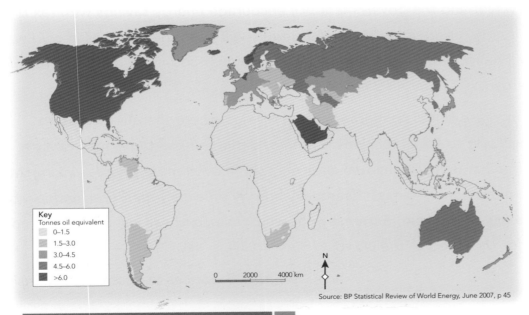

Key
Tonnes oil equivalent
- 0–1.5
- 1.5–3.0
- 3.0–4.5
- 4.5–6.0
- >6.0

0   2000   4000 km

N

Source: BP Statistical Review of World Energy, June 2007, p 45

Figure 7.1 Energy consumption per capita 2006

◆ *Oil* Nowhere is the contribution of oil less than 30 per cent and it is the main source of energy in four of the six regions shown in Figure 7.3. In the Middle East where most easily accessible oil reserves can be found it accounts for approximately 50 per cent of consumption.

◆ *Coal* Massively depleted coal reserves means that only in the Asia Pacific region is coal the main source of energy. In contrast it accounts for less than 5 per cent of consumption in the Middle East and South and Central America.

◆ *Natural gas* Natural gas is the main source of energy in Europe and Eurasia where there are still large gas fields in operation, and it is a close second to oil in the Middle East. Its lowest share of the energy mix is 11 per cent in Asia Pacific.

◆ *Nuclear energy* Nuclear energy is not presently available in the Middle East and it makes the smallest contribution of the five energy sources in Asia Pacific, Africa and South and Central America. It is most important in Europe and Eurasia and North America. France, for example, has a paucity of fossil fuels so relies on nuclear power for most of its energy needs.

◆ *Hydroelectricity* The relative importance of hydroelectricity is greatest in South and Central America (28 per cent). Elsewhere its contribution varies from 6 per cent in Africa to less than 1 per cent in the Middle East where there aren't the river systems to sustain this type of energy provision.

Other renewable sources, such as solar power, wind power, biomass and tidal power, form a negligible part of the global energy mix at present. However, this is changing and these energy sources will be considered in the final section of this chapter (page 280).

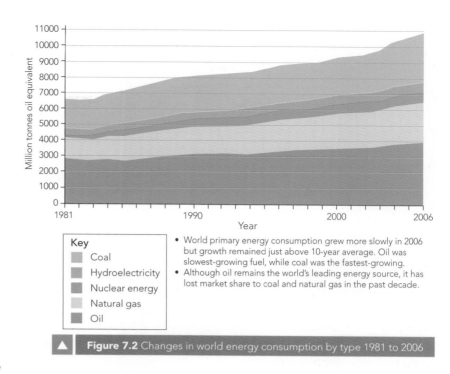

Key
- Coal
- Hydroelectricity
- Nuclear energy
- Natural gas
- Oil

• World primary energy consumption grew more slowly in 2006 but growth remained just above 10-year average. Oil was slowest-growing fuel, while coal was the fastest-growing.
• Although oil remains the world's leading energy source, it has lost market share to coal and natural gas in the past decade.

▲ **Figure 7.2** Changes in world energy consumption by type 1981 to 2006

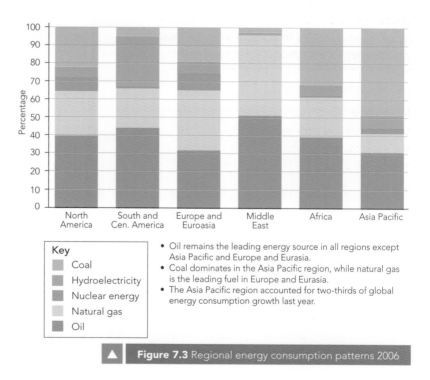

Key
- Coal
- Hydroelectricity
- Nuclear energy
- Natural gas
- Oil

• Oil remains the leading energy source in all regions except Asia Pacific and Europe and Eurasia.
• Coal dominates in the Asia Pacific region, while natural gas is the leading fuel in Europe and Eurasia.
• The Asia Pacific region accounted for two-thirds of global energy consumption growth last year.

▲ **Figure 7.3** Regional energy consumption patterns 2006

Future trends

Global primary energy demand is expected to increase by at least 50 per cent between 2006 and 2030. This would mean an average annual increase of 1.6 per cent. More than 70 per cent of this increase will be from developing countries with 30 per cent from China alone (Figure 7.4).

Figure 7.4 Traffic congestion in Beijing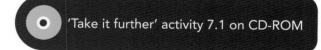

Activities

1 Describe the variations in global per capita energy consumption shown in Figure 7.1.

2 Look at Figure 7.2. Produce a table to show the contribution of the five energy sources in 1986 and 2006.

3 Look at Figure 7.3. Comment on the energy mix in each of the six regions.

⊙ 'Take it further' activity 7.1 on CD-ROM

The physical, economic and political reasons for variations in energy supply

Global variations in energy supply occur for a number of reasons. These can be broadly subdivided into physical, economic and political factors. Figure 7.5 shows examples for each of these groupings.

Variable energy patterns over time

The use of energy in all countries has changed over time because of a number of factors:

◆ *Technological development*   For example: a) nuclear electricity has only been available since 1954; b) oil and gas can now be extracted from much deeper waters than in the past; c) renewable energy technology is advancing steadily.

◆ *Increasing national wealth*   As average incomes increase, living standards improve and this involves the increasing use of energy and the use of a greater variety of energy sources.

◆ *Changes in demand*   At one time all of Britain's trains were powered by coal and most people also used coal for heating in their homes. Before natural gas was discovered in the North Sea, Britain's gas was produced from coal (coal gas).

◆ *Changes in price*   The relative prices of the different types of energy can influence demand. Electricity production in the UK has been switching from coal to gas over the past 20 years mainly because power stations are cheaper to run on natural gas.

Physical	◆ Deposits of fossil fuels are only found in a limited number of locations.
	◆ Large-scale hydroelectric development requires high precipitation, major steep-sided valleys and impermeable rock.
	◆ Large power stations require flat land and geologically stable foundations.
	◆ Solar power needs a large number of days a year with strong sunlight.
	◆ Wind power needs high average wind speeds throughout the year.
	◆ Tidal power stations require a very large tidal range.
	◆ The availability of biomass varies widely due to climatic conditions.
Economic	◆ The most accessible, and lowest cost, deposits of fossil fuels are invariably developed first.
	◆ Onshore deposits of oil and gas are usually cheaper to develop than off-shore deposits.
	◆ Potential hydroelectric sites close to major transport routes and existing electricity transmission corridors are more economical to build than those in very inaccessible locations.
	◆ In poor countries foreign direct investment is often essential for the development of energy resources.
	◆ When energy prices rise significantly, companies increase spending on exploration and development.
Political	◆ Countries wanting to develop nuclear electricity require permission from the International Atomic Energy Agency.
	◆ International agreements, such as the Kyoto Protocol, can have a considerable influence on the energy decisions of individual countries.
	◆ Potential hydroelectric power (HEP) schemes on 'international rivers' may require the agreement of other countries that share the river.
	◆ Governments may insist on energy companies producing a certain proportion of their energy from renewable sources.
	◆ Legislation regarding emissions from power stations will favour the use, for example, of low sulphur coal, as opposed to coal with a high sulphur content.

Figure 7.5 The physical, economic and political factors of energy supply

◆ *Environmental factors/public opinion*   Public opinion can influence decisions made by governments. People are much better informed about the environmental impact of energy sources today compared to the past.

Activities

1   Look at Figure 7.5. Pick two bullet points from each of the three categories to investigate further. Present your findings to your group.

2   For the UK find out when:

   a   the last steam trains (burning coal) stopped being used on Britain's general railway network

   b   nuclear electricity first came on line

   c   North Sea gas first came on line.

Discussion point

Working in groups, try to produce other examples for each of the three categories in Figure 7.5.

The traditional sources of energy: spatial variations

Oil: the most essential global energy resource?

The data quoted below is taken from the BP Statistical Review of World Energy 2007. Not everyone agrees with these data as we will see later in this section. Figure 7.6 shows the change in daily oil consumption by world region from 1981 to 2006. After a fall in demand in the early 1980s to under 60 million barrels, daily global demand rose steeply to almost 84 million barrels a day in 2006. The largest increase has been in the Asia Pacific region, which now accounts for 29.4 per cent of consumption. Only North America has a higher global share at 29.6 per cent. In contrast, Africa consumed only 3.3 per cent of global oil, behind South and Central America with 6.2 per cent.

The pattern of regional production is markedly different from that of consumption. In 2006, the Middle East accounted for 31.2 per cent of production, followed by Europe and Eurasia (21.6 per cent), North

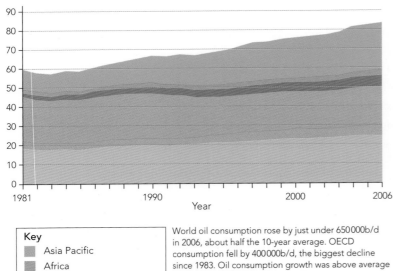

Key

- Asia Pacific
- Africa
- Middle East
- Europe and Eurasia
- South and Central America
- North America

World oil consumption rose by just under 650000b/d in 2006, about half the 10-year average. OECD consumption fell by 400000b/d, the biggest decline since 1983. Oil consumption growth was above average in China and oil-exporting countries.

Figure 7.6 Oil consumption by world region 1981–2006 ▲

the problem with demand increasing at a faster rate than proved reserves. In 2006, the Middle East accounted for almost 62 per cent of global proved reserves. The main countries contributing to the latter figure are: Saudi Arabia 21.9 per cent; Iran 11.4 per cent; Iraq 9.5 per cent; Kuwait 8.4 per cent and the United Arab Emirates 8.1 per cent. Europe and Eurasia held the second largest proved reserves with 12 per cent of the world total. The Russian Federation accounted for over half of the latter figure.

Figure 7.8 shows the reserves-to-production ratio for the world in 2006. While the R/P ratio is almost 80 years in the Middle East it is only 12 years in North America.

America (16.5 per cent) and Africa (12.1 per cent). Within the Middle East, Saudi Arabia dominates production, accounting for 13.1 per cent of the world total. The Russian Federation accounts for over half the total production of Europe and Eurasia.

Figure 7.7 illustrates the spatial distribution of proved oil reserves. In the period 1986–2006, proved reserves rose considerably but much more so in the earlier part of the period than in the latter part. And here lies

Key terms

Proved oil reserves: quantities of oil that geological and engineering information indicates with reasonable certainty can be recovered in the future from known reservoirs under existing economic and operating conditions.

Reserves-to-production (R/P) ratio: the reserves remaining at the end of any year are divided by the production in that year. The result is the length of time that those remaining reserves would last if production were to continue at that level.

Key
- Middle East
- Europe and Eurasia
- Africa
- South and Central America
- North America
- Asia Pacific

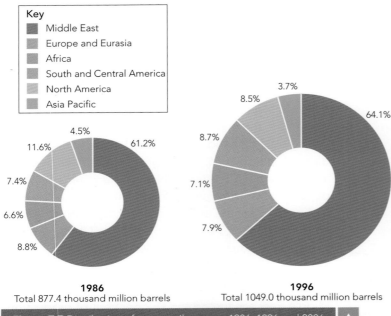

1986
Total 877.4 thousand million barrels

1996
Total 1049.0 thousand million barrels

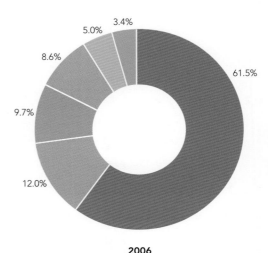

2006
Total 1208.2 thousand million barrels

Figure 7.7 Distribution of proven oil reserves 1986, 1996 and 2006 ▲

Region	Reserves/Production ratio (years)
North America	12.0
South and Central America	41.2
Europe and Eurasia	22.5
Middle East	79.5
Africa	32.1
Asia Pacific	14.0
World	40.5

Source: BP Statistical Review of World Energy 2007

Figure 7.8 Reserves-to-production ratio

The US government's Energy Information Agency predicts that the demand for oil will rise by 54 per cent in the first quarter of the twenty-first century. This amounts to an extra 44 million barrels of oil each day by 2025. Much of this extra demand will come from Asia. All estimates indicate that the Persian Gulf's share of the oil trade will rise steadily over the next two decades and along with it the risk of terrorist attack and embargo by the key producing countries.

Recent price rises

The price of oil rose from $10 a barrel in 1998 to more than $65 a barrel in late 2005. In October 2005 prices in some petrol stations in the UK reached £1 a litre. With some variations the price of oil remained high thereafter and this pattern is likely to continue.

The price of oil affects the cost of almost all other products and activities. The previous three global economic recessions were all strongly linked to a substantial increase in oil prices. However, up to early 2007 the global economy had managed to ride out the rise in the price of oil but some economists doubted that this could continue. Previous sharp and significant increases in the price of oil have been mainly the result of supply shocks (the OPEC oil embargo in 1973–74, the Iranian revolution in 1979, Iraq's invasion of Kuwait in 1990). The current high price situation is mainly due to rising demand. This means that prices are more likely to remain high, although the level may fall back to a certain extent.

> **Key terms**
>
> **Supply shock**: a significant interruption to supply due to an environmental, economic or political event.
>
> **OPEC:** the Organisation of Petroleum Exporting Countries.

Figure 7.9 Oil installations in Saudi Arabia

The following are the main reasons for the recent rapid rise in the price of oil.

◆ The significant increase in the demand for oil is the largest for almost 30 years. Very high growth in demand in China, India, the USA and some other economies has impacted heavily on the rest of the world. Many emerging economies such as India and China subsidise oil, which encourages consumption.

◆ Insufficient investment in exploration and development over the last two decades. The low oil prices for most of this period did not provide enough incentive for investment.

◆ Problems in the Middle East centred on Iraq. Exports of oil from Iraq are well below potential because of terrorist attacks and the slow pace of reconstruction after the war.

◆ Major buyers, particularly governments, stocking up on oil to guard against disruptions to supply.

◆ The impact of hurricanes, particularly on US oil production in the Gulf of Mexico.

◆ Limited US refining capacity due to inadequate investment in recent decades. US refineries are ageing and thus require more maintenance, which in turn reduces capacity.

◆ A lack of spare oil production capacity. In the past Saudi Arabia has maintained a significant amount of spare capacity (wells it was not pumping oil from) to prevent global supply problems when supplies were interrupted in other producing countries. This spare capacity has declined to a 20-year low.

America and China

The USA consumes almost one-quarter of global oil output but has only 2.5 per cent of its proven reserves. Although petrol prices have risen significantly in the USA in recent years, they remain less than half the price in the UK. Of the 20 million barrels a day consumed in the USA, 25 per cent is used for transportation. However, the oil efficiency of US vehicles is at a 20-year low – a result of complacency in the period of low energy prices.

The USA's high dependence on oil leaves it vulnerable to supply shocks and also pushes prices higher for the rest of the world. The only realistic way to limit the demand for oil in the USA is to increase the tax on petrol substantially. However, it is unlikely that any American president would take such a big political risk.

China alone has accounted for one-third of the growth in global oil demand since 2000. China passed Japan as the world's second largest user of oil in 2004. Its average daily consumption of 6.63 million barrels is about twice its domestic production. Because of this situation its oil imports doubled between 1999 and 2004. However, oil consumption per person is still only one-fifteenth of that in the USA. As this gap narrows, it will have a considerable impact on global demand. The demand for oil in China is expected to increase by 5 to 7 per cent a year. If this occurs, China will take over from the USA as the world's largest consumer of oil by 2025.

When will global peak oil production occur?

There has been growing concern about when global oil production will peak and how fast it will decline thereafter. For example, in the USA oil production peaked in 1970. There are concerns that there are not enough large-scale projects underway to offset declining production in well-established oil production areas. The rate of major new oil field discoveries has fallen sharply in recent years. It takes six years on average from first discovery for a very large-scale project to start producing oil. The International Energy Agency expects peak oil production somewhere between 2013 and 2037, with a fall by 3 per cent a year after the peak. The United States Geological Survey predicts that the peak is 50 years or more away.

> **Key term**
>
> **Peak oil production:** the year in which the world or an individual oil producing country reaches its highest level of production, with production declining thereafter.

However, in total contrast, the Association for the Study of Peak Oil and Gas (ASPO) predicts that the peak of global oil production will come as early as 2011, stating 'Fifty years ago the world was consuming

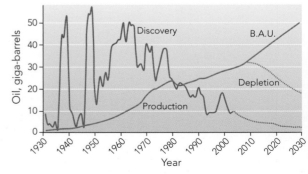

B.A.U. – 'Business as usual' prediction of those who do not accept that world oil production will peak soon

Depletion – the ASPO prediction

Figure 7.10 Global projections for peak oil ▲

4 billion barrels of oil per year and the average discovery was around 30 billion. Today we consume 30 billion barrels per year and the discovery rate is now approaching 4 billion barrels of crude oil per year.' If ASPO is correct and the oil peak is imminent, it will not allow time to shift energy use to alternative sources (Figure 7.10).

Activities

1 Explain why the locations of global oil production and consumption vary so widely.

2 Define the 'reserves/production ratio'. Describe how this varies around the world.

3 a Why is the prediction of peak oil production so important?

   b Suggest why the predictions of when peak oil production will occur vary so widely.

○ 'Take it further' activity 7.2 on CD-ROM

Coal: the number 1 global energy polluter

Figure 7.11 shows that the USA has the largest reserves of coal, enough to last 250 years at its present rate of consumption. Thus, it is not surprising that coal accounts for 50 per cent of US electricity generation and 83 per cent of power plant CO_2 emissions. However, China is by far the largest consumer of coal and the gap between China and the rest of the world will steadily increase in the future. China is expected to need 3242 million tons of coal a year by 2025. It is likely that China will build several hundred new coal-fired power stations to satisfy its demand for energy. This will have a huge impact on greenhouse gas emissions.

Coal is the most polluting source of energy (Figure 7.11). Environmental legislation in a number of countries has required coal-burning power plants to reduce pollutants such as nitrogen oxides and sulphur dioxide by installing building-size scrubbers and catalytic units. However, at present, all the carbon dioxide produced is still released into the atmosphere. This amounts to nearly two billion tons each year from US coal power plants alone.

Coal gasification is the technology that could transform the situation. At the Wabash River plant in the USA the coal is mixed with water and pure oxygen to produce a flammable gas. Sulphur and other contaminants are then removed from the syngas, as it is called, before it is burned in a gas turbine to produce electricity. The Wabash River plant does not take out the carbon dioxide but it is technically possible. The carbon dioxide could then be pumped underground into depleted oil fields. However, scientists say that they need to know more about how buried carbon dioxide behaves to be certain that it will not leak back to the surface.

Key term

Coal gasification: a process which converts solid coal into a gas that can be used for power generation.

Figure 7.11 Coal factfile ▼

WHO HAS COAL? The world has more than a trillion tons of readily available coal. The US has the largest share, but other energy-hungry countries, such as China and India, are richly endowed as well.

27%	17%	13%	10%	9%	5%	19%
USA	RUSSIA	CHINA	INDIA	AUSTRALIA	S.AFRICA	OTHER

WHO USES COAL NOW? Gloabl coal consumption is roughly five billion tons a year, with China burning the most. Western Europe has cut coal use by 36% since 1990 by using natural gas from the North Sea and Russia.

Millions of tons

1531	1117	1094	431	251	1016
CHINA	EUROPE*	USA	INDIA	RUSSIA	OTHER

WHO WILL USE IT TOMORROW? China's coal needs will more than double by 2025 to satisfy factories and consumers. The country also plans to convert coal to liquid motor fuels. Worldwide, consumption will rise by 56%.

Millions of tons

3242	1505	853	736	288	1602
CHINA	USA	EUROPE*	INDIA	RUSSIA	OTHER

* Excluding Russia

WHAT'S IN COAL SMOKE?

SULPHUS DIOXIDE The sulfur in coal forms this gas, which gives rise to acid rain when it reacts with water in clouds. Many plants control sulfur emissions by burning low-sulfur coal and passing the exhaust through scrubbers, which capture sulfur dioxide.

NITROGEN OXIDE The heat of power-plant burners turns nitrogen from the air into nitrogen oxides, which can contribute to acid rain and ground-level ozone. Pollution controls on many plants limit nitrogen oxide emissions.

MERCURY The traces of mercury in coal escape in power-plant exhaust. Falling hundreds of miles away in rain or snow, the mercury builds up in fish, making some species unsafe for children and pregnant women to eat.

CARBON DIOXIDE Coal produces more CO_2 per energy unit than any other fossil fuel. CO_2 is a green-house gas, affecting climate by trapping heat that would otheriwise escape to space. Power plants today release all their CO_2 into the atmosphere.

PARTICULATES Particles from coal-burning plants can harm people who have heart and breathing disorders. Soot and ash are captured before they go up the stacks, but finer particles can form later, from oxides of sulphur and nitrogen.

Source: National Geographic

Figure 7.12 A long coal train in USA ▲

Coal gasification has been around for many years. Before the UK discovered natural gas in the North Sea, gas supplies to homes and industry were produced from coal. However, what is relatively new is the use of coal gas to produce electricity. Currently there are 117 modern gasification plants operational worldwide. About half of these use coal. The others use petcoke, refinery bottom waste and biomass.

At present electricity from coal gasification is more expensive than from traditional power plants but if tougher pollution laws are passed in the future, this situation could change significantly.

Activities

Look at Figure 7.11.

1 Which countries are the main producers and consumers of coal?

2 Discuss the pollution problems associated with coal.

Natural gas: the world's favourite fossil fuel

Natural gas, composed mainly of methane, is the least polluting of all the fossil fuels (Figure 7.13). Production is dominated by Russia and the USA, together accounting for almost 40 per cent of the world total (Figure 7.14). However, the distribution of proven reserves paints a different picture (Figure 7.15). The

Fossil fuel emission levels pounds per billion btu of energy input			
Pollutant	Natural gas	Oil	Coal
Carbon dioxide	117000	164000	208000
Carbon monoxide	40	33	208
Nitrogen oxides	92	448	457
Sulfur dioxide	1	1122	2591
Particulates	7	84	2744
Mercury	0.000	0.007	0.016

Source: EIA – Natural Gas Issues and Trends 1998

Figure 7.13 Fossil fuel emission levels ▲

Country	Production 2006 million tonnes of oil equivalent	2006 share of total %
Russia	550.9	21.3
USA	479.3	18.5
Canada	168.3	6.5
Iran	94.5	3.7
Norway	78.9	3.0
Algeria	76.0	2.9
UK	72.0	2.8
Indonesia	66.6	2.6
Saudi Arabia	66.3	2.6

Source: BP Statistics Review of World Energy

Figure 7.14 Natural gas production 2006

Middle East now holds the largest reserves, a position held by Europe and Eurasia previously.

The relatively low production of natural gas in the Middle East is down to limited markets for the product in the region where in the past the emphasis has been firmly on oil. Much of the natural gas that has been brought to the surface was 'flared off' because there were no pipeline networks to take it to consumers. However, this position is beginning to change.

'Conventional' natural gas, which is generally found within a few thousand metres or so below the surface, has accounted for most of the global supply to date. However, in recent years 'unconventional' deposits

have begun to contribute more to supply. The main categories of unconventional natural gas are:

◆ deep gas

◆ tight gas

◆ gas-containing shales

◆ coalbed methane

◆ geopressurised zones

◆ Arctic and sub-sea hydrates.

Key term

Unconventional natural gas: natural gas that is more difficult and therefore more expensive to extract than 'conventional' reserves.

Unconventional deposits are clearly more costly to extract but as energy prices rise and technology advances, more and more of these deposits are attracting the interest of governments and energy companies.

'Take it further' activity 7.3 on CD-ROM

Nuclear power: a global renaissance?

No other source of energy creates such heated discussion as nuclear power. The main concerns about nuclear power are:

◆ power plant accidents, which could release radiation into air, land and sea

Key
- Middle East
- Europe and Eurasia
- Africa
- South and Central America
- North America
- Asia Pacific

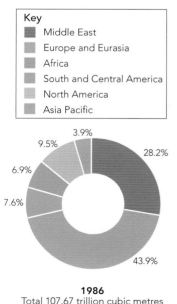

1986
Total 107.67 trillion cubic metres

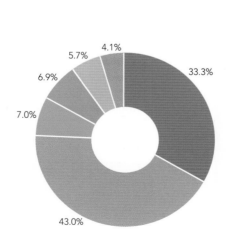

1996
Total 147.89 trillion cubic metres

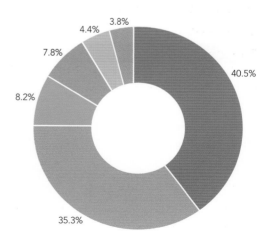

2006
Total 181.46 trillion cubic metres

Figure 7.15 Distribution of proved reserves for natural gas in 1986, 1996 and 2006

- radioactive waste storage/disposal
- rogue state or terrorist use of nuclear fuel for weapons
- high construction costs
- the possible increase in certain types of cancer near nuclear plants.

With 103 operating reactors, the USA leads the world in the use of nuclear electricity. This amounts to a quarter of the world's total, producing 20 per cent of the USA's electricity. At one time the rise of nuclear power looked unstoppable. However, a serious incident at the Three Mile Island nuclear power plant in Pennsylvania in 1979 and the much more serious Chernobyl disaster in the Ukraine in1986 brought any growth in the industry to a virtual halt as Figure 7.16 shows. No new nuclear power plants have been ordered in the USA since then, although public opinion has become more favourable in recent years as: a) Three Mile Island and Chernobyl recede into the past; and b) worries about polluting fossil fuels increase.

The big advantages of nuclear power are:

- zero emissions of greenhouse gases
- reduced reliance on imported fossil fuels.

Other countries, deeply concerned about their ability to satisfy demand, are going ahead with plans for new nuclear power plants (Figure 7.17). China currently produces 6600 megawatts of power from nine reactors. It aims to increase this to 40000 megawatts. India

A sequence of events that would lead to the worst accident possible – the 'China Syndrome': A break on the coolant pipe (1) and failure of emergency cooling (2) would cause the reactor vessel (3) to overheat, melting the fuel (4). This would escape through the concrete slab (5). The containment building (6) would crack, allowing more escapes.

Figure 7.18 Diagram showing the 'China Syndrome'

already has 15 operating nuclear power plants with eight more under construction. France obtains 78 per cent of its electricity from nuclear power and is thinking about replacing its older plants with new ones. But it has yet to decide on this course.

A few countries have developed **fast breeder reactor** technology. These reactors are very efficient at manufacturing plutonium fuel from their original uranium fuel load. This greatly increases energy production but it could prove disastrous if there was an accident (Figure 7.18) or if the plutonium got into the wrong hands as plutonium is the key ingredient for nuclear weapons.

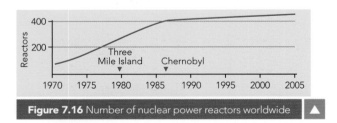

Figure 7.16 Number of nuclear power reactors worldwide

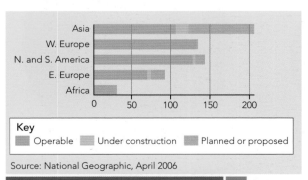

Source: National Geographic, April 2006

Figure 7.17 Number of nuclear power reactors by region (as of December 2005)

> ## Key term
>
> **Fast-breeder reactor:** a nuclear reactor in which the chain reaction is maintained mainly by fast neutrons. It is capable of producing more fissionable material than it consumes.

Activities

1 Describe the information shown in Figure 7.17.

2 Explain the sequence of events that would occur during a nuclear meltdown (the 'China Syndrome' – Figure 7.18).

Figure 7.19 Chernobyl after the disaster in 1986 ▲

Country	Million Tonnes Oil Equivalent	2006 Share of World Total %
China	94.3	13.7
Canada	79.3	11.5
Brazil	79.2	11.5
USA	65.9	9.6
Russia	39.6	5.8
Norway	27.1	3.9
India	25.4	3.7
Japan	21.5	3.1

Source: BP Statistical Review of World Energy

Figure 7.20 HEP consumption 2006 ▲

Discussion point

Divide into two groups. Research the pros and cons of nuclear power in more detail and present your findings to the whole group.

Hydroelectric power (HEP): the largest renewable

The HEP figures produced by the BP Statistical Review (Figure 7.20) are for consumption rather than production but the trade in HEP between countries is extremely limited. The 'big four' HEP nations of China, Canada, Brazil and the USA account for over 46 per cent of the global total.

Of the traditional five major sources of energy, HEP is the only one which is renewable. However, most of the best HEP locations are already in use so the scope for more large-scale development is limited. However, in many countries there is scope for small-scale HEP plants to supply local communities.

Although HEP is generally seen as a clean form of energy, it is not without its problems. Large dams and power plants:

◆ can have a huge negative impact on the environment (Figure 7.20)

◆ obstruct rivers causing problems for aquatic life

◆ may cause deterioration in water quality

◆ may cause large areas of land to be flooded

◆ may cause the release of significant quantities of methane, a greenhouse gas if large forests are submerged without prior clearance.

 'Take it further' activity 7.4 on CD-ROM

U2

7

The energy issue

Figure 7.21 HEP plant and dam ▲

7.2 What is the relationship between energy use and economic development?

There is a very strong correlation between energy use and economic development. As an economy develops greater amounts of energy are required for many reasons including:

- mechanisation of agriculture
- expansion of manufacturing industry
- demands of a growing service sector
- increased freight and general business transportation
- increasing car ownership
- higher levels of domestic use
- expansion of power networks to isolated areas.

The relationship between energy and poverty

Figure 7.22 shows the relationship between GNP per capita and energy use for a large number of countries in 2000. There is a strong positive correlation. However, anomalies do occur for a number of reasons.

Activity

Think of at least one explanation for why such anomalies occur.

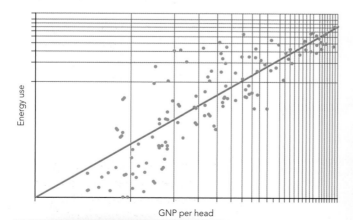

Figure 7.22 Relationship between energy use and GNP per head

Around the world two billion people lack access to household electricity. It has been estimated that connecting these people to electricity services would add only 1 per cent extra to emissions of greenhouse gases. In the poorest countries of the world, traditional biomass often accounts for 90 per cent or more of total energy consumption.

Access to energy services brings important benefits, which include:

- *a better quality of life*   Many poor people in LEDCs meet their energy needs by: a) collecting biomass (fuel wood (Figure 7.23), dung, agricultural residues) for cooking and heating; and b) buying kerosene and batteries for lighting or radio. Collecting biomass takes women and children

Figure 7.23 Use of biomass in LEDC

between two and seven hours a day while kerosene, candles and batteries are usually more expensive than modern energy services.

◆ *health improvements*   Inefficient cooking and heating with biomass fuels causes indoor air pollution, which is a significant cause of poor health and death. Indoor air pollution causes about 1.8 million deaths a year. Access to electricity allows clinics to keep vital drugs under refrigeration, which can make a big difference to health and mortality levels.

◆ *empowerment of the poor*   Access to modern communications helps poor people to become better informed and more independent. It allows people to educate themselves to an extent. For example, farmers become more aware of agricultural prices and how they vary over time and space.

◆ *increased productivity and income*   Electricity is vital for the success of many small-scale businesses. Productivity can be increased by extending the working day and by mechanisation.

◆ *improving the environment*   Direct use of biomass is often very inefficient, causing emissions of toxic materials and greenhouse gases. In addition, the poorest people often live in ecologically sensitive areas which are vulnerable to desertification and deforestation.

Activities

1   Explain why energy use increases with economic development.

2   Look at the five bullet points describing the benefits of access to energy services. Which do you think is the most important benefit?

Case study | The energy situation in the UK

The UK is rapidly running out of the significant reserves of oil and gas that made it a leading producer over the last three decades. In 2005 Britain became a net importer of natural gas and is expected to lose its self-sufficiency in oil by 2009. To add to this the UK coal industry has continued its considerable decline in recent decades. By 2020 Britain will be importing about three-quarters of its primary energy needs. This is at a time when there are growing fears that energy could become a political weapon.

Many of the country's coal and nuclear power stations have been in service for a long time and will need to be closed in the next decade or so. New development will

be costly. The government is also under considerable pressure to reduce the amount of pollution caused by energy production and consumption. It has set the country the optimistic target of reducing greenhouse emissions by 60 per cent by 2050.

Figure 7.24 shows how the UK's energy consumption by source has changed since 1990. The main changes are:

◆ a modest decline in the share of petroleum

◆ a very considerable rise in the relative importance of gas

◆ an equally significant decline in the share of coal.

The contribution of nuclear power and other sources of energy (mainly HEP) changed little during this period.

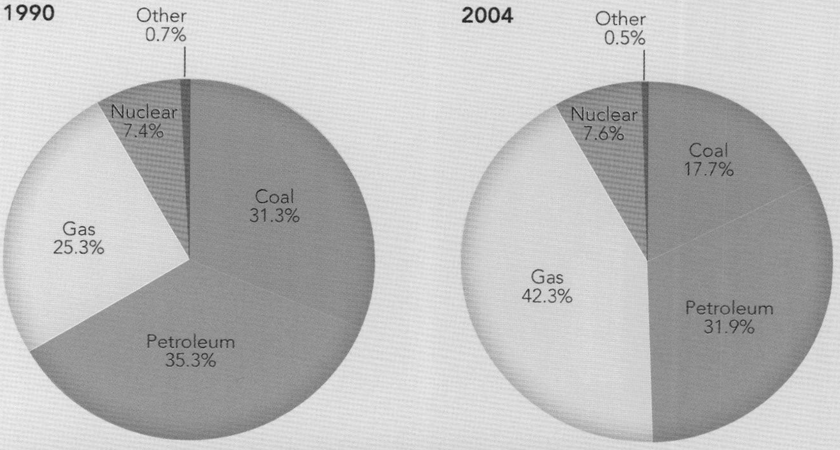

Oil and gas

The UK has already taken between half and three-quarters of the oil and gas in its territorial waters. Much of the remaining North Sea reserves are in small and remote fields. UK oil production peaked at the end

1990

Other 0.7%
Nuclear 7.4%
Coal 31.3%
Gas 25.3%
Petroleum 35.3%

2004

Other 0.5%
Nuclear 7.6%
Coal 17.7%
Gas 42.3%
Petroleum 31.9%

Figure 7.24 UK energy consumption by source 1990 and 2004 ▲

of the 1990s and it has now fallen by about 30 per cent to around two million barrels a day. By 2010 it could be down to 1.2 million barrels a day. During the same period it has been estimated that natural gas production will fall from 9400 million cubic feet to about 6000 million cubic feet.

More than 90 per cent of Britain's gas comes from the North Sea. Approximately 10 per cent is imported via the European gas network. Only 2 per cent of Britain's gas supply comes from Russia, a figure well below that of many mainland European countries. The importation of liquefied natural gas (LNG) to a plant on the Isle of Grain, in Kent, began in 2005.

Clearly, gas imports will rise in the future. There is no other choice as gas is projected to account for an increasing share of electricity generation (Figure 7.25).

The government is trying to encourage development of the remaining reserves in the North Sea. One way is a policy under which oil fields left fallow by their owners may be given to other companies with plans to develop. Also, as many of the largest fields decline in production and become less profitable to the large oil companies, smaller companies are taking their place.

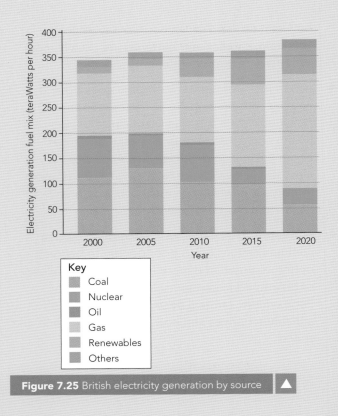

Key
Coal
Nuclear
Oil
Gas
Renewables
Others

Figure 7.25 British electricity generation by source ▲

▼ **Figure 7.26** Oil exploration rig on the Cromarty Firth

New techniques are being developed to extract more oil than previously possible from the North Sea. An example applies to the Miller oil field 240km north-east of Peterhead in Scotland. Production peaked in 1995 and the oil field was due to shut down by 2007. However, by injecting carbon dioxide from a mainland power station into the oil field it has been estimated that an additional 40 million barrels of oil could be extracted. This would increase the life of the Miller field by 15 to 20 years.

Nuclear power

Until mid-2005 it seemed unlikely that the UK would consider building a new generation of nuclear power plants. However, with falling North Sea energy production and other energy concerns, nuclear appears to be back on the agenda. The government is faced with the difficult decision of either allowing the industry to gradually run down as old plants have to be closed or to build new plants. A significant problem is that it takes at least ten years to plan and build a nuclear reactor. Environmental organisations such as Greenpeace are absolutely opposed to nuclear power.

The supporters of nuclear power argue that it is the only way that the UK can avoid electricity shortages and meet its climate change obligations at a reasonable cost. Without the construction of new power plants, the share of nuclear generated electricity will decline from 23 per cent in 2005 to 7 per cent by 2020. Of the country's 12 nuclear plants, nine are due to be closed in the next ten years.

Coal

At the beginning of the twentieth century, coalmining was the country's biggest employer. At its peak the industry employed over one million workers. However, at the end of March 2005 there were only 42 opencast sites and eight major deep mines in production in the UK, employing 9300 people. In 2004, total UK coal production was 25.1 million tonnes. Coal has declined because it is the dirtiest and the most inflexible of the fossil fuels, and because virtually all of the country's easily accessible coal has already been mined.

Total UK coal consumption in 2004 was 60.6 million tonnes with 33 per cent of electricity generated in 2004 coming from coal. The country has 19 coal-fired power stations. Very little UK coal is exported. However, imports are very significant, amounting to 36.2 million tonnes in 2004.

Nevertheless, the coal industry in the UK may be on the point of a limited comeback with the development of 'clean coal technologies'. This new technology has developed forms of coal that burn with greater efficiency and capture coal's pollutants before they are emitted into the atmosphere. The latest 'supercritical' coal-fired power stations, operating at higher pressures and temperatures than their predecessors, can operate at efficiency levels 20 per cent above those of coal-fired power stations constructed in the 1960s. Existing power stations can be upgraded to use clean coal technology.

> **Key term**
>
> **Clean coal technology:** power plant processes that both increase the efficiency of coal burning and significantly reduce emissions.

Hydroelectric power

The UK generates only about 0.8 per cent of its electricity from HEP. Most of the large-scale plants (producing more than 20 megawatts) are located in the Scottish Highlands. There are very few opportunities to increase large-scale HEP production in the UK as most commercially attractive and environmentally acceptable sites are already in use.

However, in July 2005 Scottish ministers approved plans to build a new HEP generating station at Glendoe in Inverness-shire. The power station will be built underground at the side of Loch Ness. The new plant will generate up to 100 megawatts of electricity, sufficient to meet the power requirements of 37 000 homes.

It has been estimated that if small-scale HEP from all of the streams and rivers in the UK could be tapped, it would be possible to meet just over 3 per cent of the country's total electricity needs.

Other renewable forms of energy

The government has set a target of 10 per cent of electricity from renewable sources by 2010. In 2003, biomass used for both heat and electricity generation accounted for 87 per cent of renewable energy in the UK. The majority came from landfill gas (33 per cent) and waste combustion (14 per cent). Electricity production from biomass accounted for 1.55 per cent of total electricity supply in 2003.

Of the other available sources of renewable energy, wind seems to be the only new renewable source of energy available to the UK in any significant quantity. Nevertheless, there has been some progress with other forms of renewable energy. Listed below are some examples.

◆ A small geothermal power plant is in operation in Southampton. Opened in 1986 it provides heating and cooling systems for a number of domestic and commercial consumers. The plant uses hot water from deep below the city.

◆ In 2003 the total capacity for solar photovoltaics in the UK was only 6 megawatts.

◆ There are two wave power devices operating in the UK, both in Scotland. The total capacity amounts to 1.25 megawatts.

Wind energy

Scroby Sands, Britain's newest (2005) and largest wind farm, is located on a sand bank 3km off the coast of Great Yarmouth (Figure 7.27). Its 30 turbines can produce sufficient electricity for 41 000 homes.

> **Key term**
>
> **Wind farm:** a number of wind turbines grouped together at a particular location.

The government promotes wind power at least partly to help meet its commitment to reduce carbon dioxide emissions. Under the 'renewables obligation certificates', energy companies are obliged to generate part of their electricity through renewable sources. At present the requirement is 4.3 per cent but this will rise to 15 per cent by 2015. Power companies without renewable sources of energy can meet their obligation by buying credits from other companies that operate renewable energy facilities. The credit system has resulted in significant payments from conventional power companies to 'green' operators. Conventional power companies are also subject to a climate change tax.

In 2005 the National Audit Office estimated that total financial assistance to the renewable energy industry (mainly wind farms) amounted to £700 million a year. This is expected to rise to £1 billion by 2010.

A recent estimate by the trade magazine *Platts Power UK* was that renewable energy capacity would rise 21-fold between 2005

and 2010. If this materialises, 7 per cent of the UK's electricity supply will come from wind. It takes about three years from planning consent for a wind farm to become operational. A recent report by the Royal Academy of Engineering stated that the only forms of electricity more costly than wind power are wave power and poultry-litter power. It has been estimated that a wind farm the size of Dartmoor would be needed to produce the same output as an average conventional power station.

Figure 7.27 Wind energy locations in the UK ▼

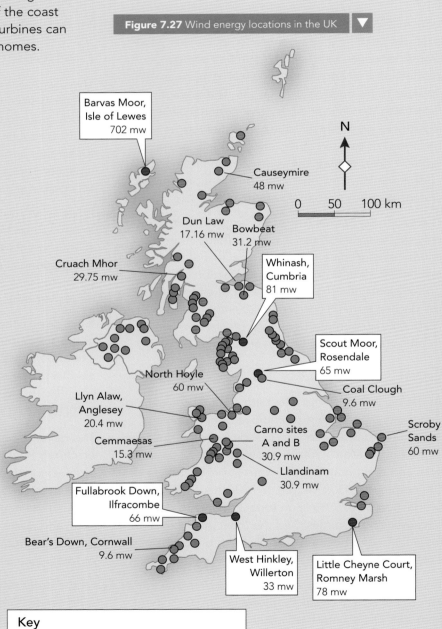

Barvas Moor, Isle of Lewes 702 mw

Causeymire 48 mw

Dun Law 17.16 mw

Bowbeat 31.2 mw

Cruach Mhor 29.75 mw

Whinash, Cumbria 81 mw

Scout Moor, Rosendale 65 mw

North Hoyle 60 mw

Coal Clough 9.6 mw

Llyn Alaw, Anglesey 20.4 mw

Scroby Sands 60 mw

Cemmaesas 15.3 mw

Carno sites A and B 30.9 mw

Llandinam 30.9 mw

Fullabrook Down, Ilfracombe 66 mw

Bear's Down, Cornwall 9.6 mw

West Hinkley, Willerton 33 mw

Little Cheyne Court, Romney Marsh 78 mw

N

0   50   100 km

Key

● Under construction or given consent

● Operating

● Controversial

mw Megawatts

Microgeneration

There has been a developing interest in new small-scale energy generators. In the UK, microgenerators are generators with an output of less than 50 kilowatts. Photovoltaic tiles and small wind turbines on roofs are no longer a rarity. If these were installed in large enough numbers they could take a considerable strain off overloaded distribution grids. The government-sponsored Energy Saving Trust (EST) estimates that home-powered generators of various types could provide 30 to 40 per cent of the UK's electricity needs by 2050.

> **Key term**
>
> **Microgeneration:** generators producing electricity with an output of less than 50kW.

Activities

1 Look at Figure 7.24.

  a Explain the composition of Britain's energy supply in 2004.

  b Suggest reasons for the changes between 1990 and 2004.

2 Describe the trends shown in Figure 7.25.

'Take it further' activity 7.5 on CD-ROM

Case study | Energy issues in Mali: an LEDC

Mali in West Africa is a huge landlocked and extremely poor country (Figure 7.28). It is in the Sahel, a region threatened by drought and desertification. The northern 65 per cent of the country is desert or semi-desert. Most people depend on the environment, in terms of farming, herding or fishing for their livelihoods. The population of 12 million is growing at about 3 per cent a year.

Energy is a big issue in Mali. The country has no fossil fuel resources of its own. This means that fossil fuels have to be imported through neighbouring coastal countries, increasing costs considerably. For example, generating costs for grid electricity are twice as high as those for the Ivory Coast. Imported petroleum accounts for 8 per cent of the country's trade balance. This is a major financial cost for a very poor country.

In rural areas 80 per cent of energy needs are supplied by firewood and charcoal. This uses over 50 million tons of national forest reserves every year. Kerosene lamps, torches and rechargeable car batteries are used for lighting. The latter also facilitate TV and radio. A small minority have a generator or solar panels.

Woodcutting is a rural industry in itself and provides the only source of employment for many people. Areas close to the main urban areas are particularly vulnerable to this activity. In 1994, there were 600 000

▼ **Figure 7.28** The location of Mali

tonnes of wood cut for use in the capital Bamako. By 2006 this had increased to almost 900 000 tonnes. The government predicts that if nothing is done to reverse this trend, the demand for wood will be greater than supply by 2010.

Less than 12 per cent of the population has access to formal electricity (Figure 7.29). The contrast between urban and rural areas is huge. Most people connected to mains electricity live in the capital Bamako and the main towns. In rural areas less than 1 per cent of the population has access to mains electricity.

The lack of access to electricity and energy services has a big impact on the quality of life in general. For example:

◆ the electrification of rural health centres can bring significant improvements in life expectancy and infant mortality

◆ the electrification of schools not only improves children's education but also allows evening educational opportunities for the adult population

◆ the lack of electricity severely hinders the development of income-generating activities.

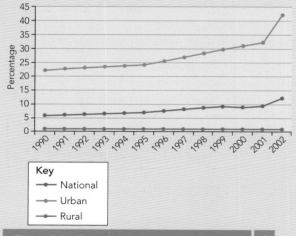

Figure 7.29 Electricity access in Mali 1990–2002 ▲

Renewable possibilities

In 1990, the government formulated a new National Domestic Energy Strategy. Its stress on renewable energy gained important international financial backing.

With an average of 5–6 hours of sunshine a day, there are few countries with a better solar resource. The Mali Folke Centre (MFC), a Christian Aid partner, has installed solar panels on the roofs of 30 schools to provide lighting. Local people have been trained to maintain them.

MFC has also helped to develop plantations of jatropha. This plant not only helps to stabilise areas close to desertification, but it also produces seeds which are made into a biofuel substitute for diesel oil.

The African Rural Energy Enterprise Development (AREED) is a United Nations programme designed to develop new sustainable energy enterprises that use clean, efficient and renewable energy technologies. The aim is to meet the energy needs of underserved populations while cutting the environmental and health consequences of existing energy use, particularly low-quality biomass fuels such as dung and wood. AREED provides early-stage funding and advice to entrepreneurs. One project has involved the production of fuel briquettes made from agricultural by-products. The briquettes burn for longer and are 25 per cent cheaper than traditional wood-based products.

In June 2004 a law was passed to protect certain forest species against excessive cutting. The government has also suspended the export of wood obtained from living trees since 2004. The government and NGOs are trying to encourage women to make greater use of energy-efficient stoves, which use 4–5 times less wood than traditional stoves.

Activities

1   Produce a series of six or more bullet points to explain Mali's energy problems.

2   Describe the contrasts and trends shown in Figure 7.29.

3   Describe and explain the differences in energy production and consumption in the UK and Mali.

7.3 What are the social, economic and environmental issues associated with the increasing demand for energy?

The increasing production and use of energy impacts in a variety of ways on people and the environment. These effects can be broadly grouped into economic, social and environmental (Figure 7.30).

Economic	<ul><li>Building more power stations and extending electricity transmission corridors.</li><li>New sources of power may be required if existing sources reach their upper limits.</li><li>High levels of investment are needed to install a modern energy infrastructure. Foreign direct investment may be required.</li><li>Skilled energy workers may have to be brought in from abroad initially. However, over time the skills base in the domestic energy sector should develop, providing considerable employment.</li><li>As the energy infrastructure expands it may encourage the development of other industries.</li><li>Some countries gain considerable wealth from exporting surplus energy.</li></ul>
Social	<ul><li>Connection to an electricity system can bring considerable social benefits, particularly in health and education.</li><li>There are debatable health concerns about living close to nuclear and coal-fired power stations in particular and also about living in proximity to electricity pylons.</li><li>A modern energy economy allows people to communicate in a wide variety of ways but face-to-face contact is often reduced as a result.</li></ul>
Environmental	<ul><li>The energy infrastructure has a significant visual impact.</li><li>Fossil fuel combustion is the main source of three major pollution problems: climate change, acid deposition and urban smog.</li><li>Flooding forested areas behind HEP dams results in the release of methane.</li><li>Oil spills can occur from wells, pipelines and along tanker routes.</li><li>The increasing demand for firewood can contribute to desertification in semi-arid areas.</li><li>Emissions from transportation increase rapidly as an economy develops.</li></ul>

Figure 7.30 The impact of increased energy demand

> ### Key term
>
> **Foreign direct investment:** overseas investment in physical capital by transnational corporations.

The exploitation of energy resources brings both opportunities and problems for people and the environment.

Case study 1 | Norway: the social and economic opportunities created by the exploitation of energy resources

The discovery of oil and gas in Norwegian waters in the late 1960s has brought major economic and social opportunities to Norway but also some problems, mainly environmental, to overcome. Because the quantities of oil and gas have been large and Norway's population is relatively small (4.5 million), the benefits per person have been considerable. Oil and gas account for one-third of Norway's export earnings. Norway is the seventh largest producer of crude oil in the world and the third largest exporter of oil behind Saudi Arabia and Russia. No other country produces more offshore oil than Norway. In terms of total energy production, Norway ranks eight in the world. Less than a third of Norway's reserves have been extracted.

Oil and gas is Norway's most important industry. It is vital to the economy. Around 80 000 people are directly employed by oil-related businesses, with considerable knock-on effects. Almost 250 000 jobs are attributed directly or indirectly to the oil and gas industry. In the early years Norway largely relied on the expertise of foreign companies but now its own oil and gas industry is very well developed (Figure 7.31). Such expertise is in demand in other parts of the world. Norway is acknowledged as a global leader in sub-sea technology. In addition, the oil and gas industry has boosted innovation and technological development in other industrial sectors.

▲ **Figure 7.31** Norwegian offshore oil and gas

Norway has also been richly endowed with the physical characteristics to develop larger-scale hydroelectricity. Hydroelectric power accounts for 99 per cent of electricity generation in Norway. The small remaining amount comes from electricity imports. Thus, it is not surprising that Norway is one of Europe's cleanest nations in terms of energy use. Norway has about 850 hydroelectric plants giving a total installed capacity of over 27 000MW. However, as all the most obvious hydroelectric locations are in use, there is only limited capacity for further development.

Norway's relatively cheap hydroelectricity has attracted heavy industries requiring large power inputs. Such foreign direct investment has provided a significant number of jobs, creating a cycle of cumulative causation in the regions affected. Southern Norway has the largest concentration of electro-metallurgical and electro-chemical industries in Western Europe. More than 30 plants produce aluminium, zinc, ferro-alloys, fertilisers and ammonia.

> **Key term**
>
> **Cumulative causation:** the process where a significant increase in economic growth can lead to even more growth as more money circulates in the economy.

The standard of living in Norway is much higher than it would be without oil and gas revenues. Incomes are very high and the country is able to afford a welfare system which is one of the best in the world.

Community development is a top priority on Norway's social agenda. Revenue from oil and gas has allowed a high per capita spending on sports, youth, transport and general community facilities, not just in urban areas but also in smaller more isolated communities. This has helped to sustain population in the less accessible parts of the country.

Environmental and economic concerns have been raised; the large number of hydroelectric plants are bringing an increasing number of heavy industries into Norway, attracted by the cheap HEP.

Norway's oil and gas sector works to high standards in terms of both operation and inspection, resulting in a very limited environmental impact. However, with the industry concentrated offshore, the danger of a major maritime incident is always possible.

Norway has a Renewable Energy Programme. Its main objective is to reduce hydropower transmission line loss and to develop undersea cable technology. It is also looking at solar, wind and biomass possibilities.

Both people and politicians in Norway realise that their oil and gas resources will not last forever. Thus, for a number of years, surplus oil revenues have been placed in a Government Petroleum Fund. This money is invested abroad and is now valued at more than $150 billion.

Activities

1 Research a map which shows the location of Norway's oil and gas fields and its major HEP plants.

2 How have such abundant energy resources improved the quality of life in Norway?

Case study 2 | Nigeria: limited benefits from oil wealth

> Oil fouls everything in southern Nigeria. It spills from the pipelines, poisoning soil and water. It stains the hands of politicians and generals, who siphon off its profits. It taints the ambitions of the young, who will try anything to scoop up a share of the liquid riches – fire a gun, sabotage a pipeline, kidnap a foreigner.
>
> *National Geographic, February 2007*

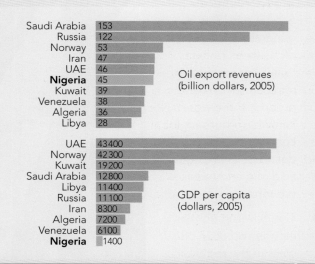

Figure 7.32 Nigeria's paradox hardship persists as oil flows ▲

Potential opportunities

Oil was first extracted from the Niger Delta in 1956. The low-sulphur oil quickly gained important export markets around the world and by the mid-1970s Nigeria had joined OPEC, the influential Organisation of Petroleum Exporting Countries.

Nigeria is the world's sixth largest oil exporter. The destinations of Nigeria's oil exports are:

◆ USA 40 per cent

◆ Western Europe 21 per cent

◆ Asia and Pacific 16 per cent

◆ Latin America 10 per cent

◆ Africa 9 per cent

◆ Other destinations 4 per cent.

Production is dominated by five major transnational corporations (TNCs): Shell, Total, Agip, ExxonMobil and Chevron. Together they operate 159 oil fields, 275 flow stations and 4500 miles of pipelines. Many of the new oilfields are offshore.

Oil related problems

In 1960, farm products such as palm oil and cacao beans accounted for nearly all Nigeria's exports. Today, oil makes up 90 per cent of export earnings and 80 per cent of its revenue. The most populous country in Africa with 130 million people has gone from being self-sufficient in food to importing more than it produces. This is because both government and many communities have neglected agriculture in pursuit of oil wealth (Figure 7.32).

Although Nigeria produces a large quantity of oil, its own refineries are old and poorly run, resulting in frequent breakdown. Thus, the country also imports the bulk of its fuel. A recent United Nations report put the quality of life in Nigeria below all the other major oil producing countries. The World Bank sees Nigeria as a 'fragile state' because of failed governance, epidemic disease and the risk of armed conflict.

It has been estimated that corruption siphons off as much as 70 per cent of annual oil revenues. Most Nigerians live on less than a dollar a day (Figure 7.33). Fifty years of oil extraction has failed to improve the lives of the majority of people. Critics blame the government and the transnational oil companies for this terrible state of affairs.

Nigeria nationalised its oil industry in 1971. In a joint venture arrangement, the Nigerian National Petroleum Corporation (owned by the government), owns 55–60 per cent of TNC oil operations. The question is – where has the money gone? A recent international report used the terms 'the cancer of corruption' and 'the institutionalised looting of national wealth'. Nigeria's anticorruption agency estimated that in 2003, 70 per cent of oil revenues, more than $14 billion, was stolen or wasted. Each one of Nigeria's 36 state governments receives a share of oil money but 'trickle down' to the country's poor is extremely limited.

One of the world's largest wetlands and Africa's largest remaining mangrove forest has suffered an environmental disaster.

◆ Oil spills, acid rain from gas flares and the stripping away of mangroves for pipeline routes have killed off fish.

◆ Between 1986 and 2003, more than 50 000 acres of mangroves disappeared from the coast mainly due to land clearing and canal dredging for oil and gas exploration.

- The oil fields contain large amounts of natural gas. This is generally burned off as flares rather than being stored or reinjected into the ground. Hundreds of flares have burned continuously for decades. This causes acid rain and releases greenhouse gases.

- The government has recognised 6817 oil spills in the region since the beginning of oil production. Critics say the number is much higher.

- Construction and increased ship traffic has changed local wave patterns causing shore erosion and the migration of fish into deeper water.

- Various types of construction have taken place without adequate environmental impact studies.

The federal environmental protection agency has only existed since 1988 and Environmental Impact Assessments were not compulsory until 1992.

▲ **Figure 7.33** Quality of life in the Niger delta

> ### Key term
>
> **Environmental Impact Assessment:** a document required by law detailing all the impacts on the environment of an energy or other project above a certain size.

Local people who have been forced to give up fishing because of reduced fish stocks often find it difficult to get alternative employment. Many local people feel that most jobs go to members of the country's majority ethnic groups – the Igbo, Yoruba, Hausa and Fulani who traditionally come from elsewhere in Nigeria. An added problem is the history of ethnic rivalry in an area inhabited by over 20 ethnic groups. The people of the Niger Delta also accuse the government of inadequate investment in the region in terms of schools, hospitals, housing and other forms of infrastructure.

The largest new development in the Delta is the Gbaran Integrated Oil and Gas Project operated by Shell. It is located along the Nun River, a tributary of the Niger. Encompassing 15 new oil and gas fields, it should begin production in 2008. Shell hope to avoid the mistakes of the past with this project which affects 90 villages by implementing greater environmental controls and conducting discussions with the local community.

From time to time local rebel groups have attacked the oil industry either out of frustration at the paucity of benefits accruing to the region or in an attempt to gain payouts. In 2006, an armed rebel group known as the Movement for the Emancipation of the Niger Delta (MEND) intensified attacks on oil platforms and pumping stations. A rising tide of violence has affected the country's financial stability and its ability to supply crude oil to the outside world.

Activities

1 Why are foreign companies so active in Nigeria's oilfields?

2 Discuss the assertion that 'oil wealth has brought more disadvantages than advantages to the average person in Nigeria'.

7.4 How can energy supply be managed to ensure sustainability?

Energy management and conservation

The improved management of energy supplies is vitally important because of:

◆ concerns about the exhaustion of fossil fuels

◆ the impact of emissions on the environment

◆ the high cost of building large energy installations.

Managing energy supply is often about balancing socio-economic and environmental needs. We have all become increasingly aware that this requires detailed planning and management. Carbon trading is an important part of the EU's environment and energy policies. Under the EU's emissions trading scheme, heavy industrial plants have to buy permits to emit greenhouse gases over the limit they are allowed by government. However, this could be extended to other organisations such as banks and supermarkets. From 2008 the British government is offering the free provision of visual display electricity meters so that people can see exactly how much energy they are using at any time. Many countries are looking increasingly at the concept of community energy. Much energy is lost in transmission if the source of supply is a long way away. Energy produced locally is much more efficient.

> **Key terms**
>
> **Carbon trading**: a company that does not use up the level of emissions it is entitled to can sell the remainder of its entitlement to another company that pollutes above its entitlement.
>
> **Community energy:** energy produced close to the point of consumption.

In March 2006, J M Barrosa took on the European Commission's first attempt at creating an EU energy policy. He argued that the EU can no longer afford 25 different and uncoordinated energy policies.

Figure 7.34 summarises some of the measures governments and individuals can undertake to reduce the demand for energy and thus move towards a more sustainable situation.

> **Activity**
>
> 1 Think of some other measures to add to Figure 7.34.
>
> 2 Which of the 'government' measures would have the greatest impact on your life? Explain why.

Government	Individuals
◆ Improve public transport to encourage higher levels of usage	*Transport*
◆ Set a high level of tax on petrol, aviation fuel etc	◆ Walk rather than drive for short local journeys
◆ Ensure that public utility vehicles are energy efficient.	◆ Use a bicycle for short to moderate distance journeys
◆ Set minimum fuel consumption requirements for cars and commercial vehicles	◆ Buy low fuel consumption/low emission cars
◆ Congestion charging to deter non-essential car use in city centres	◆ Reduce car usage by planning more 'multi-purpose' trips
	◆ Use public rather than private transport
◆ Offer subsidies/grants to households to improve energy efficiency	◆ Car pooling
	In the home
◆ Encourage business to monitor and reduce its energy usage	◆ Use low-energy light bulbs
◆ Encourage recycling	◆ Install cavity wall insulation
◆ Promote investment in renewable forms of energy	◆ Improve loft insulation
◆ Pass laws to compel manufacturers to produce higher efficiency electrical products	◆ Turn boiler and radiator setting down slightly
	◆ Wash clothes at lower temperatures
	◆ Purchase high-energy efficiency appliances
	◆ Do not leave appliances on standby

Figure 7.34 Examples of energy conservation measures ▲

AS Geography for OCR

Theory into practice

Electrical appliances left on standby use about 7 per cent of all electricity used in UK homes. Complete an audit of your domestic appliances. How many have a standby function? How many are regularly left on standby?

Alternative sources of energy: a global summary

The first major wave of interest in new alternative energy sources resulted from the energy crisis of the early 1970s. However, the relatively low price of oil in the 1980s, 1990s and the opening years of the present century dampened down interest in these energy sources. However, renewed concerns about energy in recent years and corresponding price increases have kick-started the alternative energy industry again. Increasing amounts of venture capital are being invested in new green energy initiatives and companies. Renewable energy is becoming an important industry in a number of countries.

The main drawback to the new alternative energy sources is that they invariably produce higher cost electricity than traditional sources. However, the cost gap is narrowing as:

◆ alternative energy technology improves

◆ traditional electricity supply becomes more expensive

◆ governments legislate in favour of alternative energy.

Solar power

◆ *Photovoltaic systems*   These are solar panels (Figure 7.35) that convert sunlight directly into electricity. By the end of 2002, there have been 1500MW installed globally. The leading countries were Japan (627MW), Germany (295MW) and the USA (212MW).

◆ *Thermal power plants*   Total global installed capacity at the end of 2002 was 364MW, most in the form of nine power plants in the Mohave Desert in Southern California. Located on three sites, the plants vary in size from 14 to 80MW.

Figure 7.35 Solar power ▼

Parabolic trough technology is used to collect the sun's rays. Steam is generated at 400°C to drive the turbines which produce 345MW. This is enough to meet the demand from more than half a million people.

◆ *Solar towers*   Another idea being considered is to build a large area of 'greenhouse' with a very tall tower in the middle. The hot air in the greenhouse would rise rapidly up the tower, driving turbines along the way.

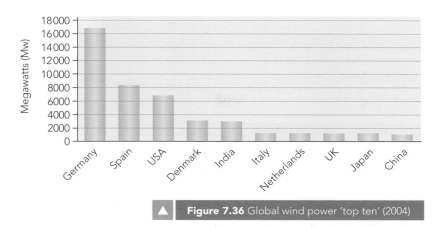

▲ Figure 7.36 Global wind power 'top ten' (2004)

Wind power

Global wind-generated electricity capacity totalled 47 300MW at the end of 2004. Figure 7.36 shows that almost 67 per cent of global wind power is concentrated in just three countries with Germany leading the way. Only seven countries produced more than 1000MW in 2004.

Apart from establishing new wind energy sites, repowering could also play an important role. This means replacing old wind turbines with new engines which give a better performance.

> **Key term**
>
> **Repowering:** replacing old wind turbines with new engines which give a better performance.

Biomass

Biomass is organic matter from which energy can be produced. Living plants generate ten times as much energy each year as people consume. The direct domestic use of biomass, such as burning firewood,

Figure 7.37 Wind farm in California ▼

2005	USA	Brazil	EU	India	China	World
Ethanol	7.50	8.17	0.48	0.15	0.51	17.07
Biodiesel	0.22	0.05	2.53	0.00	0.00	2.91
Total	7.72	8.22	3.01	0.15	0.51	19.98

Source: FT 20/6/07

Figure 7.38 Leading biofuel countries (million tonnes of oil equivalent)

is a major source of energy in many developing countries. But in terms of the future, it is biomass schemes that a) generate electricity and b) act as a substitute for oil that are the centre of attention.

Due to the high price of oil, biofuels are increasing in popularity as sources of power for trucks, cars and planes. In Brazil, all fuel sold to motorists is 22–26 per cent ethanol. Brazil is the world's largest ethanol producer, which it distills from sugar cane (Figure 7.39). However, a new generation of cars (flex-fuel vehicles) can run on ethanol alone, which is about half the price of the alternative at the petrol pumps. Brazil exports ethanol to countries such as Japan and South Korea as they start to diversify energy consumption away from oil. The USA is just behind Brazil in ethanol production, with corn being the crop source. In Germany, 'biodiesel' (from rapeseed) is increasing in production. In over 30 countries a variety of crops are now being grown for fuel. A recent Canadian report stated that biofuels have begun to pose the first serious challenge

to petroleum-based fuel in a century. China expects biofuels to contribute 15 per cent of its transportation energy needs by 2020. With rising grain prices, food and fuels may be in competition in a number of countries.

The more advanced second generation biofuels, such as cellulosic ethanol made from plant waste are still a number of years away from commercial production. The US government is currently funding the development of six pilot plants.

Geothermal power

Geothermal energy is the natural heat found in the earth's crust in the form of steam, hot water and hot rock. Rainwater may percolate several kilometres in permeable rocks where it is heated due to the earth's geothermal gradient. This is the rate at which temperature rises as depth below the surface increases. The average rise in temperature is about 30°C per km, but the gradient can reach 80°C near plate boundaries.

This source of energy can be used to produce electricity or its hot water can be used directly for industry, agriculture, bathing and cleansing. For example, in Iceland, hot springs supply water at 86°C to 95 per cent of the buildings in and around Reykjavik. At present, virtually all the geothermal power plants in the world operate on steam resources (Figure 7.39), having an extremely low environmental impact.

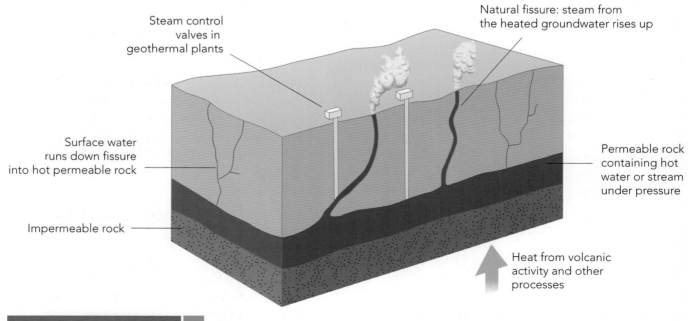

Figure 7.39 Geothermal power

Figure 7.40 Geothermal power plant ▲

By 2005 the global capacity of geothermal electricity had reached 8900MW, with considerable new development underway. The number of countries producing power from this source could rise from 21 in 2000 to 46 in 2010, with global capacity reaching 13500MV.

The USA is the world leader in geothermal electricity with plants in Alaska, California, Hawaii, Nevada and Utah. Total production accounts for 0.37 per cent of the electricity used in the USA. New federal tax incentives introduced in 2005 have encouraged new development. A survey in November 2006 identified 61 projects in various stages of development.

Other leading geothermal electricity countries are the Philippines, Italy, Mexico, Indonesia, Japan, New Zealand and Iceland.

Tidal power

There are two possible ways of exploiting tidal energy.

◆ Building barrages across estuaries and bays with a large tidal range.

◆ Harnessing local tidal currents with moveable, non-permanent installations.

The world's first tidal barrage plant, at La Rance in Brittany, France, opened in 1966. It has a capacity of 240kW and it carries a road crossing the estuary. Other barrages are located at:

◆ Annapolis Royal, Nova Scotia, Canada (20MW) (Figure 7.41)

◆ The Bay of Kislaya, Murmansk, Russia (400kW)

◆ Jiangxia Creek, China (3.2MW).

A number of other larger-scale schemes have been proposed from time to time but there have been major concerns over large capital costs, long construction times and worries about the environmental impact. In Britain there have been proposals to construct barrages across the Severn Estuary and the Solway Firth.

Ocean thermal energy conversion (OTEC)

OTEC uses the temperature difference between the ocean's warm surface water and the cold deep waters to generate electricity. The system can only work in the tropics where the temperature difference between the surface and deep water is at least 20°C.

Figure 7.41 Annapolis Royal tidal power station, Nova Scotia, Canada ▲

The USA developed the world's first at-sea power plant off the coast of Hawaii. Here the warm surface water vaporises ammonia, which is used to drive a turbine. The ammonia is then condensed by deep cold water and the cycle begins all over again.

Unlike some other renewable energy sources, OTEC provides 'base load' power (a continuous electricity flow). However, at present the costs of OTEC are high and the amount of electricity provided is very small.

Case study | Germany: satisfying energy demand in an increasingly sustainable way

◆ As a major global economy, Germany is the fifth largest consumer of energy in the world. The main components of primary energy consumption in 2005 were: oil (36%), natural gas (22.7%), coal (12.9%) and lignite (11.2%). Apart from coal, Germany does not possess any large hydrocarbon reserves, which means a large energy import bill. Developing renewable energy is important to Germany's energy security as well as its ambitions to improve the quality of the environment.

◆ Germany is one of the leading countries in the world in promoting renewable energy. In 2005, renewable energy accounted for 4.6 per cent of

Germany's primary energy supply and 10.2 per cent of its total electricity consumption. Renewable energy is now an important industrial sector in Germany.

◆ Germany meets 5 per cent of its electricity needs from hydroelectricity. Its installed output of 4600MW is concentrated mainly in the pre-Alpine region where physical and precipitation characteristics are most favourable.

◆ By early 2007, Germany had over 18000 wind turbines on line, producing more than 20000MW. Germany now obtains 5.7 per cent of its electricity from wind, employing over 64000 people in the process. Germany is now looking to build offshore windfarms to accompany its land-based turbines. The German government has legislated to promote wind energy since 1991 with the most important impetus coming from the Renewable Energy Sources Act of 2000 and its amendment in 2004. The central element is a minimum price, guaranteed for more than 20 years, which will be paid for electricity fed into the grid that has been generated from renewable energy sources. Power companies have to take this electricity by law. The renewable energy industry also benefits from favourable tax concessions.

◆ Germany boasts a $5 billion photovoltaic industry which accounts for 52 per cent of the world's installed solar panels. The world's largest photovoltaic system is Bavaria Solarpark in Muehlhausen, Germany. It generates 10MW from 57 600 photovoltaic panels spread over three sites covering a total of 62 acres.

◆ Bioenergy (energy derived from plant and animal matter) is also a source of interest. Juhnde, near Gottingen, is Germany's first model bioenergy village.

◆ Only a few regions in Germany have potential for harnessing geothermal power. The exploitation of deep geothermal power is little more than an idea at present but geothermal sources close to the surface have provided hot water and heating to households in a few areas for 30 years.

Activities

1 Research the proposal to build a barrage across either the Severn Estuary or the Solway Firth.

2 Considerable investment in biofuels is occurring. Discuss the advantages and disadvantages of this source of energy.

 'Take it further' activity 7.6 on CD-ROM

Knowledge check

1 What do you understand by the term 'energy crisis'?

2 Why is the world still so reliant on fossil fuels?

3 Explain the physical factors that cause variations on energy supply.

4 Describe the global distribution of:

a oil production

b proved oil reserves.

5 Why does coal remain such an important source of energy in some countries despite the pollution problems associated with this form of energy?

6 Why is natural gas sometimes described as 'the world's favourite fossil fuel'?

7 Discuss the concerns associated with nuclear electricity.

8 Why is it unlikely that there will be a significant expansion in global hydroelectricity in the future?

9 Explain the relationship between energy use and economic development.

10 Discuss the main social, economic and environmental issues associated with the increasing demand for energy.

11 Explain how energy supply can be managed to ensure sustainability.

A list of useful websites accompanying this chapter can be found in the Exam Café section on the **CD-ROM**

Exam Café
Relax, refresh, result!

Relax and prepare

What I wish I had known at the start of the year…

Shahid

"On the day of the exam I make sure I have a good leisurely meal high in slow release energy foods i.e. not burger and chips which are high in fats and make me feel drowsy. I also aim to get to the exam at least 20 minutes before the start so that I am seated well before the start time. Being one of the last ones in can be stressful. I also make sure I have gone to the toilet – you can leave the exam to go but someone has to go with you! Being nervous is natural – once I get started I feel better – even filling in the cover of the answer book helps. If things are tough, I breathe deeply and have some chocolate."

Mary

"I find it easy to understand the difference between **renewable** and **non-renewable** energy resources but I had to ask my teacher to explain what **semi-renewable** resources are. She explained that some renewable resources are not renewable forever or tend to stop being renewable once the level of use is faster than it can be naturally replaced. The most obvious examples are biofuels. If you cut down and use wood as a fuel faster than it grows back then it ceases to be renewable – instead it is semi-renewable."

Common mistakes – Zara

"It is all too easy to confuse the meaning of words in the heat of the exam. The ones I regularly confuse are 'describe' and 'explain'. Describe means saying what something is like, for example its shape, size, characteristics and location. This is quite straightforward and often forms the first part of a question. There should be no explanation. When describing something, quote figures e.g. 'There are 24 nuclear reactors in the UK and they are all in coastal locations.' You can then draw a map of the UK and label the reactors on the map. 'Explain' means why something is like that, or why it is there. It gives me the chance to show what I know and can apply the list of factors."

Refresh your memory

7.1 What are the sources of energy and how do these vary in their global pattern?

Non-renewable	Fossil fuels (oil, gas, coal), nuclear
Semi-renewable	Wood, biofuels, nuclear
Renewable	Wind, tidal, wave, solar, geothermal, water (water wheel and hydro)
Pattern	LEDCs – rely on wood, bio, fossil fuels + traditional e.g. water wheel
	NICs – rely on oil gas and increasingly nuclear power. Large hydro schemes are becoming more common
	MEDCs – reducing coal and oil and increasing renewables e.g. wind

7.2 What is the relationship between energy use and economic development?

Physical	Climate (wind, water, solar), relief (water), water supply (thermal), vegetation, Geology (fossil fuels, geothermal)
Economic	Capital, technology, demand, site size, transport (of fuel and energy), cost of fuels/operation, competition (imports), agriculture, forestry, waste disposal
Social	Safety, pollution (air, water, solid – acid rain), noise
Political	Cost, security, opposition from voters, impact on local economy
Development	This increases the demand for energy as: – higher incomes mean more consumer goods – increased use of transport especially petrol driven – increased demand for heating and air conditioning – mechanisation (fuel) of farming, services (computers) etc – increased demand from industry and construction

7.3 What are the social, economic and environmental issues associated with the increasing demand for energy?

Social	Inequalities in energy supply, cost etc (energy divide) Conflicts with populations living in energy producing areas Changing lifestyles e.g. use of computers
Economic	Increased costs of production and higher prices Cost of transporting – impact on networks
Environmental	Increased building of power plants impacts on environment Global warming Acid rain and other pollutants Waste disposal including impact of hot water on rivers/coasts

Refresh your memory

7.4 How can energy supply be managed to ensure sustainability?	
Supply	Expand energy production – build new plant
	New technology e.g. nuclear fusion
	Diversify energy production – new sources or new fuels
	Increase renewable e.g. fast breeder reactor, tidal
	Reduce loss in transfer
	Reduce waste e.g. flaring off gas
Demand	Ration energy – raise price e.g. petrol tax
	Reduce consumption e.g. insulation
	Make machinery more energy efficient e.g. long life bulbs
	Increase public's awareness of energy use e.g. education
Sustainability	Use of renewable sources and fuels
	Reduction in demand via price, better insulation, low energy ideas
	Increased efficiency so less used and higher yield in production
	Advanced technology e.g. nuclear fusion, space mirrors etc

Top tips . . .

'Unfortunately most students in exams provide examples from within the British Isles, usually quoting the same examples. Examiners do get fed up with the same case studies. Why not provide one example from an area in a developing country or from a place you have visited.

Get the result!

Sample question

Outline the main forms of renewable energy resources. [6 marks]

Examiner says

This answer would gain maximum marks because (1) and (2) it provides a basic definition with an example, (3) it gives a form of renewable energy resource with example, (4) it gives another form, (5) elaborates on two more forms of renewable energy resources, (6) elaborates on use of wind power with appropriate example.

Student answer

A renewable resource is one that is inexhaustible e.g. solar energy (1) or one that is renewed as part of its natural life cycle e.g. water (2). In terms of energy this includes biofuels such as wood (3) as the trees will re-grow. Others do not involve the destruction of any fuel and they include geothermal (4), which uses the earth's own internal heat, wave or tidal power (5) and wind power that is produced by harnessing the power of the wind via windmills or wind turbines e.g. off the coast at Prestatyn (6).

Examiner says

Notice that this answer is quite short but quickly gets to the focus of the question.

Examiner's tips

Questions are usually marked on a level basis – level of knowledge, understanding, evaluation etc. Higher levels can be reached by showing:

- Detail with good understanding and knowledge
- Ideas developed effectively especially cause-effect
- Examples/data/evidence are clearly integrated into the answer
- Effective use of geographical terminology
- An effective use of written communication often supported with maps or diagrams
- A level of complexity in your answer – it may vary with:
  - level of development
  - time
  - location (e.g. coast v inland, highland v lowland)
  - scale (local v regional)
  - between groups or aspects (e.g. social v environmental)

Chapter 8

The growth of tourism

Over the past 50 years tourism has developed into a major global industry which is still expanding rapidly. Under one method of economic measurement it is the world's major service industry with a significant presence in every continent. Without doubt it is one of the major elements in the process of globalisation.

Questions for investigation

- In what ways has the global pattern of tourism changed?

- What is the relationship between the growth of tourism and economic development?

- What are the social, economic and environmental issues associated with the growth of tourism?

- How can tourism be managed to ensure sustainability?

Consider this

'Tourism is essentially the renting out for short-term lets of other people's environments, whether that is a coastline, a city, a mountain range or a rainforest.'
Lord Marshall, former Chief Executive Officer of British Airways.

8.1 In what ways has the global pattern of tourism changed?

The development of tourism

People have had a fascination for travel since very early times. There has always been an urge to discover the unknown and experience new environments. Thus, travel to achieve these ends is not new, but tourism, as the term is understood today, is of relatively modern origin. Tourism is distinguished by its mass character from the individual and small group travel undertaken in earlier times.

> **Key term**
>
> **Tourism:** travel away from the home environment a) for leisure, recreation and holidays, b) to visit friends and relations (VFR), and c) for business and professional reasons.

The medical profession was largely responsible for the growth of taking holidays away from home. During the seventeenth century, doctors increasingly began to recommend the benefits of mineral waters and by the end of the eighteenth century there were hundreds of spas in existence in Britain. Bath and Tunbridge Wells were among the most famous (Figure 8.1). The second stage in the development of holiday locations was the emergence of the seaside resort. Sea bathing is usually said to have begun at Scarborough in about 1730.

The annual holiday, away from work, for the masses was a product of the Industrial Revolution, which brought big social and economic changes. However, until the latter part of the nineteenth century only the very rich could afford to take a holiday away from home.

The first package tours were arranged by Thomas Cook in 1841. These took travellers from Leicester to Loughborough, 19km away, to attend temperance (abstinence from alcoholic drink) meetings. At the time it was the newly laid railway network that provided the transport infrastructure for Cook to expand his tour operations. Of equal importance was the emergence of a significant middle class with time and money to spare for extended recreation.

By far the greatest developments have occurred since the end of the Second World War, arising from the substantial growth in leisure time, affluence and mobility enjoyed in developed countries. However, it took the jet plane to herald the era of international mass tourism. In 1970, when Pan Am flew the first Boeing 747 from New York to London, scheduled planes carried 307 million passengers. By 2006 the number had reached 2.1 billion.

Travel motivators are the reasons why people travel. All the major tourism organisations recognise three major categories (Figure 8.2).

> **Key terms**
>
> **Package tour:** the most popular form of foreign holiday where travel, accommodation and meals may all be included in the price and booked in advance, usually through a travel agent. Optional extras such as car hire and special visits may also be booked at the same time.
>
> **Travel motivators:** the reasons why people travel.

Figure 8.1 The historical spa town of Bath ▲

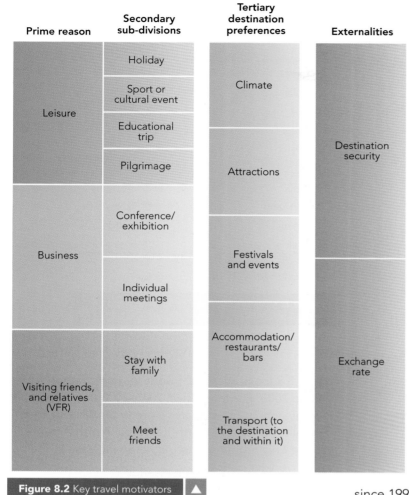

Figure 8.2 Key travel motivators ▲

The diagram columns are labelled:

Prime reason — Leisure; Business; Visiting friends and relatives (VFR)

Secondary sub-divisions — Holiday; Sport or cultural event; Educational trip; Pilgrimage; Conference/exhibition; Individual meetings; Stay with family; Meet friends

Tertiary destination preferences — Climate; Attractions; Festivals and events; Accommodation/restaurants/bars; Transport (to the destination and within it)

Externalities — Destination security; Exchange rate

Recent data

In 2005 international tourist arrivals exceeded 800 million, an all-time record (Figure 8.3). The long-term average annual growth rate has been 4.1 per cent. Europe continues to be the most important destination but the geographical share of tourist arrivals has changed significantly over the past 50 years. The World Tourism Organisation (WTO) forecasts an increase to over 1 billion arrivals in 2010 and 1.6 billion in 2020.

> **Key term**
>
> **International tourist arrivals:** tourists travelling to a country, which is not their place of residence, for more than one day but not longer than a year.

International tourism receipts totalled $680 billion in 2005 with 70 countries earning more than $1 billion from international tourism. For destination countries, receipts for international tourism count as exports. Tourism accounts for about 6 per cent of the total value of global exports of goods and services. It ranks fourth after fuels, chemicals and automotive products. According to the WTO, tourism is one of the top five export categories for as many as 83 per cent of countries and is the main source of foreign exchange for at least 38 per cent of countries. Figure 8.4 shows the trend in receipts since 1990 and the relationship between arrivals and receipts.

> **Key term**
>
> **International tourism receipts:** money spent by visitors from abroad in a destination country.

Fifty per cent of inbound tourism is for the purpose of leisure, recreation and holidays (Figure 8.5). The second most important reason was visiting friends and relatives. Air transport accounted for 45 per cent of arrivals (Figure 8.6). Air is the fastest-growing method of travel. However, road transport remains very significant.

Figure 8.7 shows the top ten tourist-destination countries by number of international arrivals and tourist receipts. Nine of the top ten destinations appear in both lists, although in a different order.

Figure 8.3 International tourist arrivals 1950–2020 ▲

Key:
- South Asia
- Middle East
- Africa
- East Asia and the Pacific
- Americas
- Europe

Source: World Tourism Organisation

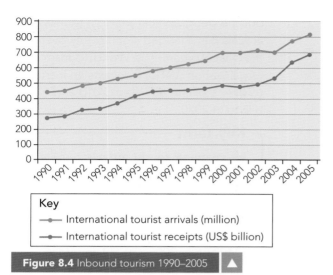

Key
— International tourist arrivals (million)
— International tourist receipts (US$ billion)

Figure 8.4 Inbound tourism 1990–2005 ▲

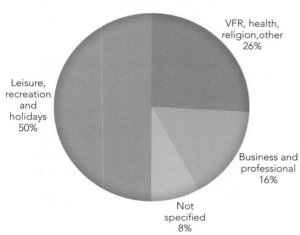

VFR, health, religion, other 26%

Leisure, recreation and holidays 50%

Business and professional 16%

Not specified 8%

Figure 8.5 Inbound tourism by purpose of visit, 2005 ▲

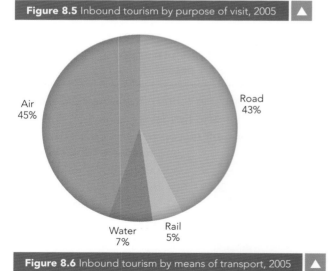

Air 45%

Road 43%

Water 7%

Rail 5%

Figure 8.6 Inbound tourism by means of transport, 2005 ▲

Key term

Tourist-destination country: a country that receives a significant number of international tourists.

a) International tourist arrivals (million)	
Rank	**Million**
1 France	76.0
2 Spain	55.6
3 USA	49.4
4 China	46.8
5 Italy	36.5
6 UK	30.0
7 Mexico	21.9
8 Germany	21.5
9 Turkey	20.3
10 Austria	20.0

b) International tourism receipts (US$ billion)	
Rank	**Billion**
1 USA	81.7
2 Spain	47.9
3 France	42.3
4 Italy	35.4
5 UK	30.7
6 China	29.3
7 Germany	29.2
8 Turkey	18.2
9 Austria	15.5
10 Australia	15.0

Source: World Tourism Organisation (UNWTO) ©

Figure 8.7 Top ten countries by a) arrivals and b) receipts (2005)

Figure 8.8 looks at global trends by world region from 1990 to 2005. For the period 2000–2005 the fastest average annual growth rates were in the Middle East, Central America and North-East Asia. North America was the only region not to record an increase in the latest five-year period.

Figure 8.9 shows the highly seasonal nature of tourism, which on a global scale is still dominated by the summer holiday period in the northern hemisphere. Seasonality is the major problem with tourism as a source of employment, having a major impact on incomes and the quality of life in the less popular times of the year.

Figure 8.10 shows recent data for inbound tourism by generating regions. About 80 per cent of tourism still takes place within the same region. People from MEDCs still dominate global tourism but many emerging economies have shown very fast growth rates in recent years. When people can afford to travel, they usually do. Tourist-generating countries have a big impact on the flow of money around the world.

Key term

Tourist-generating country: a country that supplies a significant number of international tourists.

	International tourist arrivals (million)				Market share (%)	Average annual growth (%)
	1990	1995	2000	2005	2005	00/05
World	**439**	**540**	**687**	**806**	**100**	**3.3**
Europe	265.6	315.0	395.8	441.5	54.8	2.2
Northern Europe	31.6	40.1	45.8	52.9	6.6	2.9
Western Europe	108.6	112.2	139.7	142.7	17.7	0.4
Central/Eastern Europe	31.5	60.0	69.6	87.9	10.9	4.8
Southern/Mediterranean Europe	93.9	102.7	140.8	158.0	19.6	2.3
Asia and the Pacific	56.2	82.4	110.5	155.4	19.3	7.1
North-East Asia	26.4	41.3	58.3	87.6	10.9	8.5
South-East Asia	21.5	28.8	36.9	49.3	6.1	6.0
Oceania	5.2	8.1	9.2	10.5	1.3	2.6
South Asia	3.2	4.2	6.1	8.0	1.0	5.7
Americas	92.8	109.0	128.1	133.5	16.6	0.8
North America	71.7	80.7	91.4	89.9	11.2	−0.3
Caribbean	11.4	14.0	17.1	18.9	2.3	2.0
Central America	1.9	2.6	4.3	6.5	0.8	8.5
South America	7.7	11.7	15.3	18.2	2.3	3.6
Africa	15.2	20.3	28.2	36.7	4.6	5.4
North Africa	8.4	7.3	10.2	13.7	1.7	6.0
Subsaharan Africa	6.8	13.0	17.9	23.0	2.9	5.1
Middle East	**9.6**	**13.7**	**24.2**	**39.1**	**4.8**	**10.1**

Source: World Tourism Organisation (UNWTO)©

Figure 8.8 Global trends by world region ▲

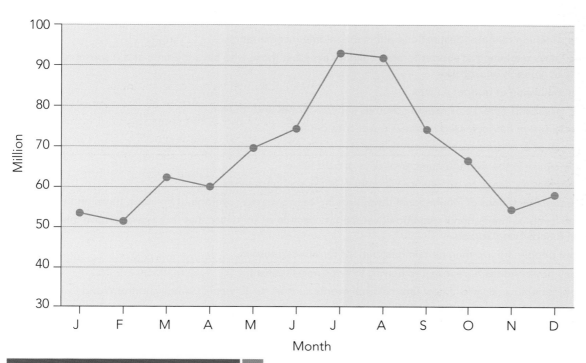

Figure 8.9 Inbound tourism by month (2005) ▲

The growth of tourism

	International tourist arrivals (millions)				Share (%)
	1990	1995	2000	2005*	2005*
World	439.4	540.5	686.8	806.3	100
From:					
Europe	252.6	309.5	396.7	449.0	55.7
Asia and the Pacific	59.1	87.4	115.5	154.3	19.1
Americas	100.3	108.9	131.5	137.1	17.0
Middle East	8.3	9.7	13.7	21.9	2.7
Africa	10.0	12.9	16.5	20.3	2.5
Origin not specified*	9.2	12.1	12.9	23.6	2.9
Same region	350.8	430.6	540.9	634.1	78.6
Other regions	79.4	97.8	133.0	148.6	18.4

Source: World Tourism Organization (UNWTO)©
(Data as collected by UNWTO 2006)

* Countries that could not be allocated to a specific region of origin. As information is derived from inbound tourism data this occurs when data on the country of origin is missing or when a category such as 'other countries of the world' is used grouping countries together that are not seperately specified.

Figure 8.10 Inbound tourism by generating regions

Activities

1 Look at Figure 8.3. Draw up a table to show the number of international tourist arrivals every ten years from 1950. Include projections to 2020.

2 Describe the trends shown in Figure 8.4.

3 Comment on the data shown in Figure 8.7.

4 Suggest why France attracts a relatively high share of short-stay tourists compared to the USA.

5 Describe and explain the variation in international tourist arrivals shown in Figure 8.9.

Discussion point

Are their any countries which you would not visit as a matter of principle? Give reasons for your answer.

Reasons for the growth of global tourism

A range of factors have been responsible for the growth of global tourism. Figure 8.11 subdivides these factors into economic, social and political reasons and also includes factors that can reduce levels of tourism, at least in the short-term.

Key term

Globalisation: the increasing interconnectedness of the world economically, culturally and politically.

Economic	◆ Steadily rising real incomes. Tourism grows on average 1.3 times faster than GDP.
	◆ The decreasing real costs (with inflation taken into account) of holidays.
	◆ The widening range of destinations within the middle income range.
	◆ The heavy marketing of shorter foreign holidays aimed at those who have the time and disposable income to take an additional break.
	◆ The expansion of budget airlines.
	◆ 'Air miles' and other retail reward schemes aimed at travel and tourism.
	◆ Globalisation has increased business travel considerably.
	◆ Periods of economic recession can reduce levels of tourism considerably.
Social	◆ An increase in the average number of days of paid leave.
	◆ An increasing desire to experience different cultures and landscapes.
	◆ Raised expectations of international travel with increasing media coverage of holidays, travel and nature.
	◆ High levels of international migration over the last decade or so means that more people have relatives and friends living abroad.
	◆ More people are avoiding certain destinations for ethical reasons.
Political	◆ Many governments have invested heavily to encourage tourism.
	◆ Government backing for major international events such as the Olympic Games and the World Cup.
	◆ The perceived greater likelihood of terrorist attacks in certain destinations.
	◆ Government restrictions on inbound/outbound tourism.
	◆ Calls by non-governmental organisations to boycott countries such as Myanmar.

Figure 8.11 Factors affecting global tourism

Discussion point

Is it ethically acceptable to take a holiday anywhere at any time?

Variations in the level of tourism over time and space

Unfortunately, more than many other industries, tourism is vulnerable to 'external shocks'. Periods of economic recession characterised by high unemployment, modest wage rises, and high interest rates, affect the demand for tourism in most parts of the world. Because holidays are a high-cost purchase for most people, the tourist industry suffers when times are hard.

Key term

External shock: an economic, political or other trend or event in a major market that significantly reduces the demand for tourism at a particular destination or a range of destinations.

Tourism in individual countries and regions can be affected by considerable fluctuations caused by:

◆ *natural disasters* Earthquakes, volcanic eruptions, floods and other natural events can have a major impact on tourism where they occur.

◆ *natural processes* Coastal erosion and rising sea levels are threatening important tourist locations around the world.

◆ *terrorism* Terrorist attacks, or the fear of them, can deter visitors from going to certain countries, in the short term at least.

◆ *health scares* For example, the severe acute respiratory syndrome (SARS) epidemic in March 2003 had a considerable short-term impact on tourism in China and other countries in South-east Asia.

◆ *exchange rate fluctuations* For example, if the value of the dollar falls against the euro and the pound, it makes it more expensive for Americans to holiday in Europe, but less expensive for Europeans to visit the USA.

◆ *political uncertainties* Governments may advise their citizens not to travel to certain countries if the political situation is tense.

◆ *international image* A US film made in 2006 called *Turistas* has caused major concern in Brazil. It depicts a group of US backpackers whose holiday in a Brazilian resort turns into a nightmare when they are drugged and kidnapped before their organs are removed by organ traffickers.

◆ *increasing competition* As new 'more exciting' destinations increase their market share, more traditional destinations may see visitor numbers fall considerably.

New types of tourism

In the past 20 years, more specialised types of tourism have become increasingly popular (Figure 8.12). An important factor seems to be a general re-assessment of the life–work balance where an increasing number of people are determined not to let work dominate their lives. Niche market tour operators have developed to satisfy the increasing demand for specialist holidays which include:

◆ *theme parks and holiday village enclaves* Theme parks create artificial destinations from scratch. The largest are the Disney theme parks located close to Tokyo, Paris, Los Angeles and Orlando. They are the world's biggest tourist draws.

◆ *gambling destinations* Las Vegas, Nevada, is the largest gambling destination in the world. However, it has long since lost its monopoly on gambling in the USA as new locations have emerged. The scale of destination gambling has increased in many other countries too. It is often a very controversial issue.

◆ *cruising* Cruising is growing faster than any other type of holiday (Figure 8.13). North Americans take more cruises than anybody else. The appeal of this sector of the industry has spread across the age and income spectrum. A new record was set in 2006 with the launch of the 160 000-ton *Freedom of the Seas* with 15 passenger decks.

◆ *heritage and urban tourism* Visits to key historical and cultural sites attract large numbers of people. Many of these sites are in urban areas and the tourism infrastructure has been built up to cater for this demand.

◆ *wilderness and ecotourism* For some people their main holiday objective is to avoid all aspects of mass tourism by connecting with nature in its most unspoiled settings.

Figure 8.12 Factors in the growth of special interest holidays ▲

Figure 8.14 British tourists, with local guide, visiting a First World War cemetery at Ypres, Belgium ▲

◆ *medical and therapy travel* More and more people from the most affluent countries in particular are looking for cheaper general or cosmetic surgery. They undergo surgery shortly after arrival and spend the rest of the two weeks or so on aftercare and tourism. Therapeutic breaks are also increasing in popularity.

◆ *conflict/dark tourism* Dark tourism is the generic term for travel associated with death, tragedy and disaster. Examples are visits to First World War battlefields (Figure 8.14) and to Second World War concentration camps.

◆ *religious tourism* Large numbers of people may visit the holy places associated with their religion, such as Roman Catholics visiting Rome or Muslims taking part in the Haj pilgrimage to Mecca.

◆ *working holidays* Young people on low budgets are often attracted to work for short periods in an attractive environment. Grape picking in France has been a destination for generations of young Britons.

◆ *sports tourism* An increasing number of people have based a holiday around following a national sports team abroad, for example, the football and cricket world cups.

Key term

Niche markets: small markets that deal in a specialised product.

Activities

1 Discuss the economic, social and political factors responsible for the growth of tourism.

2 Explain two factors that can cause variations in the level of tourism at a destination.

3 **a** Why have theme parks become so popular in recent decades?

b Research the development and growth of a major theme park.

4 Assess the advantages and disadvantages of large-scale gambling on the residents of settlements such as Las Vegas.

5 Explain why the market for cruising has expanded so rapidly.

6 Research the expected tourism benefits to London of the 2012 Olympic Games.

Figure 8.13 Cruise ship on the River Nile ▲

8.2 What is the relationship between the growth of tourism and economic development?

As economies develop, there is an increasing demand for tourism. Most of the world's tourists originate from the developed world but their numbers will not rise much further in the future because of demographic trends. However, as the population of the developed world continues to age, the nature of demand will change. The tourist industry is increasingly turning its attention to the developing world where the growing middle classes in countries such as China, India and Brazil want to experience regular foreign travel.

The relationship between tourism and economic growth has generated considerable debate. For the population of a particular country, the growth of the economy in general can stimulate the demand for tourism as average incomes increase. The part of this demand that occurs within the country in question can generate further economic growth as spending increases on hotels, food and drink, souvenirs, entry to attractions and so on. Some businesses will expand and new businesses will form to meet rising demand. New infrastructure may also need to be built to cope with higher visitor numbers. Rising numbers of visitors from other countries will add to this effect. As more people from the country in question go abroad, the impact will be similar in the countries they visit. Tourism can create, or help create, the process of cumulative causation (Figure 8.15).

Average annual growth in arrivals, 1990–2005 (%)

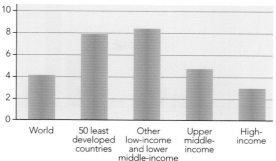

Figure 8.16 Tourism growth by income levels ▲

UNWTO and other organisations have highlighted the potential of tourism to reduce poverty in LEDCs in general and in the Least Developed Countries (LDCs) in particular. The latter are the 49 poorest countries in the world. Tourism is the primary source of foreign exchange earnings in 46 of these countries. Although LDCs accounted for only 1.2 per cent of international tourist arrivals and 0.8 per cent of receipts in 2005, the recent growth rate has been above the global average. Between 2000 and 2005, arrivals were up 48 per cent compared with 17 per cent for the world as a whole. Figure 8.16 shows the average annual growth in tourism in1990–2005 for countries at different income levels.

Figure 8.17 shows total spending by tourists from each region of the world. The gap between the developed and developing worlds is very substantial indeed. Figure 8.18 shows the ten highest countries for spending on international tourism and the ten lowest spending countries. The average spending on international tourism was $92 per tourist in 2003 so the variation around the average is huge – from $6005 to 4 US cents. There is a strong positive correlation between GNP per capita and expenditure on tourism per person by country (see statistical skills in Geographical skills, page 327).

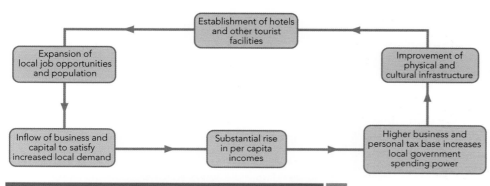

Figure 8.15 Model of cumulative causation relating to tourism ▲

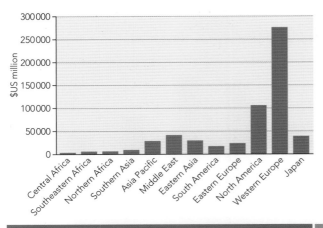

Figure 8.17 Total spending by tourists from each world region (2003) ▲

Highest and lowest spending on International Tourism 2003		
Rank	Territory	Value
1	Luxembourg	6005
2	Kuwait	1563
3	Austria	1562
4	Norway	1499
5	Iceland	1467
6	United Arab Emirates	1365
7	Bahamas	1347
8	Switzerland	1312
9	Belgium	1281
10	Iceland	1239
191	Malawi	4.03
192	Rwanda	3.98
193	Angola	3.94
194	Lao People's Dem Republic	3.09
195	Bangladesh	2.81
196	Burkina Faso	2.78
197	Niger	1.39
198	Ethiopia	0.80
199	Myanmar	0.74
200	Afghanistan	0.04

US$ tourist spending per person per year

Figure 8.18 Highest and lowest spending countries on international tourism (2003) ▲

Discussion point

Discuss the possible reasons why UNWTO considers tourism to be a good route to development for LDCs. Suggest the arguments against this view.

Case study | The UK: a mature tourism destination and tourist-generating country

Major issues in UK tourism

Tourism has had an increasing impact on the UK. With rising affluence the population have been spending more on tourism both within the UK and abroad (Figure 8.19). Likewise, as incomes have risen in other parts of the world, more people want to visit the UK. The ability to visit other places both within the country in which you live and abroad is seen by most people as an important aspect of an improving quality of life. There has been a strong relationship between the increase in average income in the UK and average spending on tourism by UK citizens.

While there are obvious benefits from tourism, particularly in terms of employment and the inflow of money from abroad, there are also significant concerns.

◆ The negative balance of payments account as more money goes out of the country than comes in through tourism. A major reason for this is the desire by UK citizens for guaranteed sunshine abroad.

◆ The regional imbalance of tourism revenues when so much foreign tourism is focused on London and a few other 'world famous' locations.

◆ The 'pressure' of tourism in the most popular 'honeypot' locations.

◆ The perception that the industry provides little reward for many communities apart from low-paid seasonal employment.

◆ The growing concern over the impact of the increasing levels of air transport due to tourism. Plans to expand airports such as Heathrow and Stansted are particularly controversial. Apart from the direct impact on populations near airports, there is growing concern about the contribution of air transport to climate change.

◆ The continued decline of the traditional British seaside resort resulting in high unemployment and urban blight in those resort areas.

Key term

Honeypot location: a place of great interest to a large number of tourists which can become extremely overcrowded at peak times.

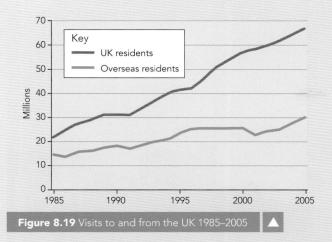

Figure 8.19 Visits to and from the UK 1985–2005 ▲

Country	Visits (000)	Country	Spend (£m)
USA	3438	USA	2384
France	3324	Germany	998
Germany	3294	Irish Republic	895
Irish Republic	2806	France	796
Spain	1786	Spain	697

Figure 8.21 Top five overseas markets for UK tourism (2005) ▲

Spending by overseas residents	£billion
Visits to the UK	14.2
Fares to UK carriers	2.8

Spending by domestic residents	£billion
Trips of 1+ nights	22.7
Day trips	44.3
Rent for second ownership	0.9

Source: VisitBritain www.visitbritain.org

Figure 8.22 Spending by overseas and domestic residents (2005) ▲

Outbound tourism

In 2005, Britons made 66.4 million visits abroad, three times the number in 1985. Two-thirds of these foreign visits were holidays, just under half of which were package holidays. Although the number of foreign holidays continues to rise, there has been a fall in the number of package holidays in recent years.

Of these journeys, 81 per cent were made by air. Spain and France accounted for 38 per cent of all destinations with 13.8 million and 11.1 million visits respectively. Of the UK's population, 15 per cent now fly abroad three or more times each year. Spending on foreign visits reached a record £32.2 billion, a fourfold rise between 1985 and 2005 in real terms.

Inbound and domestic tourism

There were 30 million visits from overseas to the UK in 2005, double the amount in 1984 and an all-time high.

Figure 8.20 The Trooping of the Colour – a major attraction for visitors to London ▼

Two-thirds of these visits were made by people going on holiday. Total spending by foreign visitors reached a record £14.2 billion. Figure 8.21 shows the top five overseas markets for the UK in 2005. Although the top four are very close in terms of number of visits, the USA is by far the largest spending country in the UK.

According to Visit Britain, total spending in 2005 was £85 billion; 80 per cent of this was by UK residents (Figure 8.22). The importance of domestic tourist spending in the UK is sometimes overlooked.

The industry is highly seasonal with the vast majority of both domestic and foreign visits occurring between April and September. July and August are by far the busiest months. Many hotels and tourist facilities close completely in the off-season.

Economic importance

Tourism is an important industry in the UK, accounting for 3.5 per cent of the UK economy. Over 2 million jobs are either directly or indirectly sustained by tourism activity. Directly related employment totals 1.4 million, about 5 per cent of all jobs in the UK. In London, tourism is the second largest (after financial services) and fastest growing sector of the economy. According to the Government Office for London it accounts for 8 per cent of its GDP and 13 per cent of

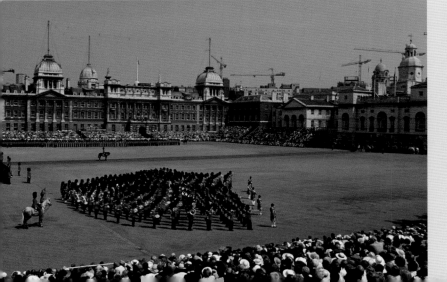

Figure 8.23 A traditional UK seaside resort ▲

employment. An increasing number of people in London and elsewhere in the UK now study for qualifications in travel and tourism.

While those parts of Britain attractive to foreign tourists have done well from the industry, many regional seaside resorts are experiencing difficulties. Of little appeal to foreigners, they have been abandoned in large numbers by British holidaymakers (Figure 8.23), who head to places such as the Mediterranean and Florida. Generally, the support given to communities of declining industries, such as coalmining and shipbuilding, has not materialised for tourism decline.

However, the south-west attracts more domestic visitors than any other region in the UK and is only overtaken by London if overseas visitors are taken into account. Tourism accounts for 29 per cent of total employment in Cornwall, the highest level in the UK. People are attracted by the beautiful coastline and the high quality of the natural environment in general. Visits to the Tate St Ives, The Eden Project and the National Maritime Museum Cornwall also bring in substantial revenue. In Cornwall between 1992 and 2003:

◆ total domestic and overseas visits increased from 3.4 million to 5.1 million

◆ total tourist spending rose from £623 million to £1216 million.

With limited opportunities elsewhere in the economy, tourism is absolutely vital to Cornwall. It is always a concern when there is such high dependency on one industry.

Discussion point

Make a list of places in the UK that people in your group have visited in the last three years. Make a note of the places your group would recommend and those it would not. Give reasons in each case.

Activities

1 Describe the trends shown in Figure 8.18.

2 Cornwall is more reliant on tourism than any other part of the country. Produce a fact file on tourism in Cornwall including a map of the most important destinations.

3 Find out which are the ten most popular individual tourist attractions (such as the London Eye) in the UK.

Case study | China: a new tourism destination and generating country

Rising prosperity and the relaxation of what were very strict limitations on foreign travel have witnessed increasing numbers of Chinese holidaying in their own country and abroad. The country is sub-divided into three broad regions – East, Middle and West (Figure 8.24). Average incomes are highest in the East and lowest in the West. It is therefore not surprising that the highest rate of involvement in both domestic and international tourism is in the East, with by far the lowest rates in the West. In much of the rural West, many families have never taken a holiday of any kind.

Outbound tourism

Chinese outbound tourism has increased rapidly over the last decade. In 2005, 31 million Chinese travelled abroad, spending $15.2 billion. This amounted to almost $500 per person. Figure 8.25 shows the rapid growth rate predicted for the future. According to the World Tourism Organisation, China will be the fourth largest source of outbound tourists at 100 million by 2020. However, at present, only 2 per cent of China's population have travelled abroad.

The factors responsible for the rapidly increasing number of Chinese travelling abroad are:

◆ rising disposable incomes

◆ longer holidays

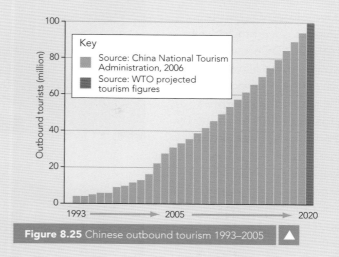

Figure 8.25 Chinese outbound tourism 1993–2005 ▲

Key
Source: China National Tourism Administration, 2006
Source: WTO projected tourism figures

◆ increasing interest in overseas travel

◆ relaxed political restrictions – the Chinese government has now given 129 countries Approved Destination Status (ADS)

◆ advertising/promotions by ADS countries anxious to secure their market share of Chinese tourists

◆ greater availability of travel products and services throughout China.

The very rapid growth in China's economy over the last 20 years has been the main factor in stimulating Chinese foreign tourism. Rising incomes have had a similar impact in China as they have had previously in more advanced economies. People are able to save more and to spend more on their homes, clothes, food, drink and transport. They are able to spend more on tourism as well. There is undoubtedly a very strong link between rising disposable incomes and the demand for tourism.

Not so long ago strict government regulations prevented people travelling abroad. A breakthrough came in 1983 when the government allowed people to participate in organised trips to visit friends and relatives in Hong Kong.

Figure 8.24 China ▼

Since the early 1990s individual countries have been given ADS by the Chinese government. Without ADS ordinary people would not be allowed to travel to a country. In 1991, the government allowed organised trips to be run by the Chinese Travel Service to Thailand, Singapore and Malaysia. In 1992, the 'market economy within socialism' was proclaimed. This opened up the pricing system within the Chinese tourist market which until then had been controlled by the government. It also allowed more scope for foreign direct investment.

Figure 8.26 An affluent Chinese tourist at ski location in northern China

Australia and New Zealand achieved ADS in 1999 and the UK in 2005, along with many other countries. ADS now applies to almost 130 countries. Notable exceptions are the USA and Canada.

Understandably, most Chinese travelling abroad so far have been very 'value-conscious' because average incomes are lower in China than in most of the countries that Chinese people visit. However, as the middle class expands and incomes rise, Chinese travel tastes have moved up the value chain. With higher incomes, people are able to afford higher-priced holidays in more exotic locations. A growing number of destinations are making considerable efforts to attract Chinese tourists because of the potentially large amounts of money involved.

China as a destination

The number of foreign visitors to China reached 22 million in 2006, excluding arrivals from Hong Kong, Macau and Taiwan (Figure 8.27). This compares with 300 000 in 1978. UNWTO predicts that China will become the world's most visited destination by 2020, if not before.

The guide books suggest that spring (March–April) and autumn (September–October) are the best times to visit China, and visitor numbers bear this out. The higher altitude areas, including Tibet, are best visited in high summer.

The 2008 Olympic Games in Beijing and the 2010 World Expo in Shanghai will be important boosts to

Country of origin	Arrivals in millions
South Korea	3.9
Japan	3.7
Russia	2.4
USA	1.7
Malaysia	0.9
Singapore	0.8
Philippines	0.7

Source: China National Tourist Office

Figure 8.27 China – foreign visitor arrivals in China 2006

tourism in China, which has the advantage of a wide variety of destinations within the country. For example, China has 31 UNESCO World Heritage Sites. Key attractions within China are:

◆ the Forbidden City and Tiananmen Square in Beijing

◆ the Terra Cotta Warriors in Xian

◆ the 6700km-long Great Wall of China

◆ the Yangtze River and the Three Gorges Dam

◆ the giant pandas of the Wolong Nature Reserve

◆ the skyscrapers, shopping and general atmosphere of Hong Kong.

The Impact on employment and GDP

The World Travel and Tourism Council (WTTC) estimates that 16.6 million people, 2.1 per cent of total

	RECEIPTS (100 million US$)	P.C.TOTAL (%)
TOTAL	292.96	100.0
Long distance transportation	82.94	28.3
Air	59.28	20.2
Rail	9.04	3.1
Motor	7.18	2.4
Sea	7.44	2.5
Sightseeing	12.27	4.2
Accommodation	37.75	12.9
Food and beverage	27.48	9.4
Shopping	63.78	21.8
Entertainment	17.02	5.8
Communication	8.44	2.9
Local transportation	10.30	3.5
Others	32.99	11.3

Source: China National Tourist Office

Figure 8.28 Breakdown of international tourism receipts (2005) ▲

employment, are employed in the travel and tourism industry. However, employment in the wider travel and tourism economy was estimated at over 72 million. Contribution to GDP is estimated at 2.5 per cent for the industry itself and 12.2 per cent for the wider travel and tourism economy. Seventy per cent of tourist revenues come from internal tourism. This is not surprising as 1.5 billion trips were taken within China in 2006. In 2007, total tourism demand in China represented 6.2 per cent of world market share. China's booming economy has made it possible for Chinese people to spend more on tourism because of higher incomes. In turn, such spending has stimulated further economic growth.

Figure 8.28 gives a breakdown of international tourism receipts for 2005. Beijing is China's top tourist destination. Here the industry contributes 8 per cent to the city's GDP.

Tibet and human rights

The most controversial issue with regard to foreign tourism in China is Tibet, although many people are also concerned about human rights in general in China. In 2006, the Free Tibet Campaign called for a tourist boycott of the new £2.3 billion Golmud–Lhasa railway. The highest passenger railway in the world now makes it possible to travel from Beijing to the Tibetan capital Lhasa in 48 hours. It is viewed by some as the completion of the 50-year-long colonisation of Tibet. The Chinese say the railway will bring big economic benefits to Tibet, the poorest part of the country. It will make Tibet more accessible not just for tourism but also for business in general. Critics fear it will speed up the influx of ethnic Chinese and further undermine the fragile Tibetan culture.

Some Tibetan organisations have urged tourists to use the services of Tibetan businesses rather than those run by the Han Chinese who have been encouraged to settle in Tibet by the Chinese government since the invasion by China in 1950.

Sanya, Hainan Island

Sanya, the second largest city on Hainan Island, is a leading destination for domestic tourists and is becoming increasingly popular with foreign tourists (Figure 8.29). It is the number one domestic choice for the rich of Beijing and Shanghai. The island is the most southerly point in China and because of its tropical climate and palm-fringed beaches it is often referred to as 'China's Hawaii'. The island has been heavily promoted by the Chinese government. In addition to its beaches, it boasts good quality golf courses, spectacular rainforest and mountain scenery, and many points of cultural and historic interest. Tourism has spurred a huge investment in infrastructure on the island, which has brought about significant changes to the natural environment.

The first charter flights from the UK began in 2007 with holidays offering ten days on the beaches of Sanya and three days in Beijing. Many of the major international hotel chains now have a presence in Sanya. These companies view Sanya as a strategic gateway into China's luxury-tourism market.

Figure 8.29 Tourist scene on Sanya ▼

Activities

1 Explain the reasons for the growth of Chinese outbound tourism.

2 Research a major tourist attraction in China. Identify the environmental problems associated with increasing numbers of visitors.

3 Write a brief summary of the information presented in Figure 8.28.

4 Find out more about tourism in Tibet. Draw up a list of the benefits and problems associated with tourism there.

Discussion points

1 Is it better for Tibet to be poor but still have its own identity or richer but be more Chinese?

2 Will the increase in both foreigners visiting China and Chinese people holidaying abroad put pressure on the Chinese government to be more democratic?

8.3 What are the social, economic and environmental issues associated with the growth of tourism?

Tourism brings both opportunities and problems for people and for the environment. This section begins by considering the social, economic and environmental issues associated with tourism. The first case study looks at Jamaica, a popular example among proponents of tourism's effectiveness at promoting economic and social prosperity. The second case study looks at Burma, probably the most controversial tourist destination in the world.

The social and cultural impact

The traditional cultures of many communities in the developing world have suffered because of the development of tourism. The adverse impact includes the following in varying degrees.

◆ The loss of locally owned land as tourism companies buy up large tracts in the most scenic and accessible locations.

◆ The abandonment of traditional values and practices.

◆ Displacement of people to make way for tourist developments.

◆ Changing community structure.

◆ Abuse of human rights by large companies and governments in the quest to maximise profits.

◆ Alcoholism and drug abuse as drink and drugs become more available to satisfy the demands of foreign tourists. It has also been suggested that the very obvious gap in wealth between locals and tourists can result in a certain 'despair' among some local people, particularly young adults, who find solace in alcohol and drugs.

◆ Crime and prostitution, sometimes involving children. 'Sex tourism' is a big issue in certain locations such as Bangkok, but it is also prevalent in some degree in most locations visited by large numbers of international tourists. The issue of crime is more complex. Clearly the wealth that foreign visitors bring with them provides more opportunities for local criminals than existed previously but visitors may also commit various crimes themselves.

◆ Visitor congestion at key locations hindering the movement of local people.

◆ Denying local people access to beaches to provide 'exclusivity' for visitors.

◆ The loss of housing for local people as visitors buy second homes in popular tourist areas (Figure 8.30).

Figure 8.31 shows how the attitudes to tourism can change over time. An industry which is usually seen as very beneficial initially can eventually become the source of considerable irritation, particularly where there is a big clash of cultures.

Figure 8.30 Entrance to National Park in Andalucia, Spain – the graffiti refers to the number of foreigners buying up houses in the nearby village of Frigiliana

1. **Euphoria**
◆ Enthusiasm for tourist development
◆ Mutual feeling of satisfaction
◆ Opportunities for local participation
◆ Flows of money and interesting contacts

2. **Apathy**
◆ Industry expands
◆ Tourists taken for granted
◆ More interest in profit making
◆ Personal contact becomes more formal

3. **Irritation**
◆ Industry nearing saturation point
◆ Expansion of facilities required
◆ Encroachment into local way of life

4. **Antagonism**
◆ Irritations become more overt
◆ The tourist is seen as the harbinger of all that is bad
◆ Mutual politeness gives way to antagonism

5. **Final level**
◆ Environment has changed irreversibly
◆ The resource base has changed and the type of tourist has also changed
◆ If the destination is large enough to cope with mass tourism it will continue to thrive

Figure 8.31 Doxey's index of irritation caused by tourism

The tourist industry and the various scales of government in host countries have become increasingly aware of these problems and are now using a range of management techniques in an attempt to mitigate such effects. Education is the most important element so that visitors are made aware of the most sensitive aspects of the host culture.

The tourist industry has a huge appetite for resources, which often impinge heavily on the needs of local people and the community structure. A long-term protest against tourism in Goa highlighted the fact that one five-star hotel consumed as much water as five local villages and the average hotel resident used 28 times more electricity per day than a local person.

Key term

Community structure: the form and development of the community in human populations.

Changing community structure

Communities that were once very close socially and economically may be weakened considerably due to a major outside influence such as tourism. The traditional hierarchy of authority within the community can be altered as those whose incomes are enhanced by employment in tourism gain higher status in the community. The age and sex structure may change as young people in particular move away to be closer to work in tourist enclaves. Changing values and attitudes can bring conflict to previously settled communities. The close ties of the extended family often diminish as the economy of the area changes and material wealth becomes more important.

Positive attributes

However, tourism can also have positive social and cultural impacts.

◆ Tourism development can increase the range of social facilities for local people.

◆ It can lead to greater understanding between people of different cultures.

- Family ties may be strengthened by visits to relatives living in other regions and countries.
- Visiting ancient sites can develop a greater appreciation of the historical legacy of host countries.
- It can help develop foreign language skills in host communities.
- It may encourage migration to major tourist generating countries.
- A multitude of cultures congregating together for major international events such as the Olympic Games can have a very positive global impact.

Activities

1. Explain the sequence of changes illustrated in Doxey's index (Figure 8.31).
2. Research the social impact of international tourism in one destination.
3. a What is meant by the term 'community structure'?
   b How can tourism impact on community structure in traditional societies.

The economic impact

Satellite accounting

The World Travel and Tourism Council (WTTC) argues that to consider tourist receipts alone, greatly underestimates the economic importance of the industry. To rectify the situation the WTCC has developed travel and tourism satellite accounting. By including all the direct and indirect economic implications of tourism it is clear that the industry has a much greater impact than most people think (Figure 8.32). Under this system of measurement, global

	Travel and tourism industry (direct impact)	Travel and tourism economy (satellite accounting)
Employment	76.1 million	231.1 million
% of world GDP	3.6%	10.4%
% of world exports	6%	12%

Source: World Travel and Tourism Council

Figure 8.33 Global travel and tourism

tourism was estimated to have generated $7060 billion of economic activity worldwide in 2007. The travel and tourism industry contributed 3.6 per cent to worldwide gross domestic product but the broader travel and tourism economy (which includes all the indirect benefits) contributed 10.4 per cent to world GDP (Figure 8.33). The travel and tourism industry employed 76.1 million people directly, while the wider travel and tourism economy employed 231.2 million, representing 8.3 per cent of total employment worldwide. This made it the world's largest service industry.

The industry, for example, creates considerable demand in the construction and manufacturing sectors. The WTTC estimates that the public and private sectors combined will spend $1155 billion on new travel and tourism capital investment worldwide in 2007.

Key terms

Travel and tourism satellite accounting: a comprehensive system of accounting that includes not only direct expenditure and receipts but also all the indirect knock-on effects.

Capital investment (in tourism): money invested in hotels, attractions, airports, roads and other aspects of infrastructure that facilitates high volume tourism.

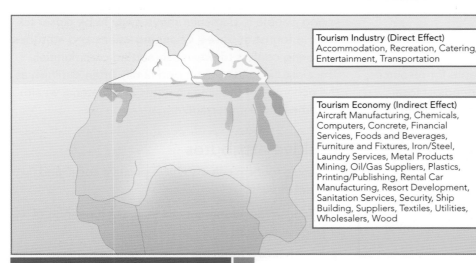

Tourism Industry (Direct Effect)
Accommodation, Recreation, Catering, Entertainment, Transportation

Tourism Economy (Indirect Effect)
Aircraft Manufacturing, Chemicals, Computers, Concrete, Financial Services, Foods and Beverages, Furniture and Fixtures, Iron/Steel, Laundry Services, Metal Products Mining, Oil/Gas Suppliers, Plastics, Printing/Publishing, Rental Car Manufacturing, Resort Development, Sanitation Services, Security, Ship Building, Suppliers, Textiles, Utilities, Wholesalers, Wood

What is commonly thought of as the "Tourism Industry" is only the tip of the iceburg

Figure 8.32 Satellite accounting 'iceberg'

AS Geography for OCR

How great are the economic benefits for developing countries?

Tourism undoubtedly brings valuable foreign currency to developing countries and a range of other obvious benefits. Foreign currency is necessary for countries to pay for the goods and services they import from abroad. However, critics argue that the value of tourism is often overrated because:

◆ economic leakages from developing to developed countries run at a rate of between 60 per cent and 75 per cent. With cheap package holidays, by far the greater part of the money paid stays in the country where the holiday was purchased

◆ tourism is labour intensive, providing a range of jobs especially for women and young people. However, most local jobs created are menial, low paid and seasonal. Overseas labour may be brought in to fill middle and senior management positions

◆ money borrowed to invest in the necessary infrastructure for tourism increases the national debt

◆ at some destinations tourists spend most of their money in their hotels with minimum benefit to the wider community

◆ tourism might not be the best use for local resources, which could in the future create a larger multiplier effect if used by a different economic sector

◆ locations can become over-dependent on tourism, which causes big problems if visitor numbers fall

◆ international trade agreements such as the General Agreement on Trade in Services (GATS) allow the global hotel giants to set up in most countries. Even if governments favour local investors there is little they can do.

Figure 8.34 Beach artist, Agadir, Morocco – an example of informal sector employment

However, supporters of the development potential of tourism argue that:

◆ tourism benefits other sectors of the economy, providing jobs and income through the supply chain – a multiplier effect

◆ it is an important factor in the balance of payments of many nations

◆ it provides governments with considerable tax revenues

◆ by providing employment in rural areas it can help to reduce rural–urban migration

◆ a major tourism development can act as a growth pole, stimulating the economy of the larger region

◆ it can create openings for small businesses

◆ it can support many jobs in the informal sector.

Key term

Growth pole: a place or region where a high concentration of investment stimulates economic growth. Such prosperity may then spread, at least to a certain extent, to the wider surrounding area.

Key terms

Economic leakages: the part of the money a tourist pays for a foreign holiday that does not benefit the destination country because it goes elsewhere.

The multiplier effect: a new or expanding economic activity in a region creates new employment and increases the amount of money circulating in the region. In turn, this attracts further economic development creating more employment, services and wealth.

The Britton and Butler models

The core–periphery enclave model of tourism (Figure 8.35) proposed by S Britton in 1981 stresses that in many developing countries the benefits/impact of tourism are very limited geographically. Most tourists come from the developed or core nations. In many developing countries (the periphery), tourists frequently stay in specially designated enclaves with all the required facilities immediately on hand. Outside of the resort enclaves there are a number of

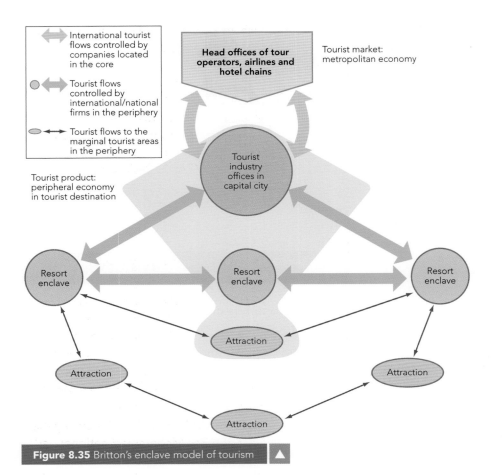

Figure 8.35 Britton's enclave model of tourism ▲

attractions (scenic, historic, cultural) at locations that can usually be reached and returned from within a day. At such locations the expected infrastructure is usually provided. Thus, the majority of the country is unaffected by tourism. As a result, most tourists have little or no contact with local people and fail to experience the reality of life in the country they have chosen to visit.

Butler's model of the evolution of tourist areas (Figure 8.36) illustrates how tourism develops and changes over time. In the first stage the location is explored independently by a small number of visitors. If visitor impressions are good and local people perceive that real benefits are to be gained, then the number of visitors will increase as the local community becomes actively involved in the promotion of tourism. In the development stage, holiday companies from the developed nations take control of organisation and management with package holidays becoming the norm. Eventually growth ceases as the location loses some of its former attraction. At this stage local people have become all too aware of the problems created by tourism. Finally, decline sets in, but because of the perceived economic importance of the industry efforts will be made to re-package the location which, if successful, may either stabilise the situation or result in renewed growth (rejuvenation).

Much more attention is now being paid to the impact of tourism on the local economy (Figure 8.37) so that the communities in and near tourist areas reap real benefits.

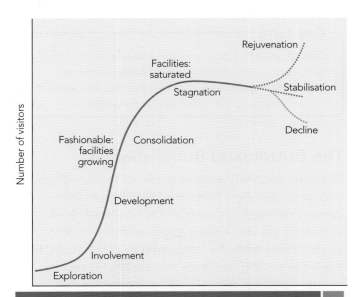

Figure 8.36 Butler's model of the evolution of tourism in a region ▲

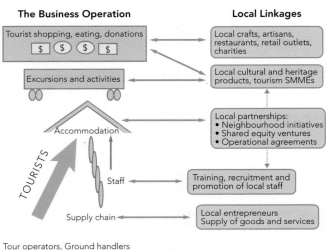

Figure 8.37 Tourism linkages into the local economy

1   With the help of Figure 8.32 discuss the indirect economic benefits of tourism.

2   Discuss the merits and limitations of the Britton model.

3   Find an example of the application of the Butler model to a particular destination. Discuss the analysis you have found in your group.

Suggest how you could minimise the economic leakage of a foreign holiday by careful planning.

The environmental impact

Tourism organised in such a way that its level can be sustained in the future without creating irreparable environmental, social and economic damage to the receiving area has come to be known as sustainable tourism. The term comes from the 1987 UN Report on the

Environment which advocated the kind of development that meets present needs without compromising the prospects of future generations. Following the 1992 Earth Summit in Rio de Janeiro, the WTTC and the Earth Council drew up an environmental checklist for tourist development, which included waste minimisation, re-use and recycling, energy efficiency and water management. The WTTC has since established a more detailed programme called 'Green Globe', designed to act as an environmental blueprint for its members.

Key term

Sustainable tourism: tourism organised in such a way that its level can be sustained in the future without creating irreparable environmental, social and economic damage to the receiving area.

In so many developing countries, newly laid golf courses have taken land away from local communities while consuming large amounts of scarce freshwater. It has been estimated that the water required by a new golf course can supply a village of 5000 people. In both Belize and Costa Rica, coral reefs have been blasted to allow for unhindered watersports.

Education about the environment visited is clearly the key. Scuba divers in the Ras Mohammad national park in the Red Sea, who were made to attend a lecture on the ecology of the local reefs, were found to be eight times less likely to bump into coral (the cause of two-thirds of all damage to the reef), let alone deliberately pick a piece. However, there is huge concern about the future of many coral reefs, no more so than the Great Barrier Reef, which receives two million visitors a year (Figure 8.38).

A new form of ecotourism in which volunteers help in cultural and environmental conservation and research is developing. An example is the Earthwatch scientific research project, which invites members of the general public to join the experts as fully fledged expedition members, on a paying basis of course. Several Earthwatch projects in Australia have helped Aboriginal people to locate and document their prehistoric rock art and to preserve ancient rituals directly.

Positive environmental impacts

The environmental impact of tourism is not always negative. Landscaping and sensitive improvements to the built environment have significantly improved the overall quality of some areas. On a larger scale, tourist revenues can fund the designation and management of protected areas such as national parks and national forests.

Figure 8.38 The Great Barrier Reef with tourist divers, fish and coral

Case study | Jamaica: the opportunities created by tourism

Economic and social development

Tourism has become an increasingly vital part of Jamaica's economy in recent decades (Figure 8.39). The contribution of tourism to total employment and GDP has risen substantially. It has brought considerable opportunities to its population, although it has not been without its problems. During the 1970s the Jamaican government introduced 'Jamaicanisation' policies designed to attract much needed foreign investment in tourism. Policies included comparatively high wages and special industry taxes that went directly into social development, health care and education. The latter are often referred to by economists as 'soft infrastructure'. However, tourism has also spurred the development of vital 'hard infrastructure' too, such as roads, telecommunications, water supply and airports. Jamaica has been determined to learn from the 'mistakes' of other countries and ensure that the population would gain real benefits from the growth of tourism.

2007	Travel and tourism Industry	Travel and tourism economy
GDP % of total	9.6%	31.1%
Employment	92 000	289 000

Figure 8.39 Effect of tourism on Jamaica's economy (2007) ▲

Tourism's direct contribution to GDP in 2007 amounted to almost $1.2 billion. Adding all the indirect economic benefits increased the figure to almost $3.8 billion or 31.1 per cent of total GDP. Direct employment in the industry amounted to 92 000 but the overall figure, which includes indirect employment, is over three times as large. In the most popular tourist areas the level of reliance on the industry is extremely high.

In terms of the industry's relative contribution to the national economy, the WTTC ranks Jamaica 19 out of 176. Tourism is the largest source of foreign exchange for the country. The revenue from tourism plays a significant part in helping central and local government fund economic and social policies. Also, as attitudes within the industry itself are changing, larger hotels and other aspects of the industry have become more socially conscious. Classic examples are the funding of local social projects.

A recent paper (2007) on tourism by the People's National Party (PNP) stated that 'The momentum generated by the current round of investment in resort development has created an enormous pull factor in terms of investor confidence. This has set the stage for an even more powerful wave of investment in the next 10 years.'

The Jamaica Tourist Board (JTB) is responsible for marketing the country abroad. Recently, it used the fact that Jamaica was one of the host countries for the 2007 Cricket World Cup to good effect. The JTB also promotes the positive aspects of Jamaican culture and the Bob Marley museum in Kingston has become a popular attraction. Such attractions are an important part of Jamaica's objective of reducing seasonality.

The high or 'winter' season runs from mid-December to mid-April when hotel prices are highest. The rainy season extends from May to November. It has been estimated that 25 per cent of hotel workers are laid off during the off-season.

Figure 8.40 The location of Jamaica's national parks ▲

Environmental protection

Jamaica's government is working to reduce the environmental impact of tourism. Figure 8.40 shows the location of Jamaica's three national parks. A further six sites have been identified for future protection. The Jamaican government sees the designation of the parks as a positive environmental impact of tourism. Entry fees to the national parks pay for conservation. The desire of tourists to visit these areas and the need to conserve the environment to attract future tourism drives the designation and management process.

The two marine parks are attempting to conserve the coral reef environments off the west coast of Jamaica. They are at risk from damage from overfishing, industrial pollution and mass tourism. The Jamaica Conservation and Development Trust is responsible for the management of the national parks while the National Environmental Planning Agency has overseen the government's sustainable development strategy since 2001.

Ecotourism is a developing sector of the industry with, for example, raft trips on the River Rio Grande increasing in popularity. Tourists are taken downstream in very small groups. The rafts, which rely solely on manpower, leave singly with a significant time gap between them to minimise any disturbance to the peace of the forest.

Community tourism

Considerable efforts are being made to promote community tourism so that more money filters down to the local population and small communities. The Sustainable Communities Foundation through Tourism (SCF) programme has been particularly active in central and south-west Jamaica. Community tourism is seen as an important aspect of 'pro-poor tourism'.

> **Key terms**
>
> **Community tourism:** this aspect of the industry fosters opportunities at the community level for local people.
>
> **Pro-poor tourism:** tourism that results in increased net benefits for poor people.

The Astra Country Inn, Mandeville has been recognised as a pioneer hotel in community tourism. Its work with surrounding communities has included:

◆ promoting B&B accommodation in private homes
◆ training local guides
◆ developing community-based tourist attractions
◆ encouraging the development of local suppliers.

However, tourism has not been without its problems. The behaviour of some tourists clashes with the island's traditional morals. However, some people also have a negative image of Jamaica because of its levels of violent crime and harassment and despite the recent initiatives of the Jamaican government to protect the environment much valuable biodiversity has already been lost.

Activities

Produce a fact file on tourism in Jamaica to include a map showing the main tourist locations.

◀ **Figure 8.41** A beach fringed with palm trees in a national park area

Case study | Myanmar: the problems created by the growth of tourism

Tourism in Myanmar (formerly known as Burma), is a hugely controversial issue (Figure 8.42). Major human rights abuses have been catalogued in the development of the tourist industry and in other sectors of Myanmar's economy. The environment has also suffered significantly as the government has rushed to try to expand tourism to bring in much needed foreign currency. The military regimes running the country have maintained an oppressive dictatorship since 1962. Most recently, peaceful pro-democracy demonstrations by Myanmar monks in Yangon were violently suppressed by the military. As a result, the government of Myanmar has come under increased international criticism. The foreign currency gained from tourism has been mainly used on military spending, which accounts for nearly half the government budget. In contrast, less than 44 pence per person per year is spent on health and education. Over 60 per cent of Myanmar people live in extreme poverty. Government policy has created huge social problems.

Travel and tourism industry jobs accounted for 3 per cent of total employment in 2007 while total travel and tourism economy employment was responsible for 1.35 million jobs. The latter figure came to 6.1 per cent of total employment. In terms of GDP, the industry accounted for 3.2 per cent of the total while the wider travel and tourism economy was responsible for 6.7 per cent of GDP. Myanmar attracts between 100 000 and 200 000 foreign visitors a year.

The pressure group Tourism Concern fights exploitation in the tourist industry. Tourism Concern argues that 'By visiting Burma, tourists are accepting and validating the ongoing human rights abuses in the country.

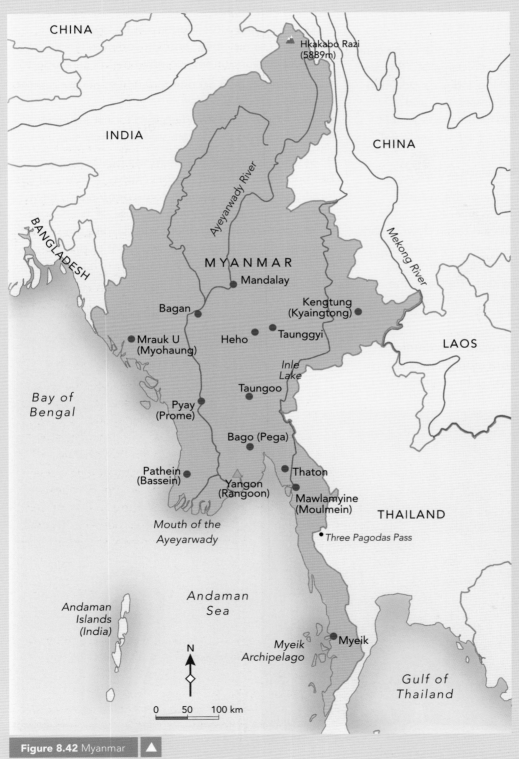

Figure 8.42 Myanmar ▲

There are well documented direct links between tourism and human rights.' Myanmars' democratically elected leader Aung San Suu Kyi is under house arrest. She was elected to power in 1990 but prevented from taking office by the military regime. She has stated 'Burma will be here for many years, so tell your friends to visit us later. Visiting now is tantamount to condoning the regime.'

The tourist boycott began after the military government announced the 'Year of the Tourist' in 1996. Its aim was to increase income from tourism considerably after the country first opened up to tourists in 1988. The boycott was started by NGOs, Burma Campaign UK and Tourism Concern. Lonely Planet has been criticised because it still publishes a guide to Myanmar. In contrast, Rough Guides have made a point of not publishing a guidebook to Myanmar. However, there are those, both inside and outside Myanmar, who feel that a blanket boycott on tourism does more harm than good.

It is very difficult to travel in Myanmar without contributing to the military government. Many hotels, transport companies and moneychangers are either owned by the government or have to pay them large fees or bribes.

Some tourist companies continue to operate in Myanmar. For example, Kuoni adopt a 'responsible tourism' policy. They follow a policy to limit interaction with the government and maximise input into the local economy by using local transport companies. They use only private hotels and air operators.

> **Key term**
>
> **Responsible tourism:** visiting a destination in such a way as to cause minimal negative impact on its environment and the culture of the host community.

Neighbouring countries are of course concerned about the situation in Myanmar. In terms of tourism, the boycott impacts on regional plans to promote this part of the world in general as a unique tourist destination. The lack of political development hinders economic and social development in Myanmar in general, with clear exemplification in terms of tourism.

Population displacement

Burma Campaign UK claims that more than one million people have been displaced from their homes 'in order to "beautify" cities, suppress dissent, and make way for tourism developments, such as hotels, airports, and golf courses'. For example, in Pagan, the 5000 people who lived in villages among the ancient pagodas were given two weeks to pack up and leave. At the Royal Palace at Mandalay, the moat and surrounding area was cleared by forced 'voluntary' labour and homes were demolished.

Displacement has caused huge changes to community structure. Established communities have often been broken up and the age structure changed. Hundreds of thousands of people have been used as forced labour to clean up existing tourist attractions and to build new facilities. In 2004, soldiers rounded up ethnic Salons or 'sea gypsies' who normally live on boats in the Mergui Archipelago. They were forced to live on land and perform traditional dances for foreign tourists. However, even where forced labour displacement has not occurred community structure has changed due to tourism, but to a lesser degree.

Environmental degradation

Myanmar boasts a huge variety of ecosystems from tropical islands, rainforests, mangroves and great rice-growing plains in the south to snow-capped peaks in the north. However, the haste to develop tourist and other aspects of infrastructure has often taken little account of the environmental repercussions. According to the *Asia Observer* 'the depletion of a renewable resource for quick commercial gain is typical of the junta's economic policies. Long-term environmental effects or their immediate impact on local residents are ignored.'

Seasonality

In terms of climate, the best season for visiting Myanmar is from November to February, when it is not too hot and rainfall is at its lowest. The south-west monsoon starts from mid-May, lasting until October. The country receives the lowest number of visitors in May, June and September. Even though July and August are not ideal in terms of weather, this is the traditional holiday period for many foreign tourists visiting Myanmar. With such low living standards and few alternative opportunities in the money economy, seasonality is a big problem for those who work in the industry.

Activities

1 Find out where the main tourist attractions in Myanmar are located. Show these locations on a map of the country.

2 Look at the Tourism Concern website (www.tourismconcern.org.uk). Which other countries and issues related to tourism is this organisation trying to bring to the attention of the general public?

8.4 How can tourism be managed to ensure sustainability?

As the level of global tourism increases rapidly, it is becoming more and more important for the industry to be responsibly planned, managed and monitored. Tourism operates in a world of finite resources where its impact is becoming of increasing concern to a growing number of people. At present, only 5 per cent of the world's population have ever travelled by plane. However, this is undoubtedly going to increase substantially.

Leo Hickman in his book *The Final Call* claims 'The net result of a widespread lack of government recognition is that tourism is currently one of the most unregulated industries in the world, largely controlled by a relatively small number of Western corporations such as hotel groups and tour operators. Are they really the best guardians of this evidently important but supremely fragile global industry?'

Environmental groups are keen to make travellers aware of their 'destination footprint'. They are urging people to:

◆ 'fly less and stay longer'

◆ carbon-offset their flights

◆ consider 'slow travel'.

> **Key term**
>
> **Destination footprint:** the environmental impact caused by an individual tourist on holiday in a particular destination.

In the latter, tourists consider the impact of their activities both for individual holidays and for the longer term. For example, they may decide that every second holiday would be in their own country (not using air transport). It could also involve using locally run guesthouses and small hotels as opposed to hotels run by international chains. This enables more money to remain in local communities.

Virtually every aspect of the industry now recognises that tourism must become more sustainable. Ecotourism is at the leading edge of this movement. An example of ecotourism in Ecuador will be considered in the following case study.

> **Key term**
>
> **Ecotourism:** a specialised form of tourism where people experience relatively untouched natural environments such as coral reefs, tropical forests and remote mountain areas, and ensure that their presence does no further damage to these environments.

Protected areas

Over the course of the past 130 years or so, more and more of the world's most spectacular and ecologically sensitive areas have been designated for protection at various levels. The world's first national park was established at Yellowstone in the USA in 1872. Now there are well over 1000 worldwide. Many countries have national forests, country parks, areas of outstanding natural beauty, world heritage sites and other designated areas which merit special status and protection. Wilderness areas with the greatest restrictions on access have the highest form of protection.

In many countries and regions there are often differences of opinion when the issue of special protection is raised. For example, in some areas jobs in mining, forestry and tourism may depend on developing presently unspoilt areas. So it is not surprising that values and attitudes can differ considerably when big decisions about the future of environmentally sensitive areas are being made. Often, a clear distinction has to be made between the objectives of preservation and conservation.

> **Key terms**
>
> **Preservation:** maintaining a location exactly as it is and not allowing development.
>
> **Conservation:** allows for developments that do not damage the character of a destination.

Tourist hubs

The concept of tourism hubs or clusters is a model that has been applied in a number of locations. The idea is to concentrate tourism and its impact in one particular area so that the majority of the region or country feels little of the negative impacts of the industry. Benidorm in Spain and Cancun in Mexico are examples where the model was adopted but both locations show how difficult it is confine tourism within pre-conceived boundaries as the number of visitors increases and people want to travel beyond tourist enclaves.

Quotas

Quotas seem to be one of the best remedies on offer. The UK Centre for Future Studies has suggested a lottery-based entrance system, an idea endorsed by Tourism Concern. Here, the number of visitors would not be allowed to exceed a sustainable level. This is an idea we are likely to hear much more about in the future.

Balancing socio-economic environmental needs

Although decision-makers in the tourist industry are more aware than ever before of environmental needs, they are also very conscious of the socio-economic needs of the communities in which tourism takes place. It is therefore not surprising that in most cases a balance has to be struck between these two objectives. This requires detailed planning and management.

Activities

1 Look at an atlas of Britain to identify the various types of protected areas and where they are located.

2 What do you think of the idea of quotas for visitor numbers at certain locations?

3 What do you understand by the concept 'slow travel'?

Case study | Sustainable tourism in Ecuador

Ecuador's (Figure 8.43) travel and tourism industry was expected to contribute 1.8 per cent to GDP in 2007, with the wider travel and tourism economy contributing 7.8 per cent. The 84 000 jobs in the industry accounted for 1.6 per cent of total employment while the 361 000 jobs in the wider travel and tourism economy made up 6.7 per cent of total employment.

International tourism is Ecuador's third largest source of foreign income after the export of oil and bananas.

The number of visitors has increased substantially, both to the mainland and to the Galapagos Islands. The majority of tourists are drawn to Ecuador by the great diversity of flora and fauna. The country contains 10 per cent of the world's plant species. Much of the country is protected by national parks and nature reserves.

As visitor numbers began to rise, Ecuador was anxious not to suffer the negative externalities of mass tourism

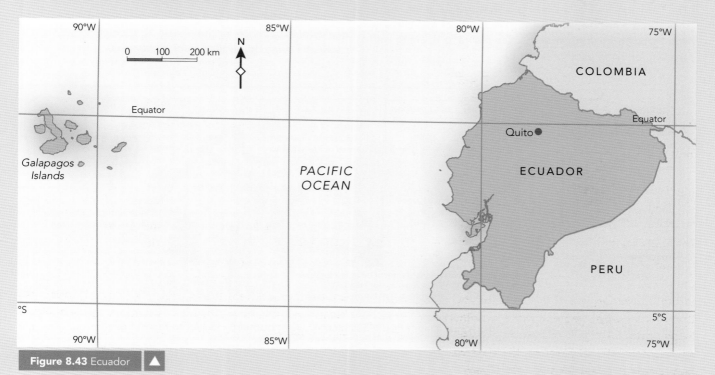

Figure 8.43 Ecuador

evident in many other countries. The country's tourism strategy has been to avoid becoming a mass-market destination but to market 'quality' and 'exclusivity' instead, in as eco-friendly a way as possible. The objective has been to strike a balance between environmental and socio-economic needs. Tourist industry leaders were all too aware that a very large influx of visitors could damage the country's most attractive ecosystems and harm its image as a 'green' destination for environmentally conscious visitors.

Ecotourism has helped to bring needed income to some of the poorest parts of the country. It has provided local people with a new alternative way of making a living. As such, it has reduced human pressure on ecologically sensitive areas.

The main geographical focus of ecotourism has been in the Amazon rainforest around Tena, which has become the main access point. The ecotourism schemes in the region are usually run by small groups of indigenous Quichua Indians (Figure 8.44).

El Pbutano Delos Buitres Nr Tena Ecuador

El Pbutano was until recently a family run rainforest farm managed in the traditional way. The farm is 300ha in size of which 80ha are cleared secondary forest, the rest is uncleared primary forest. Farming here has always followed traditional methods which protect the fragile forest ecosystem.

Over the last 8 years the farm has been developed into a sustainable ecotourism destination, run by the farmer in response to growing demand from explorer style tourists. The farm is located one and a half hours' drive from Tena and accessed by 4x4 transport along an unmade road followed by a 20 minute foot trek from the roadhead to a clearing in the forest.

The tourist capacity is small. A maximum of 16 guests can be accommodated at a time in the converted farmhouse on stilts which gives a maximum yearly capacity of 300 guests. The farm is still run as a working farm with 4 full time workers who live on site. The main outputs are bananas and fruit with small sugar and coffee plantations, along with grazing for cattle and horses.

All the farm buildings are built using local sustainable resources. The farmhouse is made of local wood, mainly bamboo, with a palm leaf roof which needs replacing every 15 years. The holidays include tours into the forest with a local guide and overnight camping to learn about local plants and to see community projects. The group size is kept small to ensure the forest is not damaged. All treks into the forest are on foot along existing paths and the routes are changed with each successive group to prevent erosion of the pathways. The cost is $40 pppn, giving an annual income of $48 000, half of which goes to the local community for improvements to local services and to pay the local guides.

Tourism is very low key and uninvasive; the local community largely ignore the tourists apart from the odd football match. The farm has been operating since 1996 as a tourist destination and its programme seems to be sustainable both in terms of the impact on the destination and the quality of experience for the tourist. The sketch shows how guests have to comply with strict regulations.

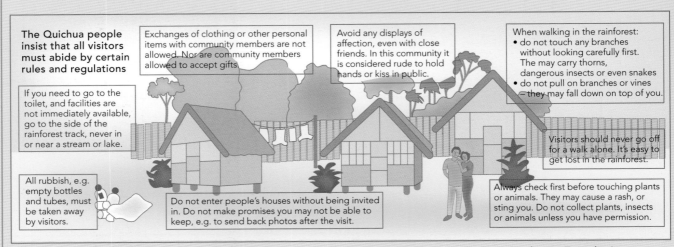

The Quichua people insist that all visitors must abide by certain rules and regulations

Exchanges of clothing or other personal items with community members are not allowed. Nor are community members allowed to accept gifts.

If you need to go to the toilet, and facilities are not immediately available, go to the side of the rainforest track, never in or near a stream or lake.

All rubbish, e.g. empty bottles and tubes, must be taken away by visitors.

Do not enter people's houses without being invited in. Do not make promises you may not be able to keep, e.g. to send back photos after the visit.

Avoid any displays of affection, even with close friends. In this community it is considered rude to hold hands or kiss in public.

When walking in the rainforest:
• do not touch any branches without looking carefully first. The may carry thorns, dangerous insects or even snakes
• do not pull on branches or vines – they may fall down on top of you.

Visitors should never go off for a walk alone. It's easy to get lost in the rainforest.

Always check first before touching plants or animals. They may cause a rash, or sting you. Do not collect plants, insects or animals unless you have permission.

Source: GEO Factsheet No. 201, page 3, Curriculum Press 'Sustainable Development – Case Studies in Ecuador.'

Figure 8.44 An ecotourism project in Ecuador

Galapagos Islands at risk

In early 2007 the government of Ecuador declared the Galapagos Islands at risk, warning that visitor permits and flights to the island could be suspended. The Galapagos Islands straddle the Equator 1000km off the coast of Ecuador. All but 3 per cent of the islands are a national park. Five of the 13 islands are inhabited. Visitor numbers are currently 100000 a year and rising.

The volcanic islands can be visited all year round but the period between November and June is the most popular. A national park entrance fee of £65 is payable on arrival. Among the many attractions are giant tortoises, marine iguanas and blue-footed boobies (Figure 8.45).

In signing the emergency decree to protect the islands, the President of Ecuador stated 'We are pushing for a series of actions to overcome the huge institutional, environmental and social crisis in the islands.'

The identified problems include:

◆ a growing population – 18000 islanders with legal status earn a living from fishing and tourism but an additional 15000 people are believed to live illegally in the islands

◆ illegal fishing of sharks and sea cucumbers is believed to be at an all-time high

◆ the number of cruise ships continues to rise

◆ internal arguments within the management structure of the national park

◆ the controversial opening of a hotel in 2006.

While the government of Ecuador has been anxious to protect this unique environment, individuals and businesses (local and external) have scope to maximise income from tourism. More careful planning and management will be required in the future to ensure the right balance between socio-economic and environmental needs.

Activities

1 Write a ten-bullet point list on ecotourism in Ecuador.

2 Why is there so much concern about the threat from tourism on the Galapagos Islands?

Figure 8.45 Giant tortoises on the Galapagos Islands

The number of people visiting Antarctica in 2007 was estimated at 33000. The annual rate of growth over the last decade has been significant and environmentalists have become increasingly concerned. The greatest worry is the increasing number of large cruise ships visiting Antarctica carrying up to 3000 passengers.

Very few tourists actually stay a night on the continent. The vast majority are brought to shore for only three or four hours (Figure 8.46). Seaborne tourism began in the late 1950s at a modest level but increased considerably in the 1980s with improvements in polar transport. The Antarctic Peninsula receives most visitors because of its relative proximity to the ports of Punta Arenas and Ushuala at the tip of South America. From here it takes only 48 hours to cross the Drake Passage. A smaller number of cruises originate in Australia and New Zealand, heading for the Ross Ice Shelf and McMurdo Sound. The alternative option is to view Antarctica from the air.

So why such concern over a relatively small number of visitors?

◆ The Antarctic ecosystem is extremely fragile. In its permafrost environment disturbances leave their imprint for a long time.

  ◆ The ecosystem dynamics on the continent are unique and are of great scientific interest.

  ◆ The summer tourist season coincides with peak wildlife breeding periods.

  ◆ Both land-based installations and wildlife agglomerate in the few ice-free locations on the continent.

  ◆ The demand for fresh water is not easy to satisfy.

  ◆ Visitor pressure is being felt on cultural heritage sites such as old whaling and sealing stations.

  ◆ The unique legal status of Antarctica makes enforcement of any code of tourist behaviour difficult.

A voluntary code has been developed by the International Association of Antarctic Tour Operators (IAATO) in conjunction with Antarctic Treaty nations. However, the Antarctic Treaty nations have yet to agree a legally binding set of tourism regulations. At present there are a limited number of tour operators in this market and they seem to be generally responsible in keeping to the guidelines and in making their clients aware of the sensitivity of the environment. The current restrictions are:

Figure 8.46 Tourism in Antarctica ▲

Case Study: Antarctica: tourism's last global frontier

U2

8

AS Geography for OCR

- the number of onshore visitors is limited

- the most sensitive areas are avoided

- visitors must keep a certain distance from wildlife, especially during the breeding season

- visitors are encouraged to take all rubbish back to their ships

- tourists are encouraged not to collect souvenirs

- tourist ships have to be able to prove that they have equipment available to clean up any oil spill they may be responsible for.

Tourism is certainly not the only aspect of development confronting Antarctica. Like the other pressure points (potential exploitation of minerals, depletion of marine resources, impact of bases), demand will undoubtedly increase. What is required is a research-based management strategy that is truly sustainable.

Activity

Describe and explain the policies you would advocate for Antarctica with regard to tourism.

Knowledge check

1 Examine the reasons for the rapid growth of tourism over the past 50 years.

2 What are the main reasons for changes in the level of tourism at particular destinations?

3 How has the location and type of tourism changed in recent decades?

4 Explain the relationship between levels of tourism and levels of development.

5 Discuss the major social issues associated with the development of tourism.

6 Suggest why there is considerable debate about the economic benefits of tourism.

7 Comment on the environmental problems linked with the growth of tourism.

8 What is sustainable tourism?

9 Why has the issue of sustainability become so important with regard to tourism?

10 In what ways will tourism change in the next 50 years?

A list of useful websites accompanying this chapter can be found in the Exam Café section on the **CD-ROM**

ExamCafé
Relax, refresh, result!

Relax and prepare

What I wish I had known at the start of the year...

Yasmin

"During the exam I try not to look around – it might look like I'm cheating or panic me if I see everyone else is working away. I always make sure I read the front cover to ensure I know how many questions I have to do from which sections.

I make sure I read each question at least twice before I start or plan my answer. I check the wording to make sure I know what the question wants and respond to the wording used."

Abebe

"I find the term 'ecotourism' confusing, so I've summarised it like this:

◆ A fairly recent phenomenon

◆ Tourists want to visit and experience environments in their wild natural states (e.g. rain forests in Columbia; coral reefs in Australia)

◆ Specialised type of tourism

◆ Basic in its amenities (to minimise impact on the fragile environment)

◆ Expensive to create funds to protect the environment

◆ Number of visitors must be rationed (to reduce impact on environment)

◆ Often it has expanded until the need to improve facilities has led to damage to the environment (so it needs managing)

I find it so much easier to remember in point form, putting my examples and reasons in brackets. I hope this question comes up in the exam!"

Hot tips

Mahmood

"I always try to work out my timings carefully and try to leave some time to check through the answers in case of errors or add an after thought where one might fit in. I use sheets of spare paper for jottings as sometimes ideas come into my head as I am working on another question."

Common mistakes – Judith

I used to get hung up on those case studies that my teacher made me learn. Often I couldn't see the meaning of the question as I was so busy getting the detailed case study down. I found it very hard to get the right amount of detail down. Too much detail and it became irrelevant, too little and I failed to show my knowledge. Now I often do a quick map to locate the example and add on only detail that is relevant to the question.

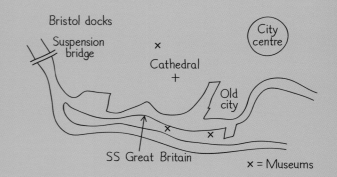

8.1. In what ways has the global pattern of tourism changed?

Increased leisure time/paid holidays

Cheaper faster types of transport especially air

Active marketing by resorts and travel firms

Increased incomes – more 'spare'

Increased psychological need to 'escape'

Rise in lifestyle expectations

Increased education about other areas and their cultures

Spread of English language

Increased media coverage

LEDCs see it as a means to develop

8.2 What is the relationship between the growth of tourism and economic development?

By Location	Increased travel to remoter areas or niche markets
	Increased internal as well as international travel
	Longer distances travelled and multi-centre holidays more popular
By Type	Purpose has increased in range e.g. active (sport) versus passive (sun)
	Duration has increased in range e.g. day trips versus longer trips
	Direction has increased in variety e.g. local, international, peak and off season trips
	Destination has increased in variety e.g. enclave, nomadism and resorts
	Impact e.g. ecotourism versus purpose built resort

8.3 What are the social, economic and environmental issues associated with the growth of tourism?

Physical	Encourages conservation e.g. National Parks
	Encourages control of pollution, dereliction etc
	Encourages coastal and river protection
	Search for and development of new resources e.g. water
Economic	Creation of employment – easy to enter industry
	Increased investment in infrastructure – roads, power etc
	Increased demand for local farm produce (move to cash farming)
	Demand for craft industries, shops, services etc
	Earns foreign currency – balance of payments
	Can lead or focus development due to multiplier effect
Social	Improved education to meet needs of tourists
	Improved health services
	Traditional cultures preserved
	Restrictive cultures made more liberal
	Introduces new ideas and expectations

Refresh your memory

8.4 How can tourism be managed to ensure sustainability?	
Physical	Construction destroys natural beauty/habitats Pollution – water, air, noise, litter, sewage Destruction of wildlife (disruption of breeding and taken as souvenirs) Water problems – loss of surface and groundwater Resource depletion e.g. building materials, fuel Soil erosion e.g. trampling
Economic	Increased imports (cost) of food etc Rise in prices e.g. food, land etc Agriculture shifts to commercial so loss of stable food crops Most profits leave area Low paid seasonal part-time and menial jobs Cost – takes money away from other areas/issues Huge drain on power and water supplies Urban coastal sprawl Cost of infrastructure e.g. roads, airport
Social	Moral corruption – vice, crime Increased inequalities – 'them and us' Loss of traditional culture, language, values Cultural colonialism

Top tips . . .

What do you do if you think your result is wrong? You should be aware of your expected grade and if you are two or more grades below it, then you might wish to think of an appeal or re-mark. First be honest with yourself over your effort and the questions you chose. Then check with friends – did they get what they expected – and with your teacher. This may indicate it wasn't just you but the whole centre. The school may seek a re-mark and so can you, but it costs and grades can go down, so be sure before following this course of action. Also remember you must appeal before the tight deadline and you can only get an upgrade if the re-marker (usually a senior examiner) thinks the mark scheme has not been followed or there is an adding up error (clerical error). You won't get an upgrade if the marker was a bit mean! You can ask for your paper to be returned – this costs but it may tell you where you went wrong and so help you in any re-sits. Check dates and costs with your examination officer and the wisdom of this action with your teacher.

Get the result!

Sample question

Suggest why some governments try to limit international tourism into their country. [6 marks]

Examiner says

This answer scores maximum marks because – (1) it provides a clear reason with an appropriate example, (2) it provides another valid reason, (3) good use of terminology, (4) and (5) clear reasons with appropriate examples, (6) another valid reason.

Student answer

Governments may want to limit foreign tourists for a variety of reasons. Politically they may be seen as a dangerous subversive element, e.g. North Korea (1), and some governments fear that visitors would undermine their culture or way of life (2) – a type of cultural colonialism (3). Others want to minimise the impact on the environment and so limit numbers, e.g. in game parks (4). Economically others fear that the profits will leave the country or tourist cash will cause inflation, e.g. Cuba (5), or the influx will lead to inequalities between regions or groups of the population (6).

Examiner says

This would score highly as the candidate tried to structure the answer rather than produce a simple list of resultant problems.

Examiner's tips

Never leave the exam room before the end – there is always something you can improve on, if only by adding another diagram.

Don't discuss the exam afterwards. It's over and you can't change your answer but also there are often a variety of ways to correctly answer a question. Go and relax until you have to prepare for your next exam.

The conclusion to an essay is key as it draws together the threads of your discussion and relates it back to the question. A conclusion can make or break an answer. For example if the question is:

"With reference to a named example assess the impact of the growth of tourism."

You could conclude it as follows:

"The recent growth of tourism in Cuba has had a mixed impact on the country (1). Benefits are largely economic whilst the resulting problems tend to be environmental and social (2). But the balance is not simple (3) as these benefits and problems vary over time, with location and between groups in the community (4)."

The conclusion above relates directly back to the question. (1) recognises that it is not a simple answer but varies between aspects (2). It goes on to highlight that the impact changes (3) with a variety of situations (4). To finally complete the conclusion some evaluation or assessment is needed – probably on the lines of the need to manage the growth carefully to reduce or minimise the negative aspects.

Conclusions are not simply afterthoughts, they should summarise the discussion and show how it has answered the initial question. A good effective conclusion can greatly lift the level of an answer.

Index

Index

index

Opposite you will find the AS Geography CD-ROM. Open up the CD, explore its contents and develop your geographical knowledge and skills further.

LiveText

On the CD you will find an electronic version of the Student Book, powered by LiveText. As well as the student book and the LiveText tools there are additional case studies, extension activities and weblinks. Within the electronic version of the Student Book, you will also find the interactive Exam Café.

Immerse yourself in our contemporary interactive Exam Café environment! With a click of your mouse you can visit 3 separate areas in the café to **Relax, Refresh your memory** or **Get the result**. You'll find a wealth of material including:

- Revision tips from students, Bytesize concepts, Common mistakes and Examiner's hints
- Language of the exam (an interactive activity)
- Revision checklists
- Sample exam questions (which you can try) with student answers and examiner comments
- Sample exam papers.

Minimum system requirements

- Windows 2000, XP Pro or Vista
- Internet Explorer 6 or Firefox 2.0
- Flash Player 8 or higher plug-in
- Pentium III 900 MHZ with 256 Mb RAM (512 Mb for Vista)
- Adobe Reader 6® or higher

To run your Exam Café CD, insert it into the CD drive of your computer. It should start automatically; if not, please go to My Computer (Computer on Vista), click on the CD drive and double-click on 'start.html'.

If you have difficulties running the CD, or if your copy is not there, please contact the helpdesk number given below.

Software support

For further software support between the hours of 8.30–5.00 (Mon–Fri), please contact:

Tel: 01865 888108

Fax: 01865 314091

Email: software.enquiries@pearson.com